# Studies in Autonomic, Data-driven and Industrial Computing

**Series Editors**

Swagatam Das, Indian Statistical Institute, Kolkata, West Bengal, India

Jagdish Chand Bansal, South Asian University, Chanakyapuri, India

The book series Studies in Autonomic, Data-driven and Industrial Computing (SADIC) aims at bringing together valuable and novel scientific contributions that address new theories and their real world applications related to autonomic, data-driven, and industrial computing. The area of research covered in the series includes theory and applications of parallel computing, cyber trust and security, grid computing, optical computing, distributed sensor networks, bioinformatics, fuzzy computing and uncertainty quantification, neurocomputing and deep learning, smart grids, data-driven power engineering, smart home informatics, machine learning, mobile computing, internet of things, privacy preserving computation, big data analytics, cloud computing, blockchain and edge computing, data-driven green computing, symbolic computing, swarm intelligence and evolutionary computing, intelligent systems for industry 4.0, as well as other pertinent methods for autonomic, data-driven, and industrial computing.

The series will publish monographs, edited volumes, textbooks and proceedings of important conferences, symposia and meetings in the field of autonomic, data-driven and industrial computing.

More information about this series at https://link.springer.com/bookseries/16624

Sudeep Tanwar

# Blockchain Technology

## From Theory to Practice

 Springer

Sudeep Tanwar
Department of Computer Science
and Engineering
Institute of Technology
Nirma University
Ahmedabad, India

ISSN 2730-6437             ISSN 2730-6445  (electronic)
Studies in Autonomic, Data-driven and Industrial Computing
ISBN 978-981-19-1490-4        ISBN 978-981-19-1488-1  (eBook)
https://doi.org/10.1007/978-981-19-1488-1

This Springer imprint is published by the registered company Springer Nature Singapore Pte Ltd.
The registered company address is: 152 Beach Road, #21-01/04 Gateway East, Singapore 189721,
Singapore

*This book is dedicated to my mom and dad, Mrs. Sneh Lata Tanwar and Late Sh Janardhan Tanwar, who gave me existence in this beautiful world; my siblings (Pradeep Tanwar and Sandeep Tanwar), who encourage me with admiration; my wife Anshul, who strengthens my soul with love and inspires me to always work toward perfection; and my sweet daughter Amodita and innocent son Divit, who enlighten me with joy every day.*

*Personally, from the core of my heart, I would like to dedicate this book to all members of ST Lab, especially Rajesh Gupta, Anuja Nair, Nilesh Jadav, Tejal Rathod, Riya Kakkar, Jigna Hethaliya, and Umesh Bodkhe, for their 24 by 7 support to convert this book into reality. Without their support, it would not have been possible for me to complete this book timely.*

*I am finally dedicating this book to Prof. (Dr.) Neeraj Kumar and Dr. Sudhanshu Tyagi for always motivating me to accelerate in all dimensions of research and development.*

*—Prof. (Dr.) Sudeep Tanwar, Ahmedabad, Gujarat, India*

# Preface

Blockchain is an emerging technology platform for developing decentralized applications and data storage, over and beyond its role as the technology underlying the cryptocurrencies. The basic tenet of this platform is that it allows one to create a distributed and replicated ledger of events, transactions, and data generated through various IT processes with strong cryptographic guarantees of tamper resistance, immutability, and verifiability. Furthermore, public blockchain platforms allow us to guarantee these properties with overwhelming probabilities even when untrusted users are participants of distributed applications with the ability to transact on the platform.

Even though blockchain technology has become popularly known because of its use in the implementation of Cryptocurrencies such as Bitcoin and Ethereum, the technology itself holds much more promise in various areas such as timestamping, logging of critical events in a system, recording of transactions, and trustworthy e-governance. Moreover, many researchers are working on use cases such as decentralized public-key infrastructure, self-sovereign identity management, registry maintenance, health record management, decentralized authentication, diamond industry, and telesurgery. Also, corporations like IBM and Microsoft are developing their applications in diverse fields such as the Internet of Things (IoT), even enabling blockchain platforms on the cloud.

Considering the need to disseminate the emerging concepts to students of UG/PG/Ph.D., researchers, and blockchain developers, I decided to write a complete reference book on blockchain technology: From Theory to Practice. The main benefits of this book to the readers are as follows:

- It introduces theoretical and implementation aspects of blockchain technology, which contains all the necessary material to become a blockchain technical expert.
- It includes an in-depth insight into the need for decentralization, smart contracts, consensus both permissioned and permissionless, and various blockchain development frameworks, tools, and platforms such as Ethereum, Bitcoin, and Hyperledger Fabric.

- It can be used as a textbook for courses related to blockchain technology and cryptocurrencies. It can also be used as a learning resource for various examinations and certifications related to cryptocurrency and blockchain technology.
- Almost every chapter covers an in-depth implementation of various concepts of blockchain technology.

This book is the first-ever 'how-to' guide addressing one of the most overlooked practical, methodological, and moral questions in any nations' journey to manage the trust among various stakeholders: What is a distributed technology, what are its properties, what it means to have consensus in a distributed system, an understanding of foundational consensus algorithms (e.g., PoW, PoB, PBFT, etc.), and why Nakamoto Consensus is a big revolution for smart applications, which have multi-party environment? How Blockchain can completely transform this landscape and what are the existing pieces of work who are working on this, what are they doing, and it gives you a sense of where we are and how some of the first production networks are coming out in this area. How to ensure scalability and computing efficiency? It differs from other published books alone on Blockchain Technology as it includes a detailed implementation on almost all aspects of Blockchain Technology and comparative case studies concerning various performance evaluation metrics, such as scalability, accessibility, reliability, heterogeneity, and Quality of Service (QoS) requirements. This book explained the nuts and bolts of blockchain technology in lucid language to make students more familiar with the implementation perspective of this much-needed technology.

This book is organized into fourteen chapters, and a brief description of each chapter is as follows:

Chapter 1 "*Introduction to Blockchain Technology*" discusses the basic fundamentals of blockchain technology that can bring trust and reliability in various business operations. It covers basic concepts of blockchain, the structure of blockchain, different types of blockchain, smart contracts, consensus mechanism, and the working of blockchain. Moreover, it briefly introduces all possible development tools and platforms to implement blockchain-based solutions.

Chapter 2 "*Blockchain Revolution from 1.0 to 5.0: Technological Perspective*" presents the historical perspective of blockchain technology ranging from 1.0 to 5.0. It discusses the different forms of blockchain with its motivation, benefits, and implementation to transform business operations in a trusted, safe, and secure environment. Finally, it shows a comparative analysis of different generations of blockchain.

Chapter 3 "*Decentralization and Architecture of Blockchain Technology*" describes the potential impact of decentralization in the current business environment and discusses decentralization methods, i.e., Disintermediation and Competition. Moreover, the adopted decentralization needs to be measured; therefore, we present a decentralization index that measures the correctness of the decentralization. Lastly, we have included some recent use cases which have incorporated decentralization into their business models.

Chapter 4 "*Basics of Cryptographic Primitives for Blockchain Development*" explores cryptography's preliminaries, which are the basic building blocks for blockchain development. It primarily covers different types of hash functions: SHA, RIPEMD, Scrypt, and Ethash, digital signatures comprising ECDSA, Schnorr, Multisignature, Ring signature, and encryption standards. Finally, this chapter shows the formal ECDSA-based secp256k1 implementation for bitcoin applications.

Chapter 5 *"Smart Contracts for Building Decentralized Applications"* discusses the basic concepts of smart contracts along with their pros and cons. The major focus of this chapter is on the implementation of smart contracts using Solidity language. This chapter covers basic primitives of Solidity language, such as arrays, functions, and structures. From the implementation point of view, this chapter presents various case studies, such as telesurgery, oil mining, banking, voting, and remote patient monitoring.

Chapter 6 *"Distributed Consensus for Permissionless Environment"* describes various consensus mechanisms with their merits and demerits. This chapter briefly discusses various consensus algorithms such as Proof of Work (PoW), Proof of Stake (PoS), Proof of Burn (PoB), Proof of Capacity (PoC), and Proof of Ownership (PoO) along with their comparative analysis, merits, and demerits. Moreover, it presents the python-based implementation of PoW, PoB, and PoO consensus algorithms.

Chapter 7 *"Mining Procedure in Distributed Consensus"* discusses the needs, technicalities, and procedures for mining a node in the blockchain network. It discusses how the incentive is distributed among the miners upon mining a block and the concept of a mining pool. From the implementation point of view, this chapter presents implementation details of the mining procedure for PoW with block mining and reward distribution.

Chapter 8 *"Distributed Consensus for Permissioned Blockchain"* focuses on identifying the practical aspect and the advantages and disadvantages of consensus protocols. The protocols are bisected into two, that is proof-based and voting-based algorithms, which aim to solve Byzantine faults. This chapter covers algorithms, such as PAXOS, RAFT, PBFT, DBFT, and FBFT. Furthermore, it shows the implementation of Leslie Lamport's algorithm for three general Byzantine problems and Proof of Authority (PoA).

Chapter 9 *"Consensus Scalability in Blockchain Network"* discusses the improvements in consensus mechanism in terms of scalability, such as the scalability of PoW consensus mechanism can be improved with Bitcoin-NG. This chapter discusses various scalable consensus mechanisms, such as Helix-based, RPoC reputation-based, and EBRC framework. Finally, this chapter highlights various bottlenecks of scaling the consensus mechanism.

Chapter 10 *"Building Trust in Blockchain Network Using Collective Signing"* discusses the implications of collective signing in the blockchain network, which eradicates the malicious to protect the authentic regulatory bodies. It highlights the architecture of collective signing (CoSi) and Byzcoin protocol in the multi-party environment to validate the digital signature and improve the blockchain's scalability.

Chapter 11 *"Adoption of Blockchain in Enterprise Computing"* discusses the basic concept of enterprise computing and also highlights the need for blockchain technology in it. From the implementation point of view, this chapter presents various case studies for enterprise computing, such as diamond supply chain, DigiLocker, transportation services, food delivery service, and restaurant management.

Chapter 12 *"Blockchain for Supply Chain Management"* describes blockchain's basic requirement and usage in supply chain management to provide security, transparency, and traceability in the network. It describes the basic research areas of adopting blockchain in supply chain management. It shows the implementation details of the food supply chain, wall mart industry, and drug logistics process utilizing the blockchain smart contract. Finally, it presents the different case studies of supply chain management with blockchain.

Chapter 13 *"Blockchain for Government Services"* presents the application of blockchain in government services such as VAT tax, payroll tax, transfer pricing, and income tax. Further, it explores the advantages of utilizing blockchain in government services. It discusses the different case studies of blockchain for taxation and land registry records and the GST process with and without blockchain. Finally, it shows the basic implementation of land registry records by implementing the blockchain smart contract.

Chapter 14 *"Impact of Blockchain on Academic Publishing"* discusses the potential of blockchain technology in open-access academic publishing. It highlights the benefits of using blockchain technology with its key technologies such as smart contracts, digital rights, cryptocurrencies, and identity management. It shows the different case studies to apply blockchain technology in open-access publishing. Finally, it shows the implementation details of blockchain using Solidity in open-access publishing and the different challenges associated with it.

The author is very thankful to all the members of Springer, especially Mr. Aninda Bose, for the opportunity to write a book focused on various implementation perspectives of Blockchain Technology.

Ahmedabad, India                                                                    Dr. Sudeep Tanwar

# Acknowledgments

Writing an authored book is more challenging than I thought and more rewarding than ever imagined. None of this would have been possible without the support of members of ST Lab; Rajesh Gupta, Anuja Nair, Nilesh Jadav, Tejal Rathod, Riya Kakkar, Jigna Hethaliya, and Umesh Bodkhe. Without their support, it would not have been possible for me to write this acknowledgment.

I would like to acknowledge and express my thanks to all M.Tech (Computer Science and Engineering/Data Science/Information and Network Security) batch 2019–2021 students for their contributions in various chapters; Avani Jain and Shubhangi Singh (Chap. 1), Samprat Bhavsar and Kshitij Deshmukh (Chap. 2), Palak Dixit and Aneri Mehta (Chap. 3), Hardikkumar Dave and Krish Shah (Chap. 4), Aarti Popat and Aneri Acharya (Chap. 5), Dakshita Reebadiya and Niyati Shah (Chap. 6), Devanshi Jain and Vishakha Jambekar (Chap. 7), Khara Parshwa and Raj Yadav (Chap. 8), Maharishi Mahadevia and Zankhana Patel (Chap. 9), Jigar Bhatt and Dharmil Shah, (Chap. 10) Farnazbanu Patel and Divya Shah (Chap. 11), Aneri Shah and Shivani Patel (Chap. 12), Kishan Vaghela and Aatrey Vyas (Chap. 13), and Yash Velankar (Chap. 14)

To everyone at Nirma University who enables me to be the Full Professor that I'm honored to be a part of, thank you for letting me serve, for being a part of our amazing University, and for showing up every day and helping more faculty members to turn their ideas into reality. A very special thanks to Prof. (Dr.) Anup K Singh, Director General, Nirma University, Ahmedabad, India, Prof. (Dr.) Madhuri Bahvsar, Head of the Department, Computer Science and Engineering, and Prof. (Dr.) R. N. Patel, Director, Institute of Technology, Nirma University, Ahmedabad, India, for providing all necessary support to complete this book.

To my family. To Anshul (my wife): for always being the person I could turn to during those dark and desperate years. She sustained me in ways that I never knew that I needed. To my big brothers, Pradeep and Sandeep, and kids, Amodita and Divit: thank you for letting me know that you had nothing but great memories of me. So thankful to have you all in my life.

I want to acknowledge my motivational mentors/friends, Prof. (Dr.) Neeraj Kumar and Dr. Sudhanshu Tyagi, who stood by me during every struggle and success. That is true friendship.

Finally, to the Springer team: Aninda Bose and Silky for giving me this opportunity to write this authored book on the much-needed title, i.e., Blockchain Technology with more focus on implementation.

Ahmedabad, Gujarat, India                                       Prof. (Dr.) Sudeep Tanwar

# Contents

# About the Author

**Sudeep Tanwar** (Senior Member, IEEE) is working as a full professor at the Nirma University, India. He is also a Visiting Professor with Jan Wyzykowski University, Poland, and the University of Pitesti, Romania. He received B. Tech in 2002 from Kurukshetra University, India, M.Tech (Honor's) in 2009 from Guru Gobind Singh Indraprastha University, Delhi, India, and Ph.D. in 2016 with specialization in Wireless Sensor Network. He has authored 04 books and edited 20 books, more than 270 technical articles, including top-cited journals and conferences, such as IEEE TNSE, IEEE TVT, IEEE TII, IEEE TGCN, IEEE TCSC, IEEE IoTJ, IEEE NETWORKS, ICC, IWCMC, GLOBECOM, CITS, and INFOCOM. He initiated the research field of blockchain technology adoption in various verticals, in 2017. His H-index is 51. His research interests include blockchain technology, wireless sensor networks, fog computing, smart grid, and the IoT. He is a member of the Technical Committee on Tactile Internet of IEEE Communication Society. He has been awarded the Best Research Paper Awards from IEEE IWCMC-2021, IEEE ICCCA-2021, IEEE GLOBECOM 2018, IEEE ICC 2019, and Springer ICRIC-2019. He has won Dr. KW Wong Annual Best Paper Prize for 2021 sponsored by Elsevier (publishers of JISA). He has served many international conferences as a member of the Organizing Committee, such as the Publication Chair for FTNCT-2020, ICCIC 2020, and WiMob2019, and a General Chair for IC4S 2019, 2020, ICCSDF 2020, FTNCT 2021. He is also serving the editorial boards of *COMCOM-Elsevier, IJCS-Wiley, Cyber Security and Applications- Elsevier, Frontiers of blockchain*, and *SPY, Wiley*. He is also leading the ST Research Laboratory, where group members are working on the latest cutting-edge technologies.

# Chapter 1
# Introduction to Blockchain Technology

**Abstract** Blockchain is a modern technology that brings trust and reliability to various business operations. This is because it has concrete characteristics such as decentralized, immutable ledger, and cryptographic solutions. Moreover, various critical business operations need security and privacy solutions and they rely on blockchain technology to provide a feasible solution to alleviate security risks. However, it is challenging to decide which type of blockchain one should adopt to encounter security threats. A blockchain learner needs to know the prerequisites and collaborative technologies involved in implementing the blockchain network. Taking care of the aforementioned objectives, this chapter convey the fundamentals of blockchain technology, where it briefly discusses the blockchain and its architecture, along with its needs and features. Additionally, this chapter has given comprehensive details on the implementation platforms used to develop blockchain-based solutions.

**Keywords** Blockchain technology · Blockchain structure · Need for blockchain · Types of blockchain

## 1.1 Introduction

In today's era, organizations are expanding their business which involves various participants such as consumers, vendors, and intermediary parties, i.e., a bank can be a mediator for the particular transaction between participants. Organizations generally use a ledger, i.e., database, to store transactions. But, a transaction in the ledger can be altered by an intruder who can easily manipulate the data, which raises security concerns [13]. Thus, blockchain as a distributed ledger technology can be used to store transactions securely with tamper resistance. Blockchain can be defined as a peer-to-peer (P2P) network, which consists of multiple numbers of nodes linked with each other in the form of blocks in a distributed manner, it can be used for carrying out transactions and storing records securely using digital signatures, distributed ledger technology, and cryptography. [2]. It has been adopted by numerous industries that makes blockchain a cutting-edge technology. Industries such as banking systems, corporate world, insurance

© The Author(s), under exclusive license to Springer Nature Singapore Pte Ltd. 2022     1
S. Tanwar, *Blockchain Technology*, Studies in Autonomic, Data-driven and Industrial Computing, https://doi.org/10.1007/978-981-19-1488-1_1

companies, healthcare [3], and other financial and non-financial sectors are under-going massive digital transformation technology by leveraging the benefits of blockchain to store transactions in the form of cryptocurrency such as bitcoin [4]. In 2008, an anonymous researcher named Satoshi Nakamoto was the first who talked about the cryptocurrency concept, i.e., bitcoin and blockchain technology. After that, the blockchain technology is evolved rapidly worldwide [5]. Cryptocurrency, i.e., bitcoin can be used to record and transfer transactions with the help of blockchain technology. Also, corporations like IBM and Microsoft develop their applications based on the Internet of things (IoT), enabling blockchain platforms on the cloud [1].

Blockchain maintains the trust between the participant nodes of the network without the involvement of any centralized authority. It protects the stored transactions from tampering so that the data can not be altered or changed in the blockchain [6]. The security among the involved participants in the blockchain network can be established using a consensus mechanism. The consensus mechanism signifies how data can be verified and added to the blockchain network. Recently, Samsung and IBM revealed that they use Proof of Concept (consensus mechanism) as a backbone for their application. As the industry continues to grow, blockchain is going to be a promising investment for future returns. Before discussing the blockchain in detail, a few concepts, such as distributed systems and centralized systems, need to be discussed.

### 1.1.1 Distributed Systems

With the growing advancement in technology, more and more computers need to communicate and connect through a high-speed network leading to the requirement of a distributed system. A distributed system, also known as distributed comput-ing, is a system in which independent computers at different locations communicate with each other and coordinate their work so that they appear as a single entity to the end-user. These autonomous components, as shown in Fig. 1.1 collaborate and make a single coherent system. Nodes of the distributed system may be computers, containers, virtual machines, or physical servers, and these nodes can interact with each other with the help of message passing [7]. The distributed system aims to make resources available to all nodes and achieve transparency by providing multi-ple nodes in front of its users while hiding their details of location, access, migration, etc. It is abstracted from the users how these computers take part in the system as a whole; it means the users do not know the internal working of the system. Distributed systems are just the opposite of centralized systems; as in distributed systems, there is no central authority to control any of the nodes; everyone collectively executes a job/process [8].

Distributed systems are categorized into four categories based on their architec-ture. These are client–server architecture, three-tier, n-tier, and peer-to-peer archi-tecture. The benefits of the distributed systems are that they are scalable. According to the requirement, more nodes can be added to the system, which further enhances

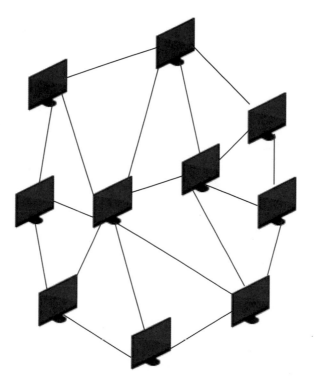

**Fig. 1.1** Distributed systems

the system's performance. It also protects the system against a single point of failure, i.e., the failure of any node does not affect the entire system's performance. But some challenges need to be focused on implementing distributed systems. One of the challenges is security, especially in public networks, where an intruder can attack to leak any confidential information. The second is coordination between nodes, as the data is being transferred between nodes simultaneously, which can overburden the system and also can be the reason for data loss in the network without any coordination [9]. So, we require a system that can overcome these security issues and tackle a huge number of nodes interacting with each other in the system without any conflict and data loss.

### 1.1.2 Decentralized Systems

Decentralized systems consist of several nodes interacting with other peers in the interconnected network. Figure 1.2 shows a decentralized system where each node makes its own decision for sending data to other nodes instead of the centralized authority in the network. It means that decision of one node does not depend on the other node, but there can be some conflict between peers in the network as nodes can send data to each other with different goals, which can affect the performance of the system. For example, users can use Bitcoin to transfer money to others without

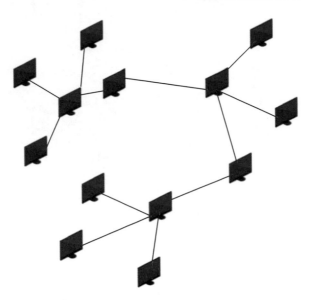

**Fig. 1.2** Decentralized systems

a centralized authority, i.e., Bank. But, in an open channel decentralized system, a user can access or leave the system at any time based on requirements. But, it can be a threat to the system's security as any malicious attacker can intrude into the system and manipulate the message that is being sent to others. Therefore, we need a system to resolve distributed and decentralized systems' security issues. That is the reason, nowadays, blockchain has been used worldwide in various industrial sectors to provide security to users in the network [11]. The following section discuss blockchain concept in detail, initiating with its basics.

## 1.2  What Is Blockchain?

Over the last few years, blockchain has completely transformed the way industries and banking sectors work by securely storing confidential data and enabling digital data transactions. As discussed earlier, digital cryptocurrency such as bitcoin utilizes blockchain technology to store and transfer transactions securely. Figure 1.3 shows the blockchain as a decentralized (distributed) open ledger network that records the transactions of nodes, i.e., a list of records called blocks linked with each other in a decentralized way [10]. The record of each transaction is transparent to all participants in the network. It means any user can request to add their data to the blockchain, ensuring transparency in the system. The primary function of blockchain is to certify the user's identity to store and share the transactions securely in a distributed network.

In traditional transaction systems, there is an involvement of centralized system such as a banking system that controls all transactions having user's confidential

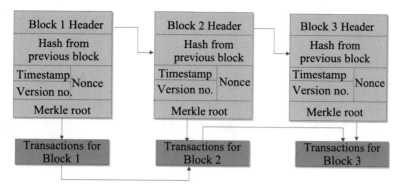

**Fig. 1.3** Decentralized systems

information. But, in blockchain technology, there is no central authority system. It works on a decentralized system where all transactions are stored in a ledger which is immutable and transparent to all. We can understand it with an example, i.e., in Google Sheets, multiple users can simultaneously edit the document. But, the environment provided for users to edit or share the document is centralized, which can be affected by a single point of failure. There is also security concern associated with the centralized systems that can be mitigated using a decentralized and immutable system. All nodes participate in the network without any involvement of centralized authority. So, to resolve these issues, we need a decentralized and distributed ledger with consistency, security, data integrity, and transparency, which is provided by blockchain technology [13].

Blockchain technology maintains the integrity of transactions in the network using cryptographic algorithms. Further, a consensus mechanism is used in the blockchain to ensure security in the system. It works on the principle that all the nodes in the network agree on the same decision regarding the data requested to add to the blockchain [14]. It means if there are some faulty nodes, also called Byzantine nodes, present in the blockchain network intruding to modify the data, then using the consensus mechanism, byzantine nodes can be removed, providing the system with Byzantine Fault Tolerance (BFT), which further lessens the probability of failure in the system. There is one more technical concept, i.e., smart contract, which can be defined as executable codes stored in the blockchain to verify and sign the transactions digitally for verifiability in the system. Unlike traditional systems using paper contracts to sign the contract between participants, blockchain uses digital contracts to verify the authenticity of transactions, which further improves the security of the system [15]. Some of the open-source blockchain tools, such as Ethereum, Geth, and Node.js can be used to execute any transaction between users with the deployment of smart contracts. In 2013, Vitalik Buterin first coined the term Ethereum after the release of Bitcoin to deploy blockchain-based applications with security due to the execution of smart contracts in the blockchain network, which has been used worldwide within many organizations and institutions [16]. So, the blockchain consists of various key technologies such as cryptographic hash functions, consensus mechanisms, and smart contracts to make the network secure, transparent, and Byzantine fault-tolerant. These key technologies will be discussed in detail in Sect. 1.4 after discussing the basic structure of blockchain and what it constitutes, discussed in Sect. 1.3.

## 1.3   Structure of Blockchain

Blockchain can be structured using a number of blocks in which nodes authenticate the transactions that need to be stored in the block. Figure 1.3 shows that the blockchain comprises blocks, which further consists of the block header and transactions for that particular block. The block header contains various fields such as hash from the previous block, Timestamp, Version no., Nonce, and Merkle root. Blockchain Technology can be defined using three crucial concepts, which are blocks, nodes, and miners. Miners are basically participants in the blockchain network whose main task is to verify the new arriving transactions, and they get the reward to validate the transaction that needs to be added to the blockchain [17, 18]. Miners solve a complex cryptographic puzzle to mine the new transactions in blocks that can be further appended to the blockchain [19]. So, these fields of the block header can be defined as follows:

- **Previous block hash function**- It contains 32 bytes of block header. This field contains the hash of the previous block, which can be computed using the block header of the previous block using the cryptographic hash function. Double SHA-256 cryptographic algorithm is used for generating hash value.
- **Timestamp**- It contains 4 bytes of block header. It indicates the time for the creation of the block, more accurately the time when the miner started hashing the header.
- **Version no.**- It contains 4 bytes of block header. Version no. denotes the version a particular block is utilizing. Blocks are linked with each other using different blockchain versions such as Blockchain version 1.0 for storing the cryptocurrencies, Blockchain version 2.0 is used to execute smart contracts, i.e., Ethereum, Blockchain version 3.0 for a decentralized entity, and Blockchain version 4.0 to enable a scalable and transparent network for industrial purpose [20].
- **Nonce**- It also contains 4 bytes of block header. It is basically defined as a number used only once. Nonce is a random number that miners are computing to validate the transactions. So miners have to randomly find the number, i.e., an integer between 0 and $2^{32}$ as 4 bytes of nonce can have a maximum of 32 bits [20].
- **Merkle root**- Figure 1.4 shows a Merkle tree built using hash values of various transactions. Hash values of all the transactions are combined to generate a Merkle root, which can be called as the root of the Merkle tree. Merkle root reduces the computation time while verifying the transactions as even a slight change in the

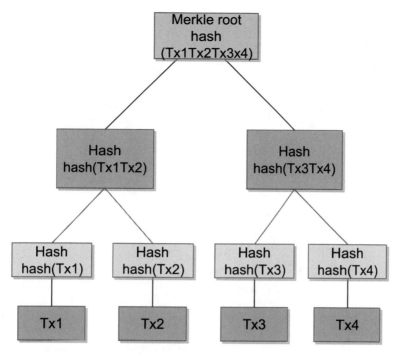

**Fig. 1.4** Merkle tree

transaction can be reflected in the Merkle root, which reduces the effort of verifying all the transactions in the blockchain.

As discussed earlier, blockchain can be classified using blocks, nodes, and miners. These terms interrelate with each other to validate the transactions in the blockchain network.

### 1.3.1   Types of Blocks in Blockchain Network

Blocks in the blockchain network can be classified as follows [21]:

– **Genesis Block**- The initial block of any blockchain network is known as the genesis block or Block 0. In 2009, an unidentified person named Satoshi Nakamoto created the Genesis Block, i.e., the first block of the blockchain, while launching the bitcoin. Genesis block initiates the blocks, which further connect to a number of blocks extending the number of transactions in the blockchain network. The Merkle root in the Merkle tree of block header ensures that all the transactions associated with their hashes remain validated throughout the network as any modification in the transaction reflects in the Merkle root, which further notifies the blockchain

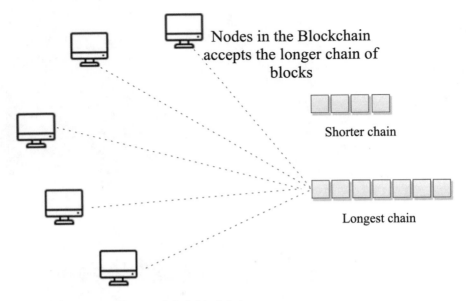

**Fig. 1.5** Nodes accept longest chain in blockchain

about the manipulation of transactions in the blockchain to ensure the security in the network.

– **Valid Block**- Valid blocks are already mined blocks by the miners added to the blocks in the blockchain. It means blocks have been appended to the blockchain by solving complex mathematical cryptographic hash algorithms. Valid blocks are added to the network by satisfying the criteria of the consensus mechanism, further improving the security of the network.

– **Orphan Block**- The consensus mechanism verifies the blocks whether they can be appended to the blockchain or not. According to the consensus, if all the nodes in the network agree to the addition of a new block, then they are the valid blocks, but the rest of the blocks are orphan blocks. There can also be a scenario where a number of miners in the network are competing to get the reward for mining the valid blocks simultaneously, or there can be the involvement of any malicious attacker to add the invalid block. Still, the blockchain only accepts the valid block or the longest chain of blocks, as shown in Fig. 1.5 making it secure and scalable for the network, and invalid blocks can be declared as orphan blocks.

### 1.3.2  Types of Blockchain Nodes

All the computers expanded across the blockchain network are known as nodes. Figure 1.6 shows different types of nodes involved in the blockchain technology.

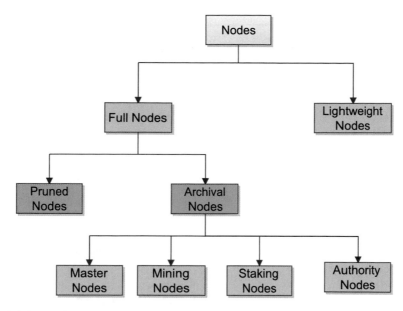

**Fig. 1.6**  Types of blockchain nodes

Every time a request to add a new transaction is issued, it must be approved by all the participating nodes by checking the validity of the transaction. Once all nodes validate the transaction, voting is done to check every node's decision about the transaction, as it may be valid or fraudulent. Each node has a copy of the digital ledger; if the majority of the nodes say that it is a valid transaction, then it is added to the blockchain network; otherwise, it is rejected [23]. There are mainly two types of nodes Full nodes and Lightweight nodes. Full nodes can be further classified into Pruned nodes and Archival nodes. The following nodes can be classified as follows [22]:

**Full Nodes**: Full nodes play an essential role in maintaining the transactions of the blockchain. It keeps track of all the transaction records, and if there is a new transaction or any modification to the blockchain, then full nodes should agree to the modifications to reflect the changes in the blockchain. But, there can be a scenario that even if the majority of the full nodes agree to any new transactions in the blockchain, then due to some conflicting nodes whole blockchain may not work according to the proposed changes. In this case, the blockchain will be split into two blockchains in which a modified blockchain can work according to the agreement of full nodes and another blockchain, i.e., due to disagreement of conflicting nodes can work in the same old manner only. Full nodes can be further classified into pruned nodes and archival nodes, which can be described as follows:

– **Pruned Nodes**- Pruned nodes are full nodes which have limited memory storage to maintain the transactions in blocks. Although, there is no limit to add blocks to the blockchain satisfying the consensus mechanism, these nodes can't hold the blocks exceeding the limit of their storage. So, if the request to add blocks to the blockchain distributed ledger exceeds, then pruned nodes consider and accept the new blocks by deleting the earliest blocks except the genesis block, but the important meta-data of the earliest blocks should not be deleted ensuring the working of blockchain in the standard manner.

– **Archival Nodes**- Archived nodes work in a similar way as full nodes that contain all the transaction records which should satisfy the criteria of the consensus mechanism. But, there is a difference that archived nodes also contain the oldest blocks, i.e., all the previous blocks of the blockchain, but these blocks are not considered in the case of pruned nodes. Therefore, archival nodes utilize more space of blockchain than pruned nodes.

1. **Master Nodes**- Master nodes can be defined as the type of archival full nodes that do not have any authority to append data blocks to the blockchain. The main task of these nodes is to validate the transactions and keep track of these transaction records.

2. **Mining Nodes**- As discussed earlier, the miners in the blockchain compete to mine or validate the transactions by solving the mathematical cryptographic puzzle fulfilling the consensus mechanism for security purposes. So, the nodes required to compute this complex puzzle by mining the blocks of valid transactions are known as mining nodes.

3. **Staking Nodes**- Similar to mining nodes, staking nodes mainly focus to satisfy the consensus mechanism in which all participating nodes should agree to the same decision for adding the new transaction to the blockchain. For staking nodes, a consensus mechanism, i.e., Proof-of-Stake is used so that authentic transactions can be appended to the blockchain, which will be discussed in further chapters.

4. **Authority Nodes**- In the open decentralized blockchain network, any users can access the network to add their transactions, but there can be some malicious or faulty nodes trying to access the network that can be a threat to the security of the network. Due to these security issues, authority nodes are required to authenticate only validated transactions for the blockchain network.

**Lightweight Nodes** Lightweight nodes accept data blocks of requisite and essential information, i.e., they only keep track of block headers of blockchain. So, if there is a request to access the block that is not considered by the lightweight nodes for faster transactions, then data can only be retrieved by considering the whole blockchain. Lightweight nodes facilitate faster transactions without considering the complete node, making it less flexible and reliable [24]. In Sect. 1.3, we have covered and explained the structure of blockchain, types of blocks, and types of blockchain nodes to give readers a better understanding of the basic constitutes of blockchain. Now, Sect. 1.4 describes the key technologies used in the blockchain network.

## 1.4  Key Technologies of Blockchain

Several technologies are associated with the blockchain to add transactions to the network securely. As we already discussed earlier, blockchain provides the secure storage of transaction records in a decentralized and secure manner. But, to achieve this immutability in the network, we need to consider several technologies associated with the blockchain, which first is cryptographic hash functions, which can be described as follows:

### *1.4.1  Cryptographic Hash Functions*

Blockchain uses cryptographic hash functions to preserve and secure transaction storage and sharing in the network. It means generated hash of the data has to pass through a cryptographic hash algorithm to yield an output that cannot be altered or tampered with by any malicious activity, which leads to a process known as hashing. Hashing takes variable length data of block as input and produces the different hashes of the fixed size. For example, M is the message considered as input for hash function $H_f$, which generates the hash value as output, i.e., $h_v = H_f(M)$ of fixed length. For blockchain, different types of cryptographic hash algorithms such as SHA-256, Message Digest algorithm (MD5), and many more can be used to generate the hash by passing the variable size data as an input [25]. Figure 1.7 shows a data block considered as input, which contains a block header and several transactions. For example, we want to validate the transactions before adding them to the blockchain network. For that, a cryptographic hash algorithm can be applied to the data block to get the output in the form of a fixed hash size for the particular transaction. So, if there is any modification in the input, change will be reflected in the hash value, which is known as the avalanche effect. Now, we need to discuss several characteristics of hash functions to ensure the security in the blockchain network, which can be explained as follows [26]:

**Fig. 1.7**  Hashing

– **Deterministic**- The deterministic property of the hash function ensures that for the same input data, the output in the form of hash value should be the same no matter how many times input data of block has been passed through the cryptographic hash function.
– **Fast Computation**- The whole process of hashing should be quick enough to get the desired hash value of data as output; otherwise, delay in getting the desired output on applying the cryptographic hash function will not be efficient for the security of the blockchain network.
– **Feasibility**- The feasibility property of the hash function states that once the hash value of data has been obtained from the input data, the user can't determine the original data using the output hash value. This property ensures the security in the network as the original data is secure after getting its hash value as the malicious attacker can't get knowledge about the original data.

We have discussed how blockchain uses cryptographic hash functions to secure the data storage in the network. But, to achieve this security, we need to focus on different applications of cryptographic hash functions to enhance and ensure the confidentiality and authenticity in the blockchain network, which can be achieved with the help of message authentication, digital signature, and one-way password file. Therefore, the following section discusses about the applications of cryptographic hash functions.

## *1.4.2   Message Authentication*

Message authentication is a mechanism used to verify the integrity of data in the blockchain network. It ensures the authenticity of data, as data received at the recipient's side should be the same as that sent from the sender's side. Data should not be manipulated or modified while sending it to the recipient. So, the cryptographic hash function facilitates message authentication in which the output hash value for the input is often termed as a message digest. Message authentication is achieved using a Message Authentication Code (MAC), which can be explained as follows:

– **Message Authentication Code**- MAC is a keyed hash function that is used between two parties, i.e., sender and receiver by sharing a secret key to authenticate information exchanges between them. MAC function works in the same way as the hash function, but it uses a secret key additionally to authenticate the message. Figure 1.8 shows the complete procedure for message authentication using MAC, which can be discussed in the following steps [27]:

  1. MAC is an algorithm that takes as input a message and a secret key and produces a fixed-sized output, i.e., MAC value, similar to the hash function, which can be later on verified to match the message.
  2. The message incorporated with the generated MAC is sent to the recipient's side to verify the authenticity of the sender's message.

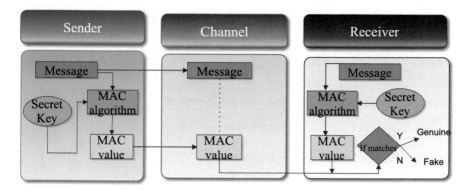

**Fig. 1.8** Procedure of MAC generation for message authentication

3. On receiving the message with MAC, the sender's message with the same shared secret key is passed into the MAC algorithm at the recipient's side to determine the MAC value. If the MAC value at the recipient's side matches with the MAC value at the sender's side, the message is authenticated, and the receiver accepts the message due to its authenticity computed using MAC.
4. But, if the computed message at the receiver's side does not match with the message at the sender's side, then there has been some modification in the original message sent by the sender, so the receiver rejects the message ensuring the authenticity in the network.
5. A good MAC provides blockchain network with a property of unforgeability, i.e., it should be infeasible to compute a pair message+MAC value which successfully verifies with a given secret key without knowing the key exactly and in its entirety.

**Digital Signature** In the real world, we use signatures to verify someone's identity; for example, if some person requests to access confidential information, access can be granted if that person's signature matches with the signature signed for that confidential information. Similarly, a digital signature associates person with a particular information digitally. A digital signature is a mathematical technique used to validate the authenticity and integrity of a message, software, or digital document. The operation of the digital signature is similar to that of the MAC. In the case of the digital signature, the hash value of a message is encrypted with a user's private key. Anyone who knows the user's public key can verify the integrity of the message that is associated with the digital signature [28]. Figure 1.9 shows the sequence to authenticate the message using digital signature based on the public-key encryption, which can be explained in the following steps [29]:

– The authentication of message exchange using a digital signature between sender and receiver involves a public key and private key.
– The message as an input is passed to the cryptographic hash function, which yields an output hash value at the sender's side.

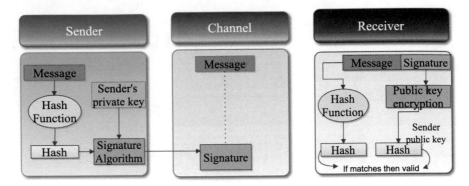

**Fig. 1.9** Message authentication using digital signature

- Then, the output hash value at the sender's side is encrypted with the sender's private key using a digital signature, so that anyone who knows the sender's public key can validate the authenticity of the message.
- At the receiver's side, the message sent from the sender is again passed through the hash function to check the authenticity of the message.
- For that, the hash generated at the receiver's side is verified with the sender's public key and if it matches with the hash value of the message, then the message is originally sent by the sender and the receiver can decrypt the message with the help of sender's private key.

### 1.4.3  Smart Contract

Traditionally, a paper contract was involved between users to initiate the transaction via a centralized party, i.e., a bank or any government institution. But, it tends to pose various security concerns for users as the transaction can be modified or altered by the third-party systems. As discussed in Sect. 1.2, a computer scientist and cryptographer, Nick Szabo, proposed the term smart contract to perform transactions between users without any intermediary or centralized authority. For that, smart contracts as shown in Fig. 1.10, which consist of self-executable code, need to be executed with the blockchain technology, i.e., on an open-source blockchain platform often called Ethereum [12]. Suppose the users fulfill predefined agreements or conditions mentioned in the code. In that case, smart contracts authenticate them to initiate the transaction and mask the user's identity providing security in the network [30].

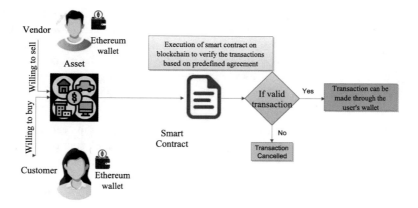

**Fig. 1.10**   Transaction between users using smart contract

## *1.4.4   Consensus Mechanism*

Blockchain is a distributed decentralized network with no central authority to check or validate transactions. It provides immutability, transparency, security, and privacy which are possible due to consensus protocol. The process through which every node present in the network comes to a common accord about which transactions are valid and will be added to the blockchain is known as consensus algorithm. If the consensus mechanisms used in blockchain are not suitable, they are at risk of various attacks. Consensus can be reached in many ways, and some of them are explained as follows [37]:

– **Proof of Work**- The beginning of the blockchain consensus mechanism was with Proof of Work (POW), used by Bitcoin. After that, many cryptocurrencies adopted this consensus mechanism. In this process, known as mining, there are miners (nodes) who find the solution to complex mathematical puzzles. A lot of computational power is required to solve these mathematical puzzles. The first node to solve this puzzle gets a chance to add their block and receives an award. These puzzles are asymmetric, i.e., takes a lot of time, but the answer can be verified easily. These puzzles are solved using the trial and error method, and to solve quickly, they need more computational power.

– **Proof of Stake**- Proof of Stake (POS) can be seen as the most common substitute for Proof of Work. In PoS, instead of investing in costly hardware to solve mathematical puzzles, the validator invests in coins. They lock some of their coins as stakes. The process determines who gets to produce the next block is randomized. Generally, the people who have the biggest stake have higher chances to create the next block. Another factor can be how much time the coin has been staked. Here, the reward is given in return for the work either all or a part of transaction fees. PoS provides enhanced privacy and preservation to users as executing an attack is expensive and also it is much more energy efficient than POW.

- **Proof of Capacity**- The process used in Proof of Capacity (PoC) consensus mechanism is known as plotting. In Proof of Work, high computing power is used to solve puzzles; however, in PoC, the solutions are prestored in the hard drive. After the hard disk is filled with solutions, that miner can take part in the process of creating the next block. The one with the fastest solution gets to create the next block. If you have more storage, then it is easy to store more solutions and your chances of creating a block are higher.

- **Proof of Burn**- In Proof of Burn (POB) consensus mechanism, instead of spending on expensive hardware, miners should show a proof that they sent some coins to a verifiably unspendable address from where the coins are irretrievable. This is known as the burning of coins. By doing this, a miner earns a privilege to mine on the system. The selection process is random. The miner can burn native coins of blockchain, or they can burn coins of an alternate chain. The more you burn coins, the more are your chances of mining the next block. POB is power efficient as you burn virtual currency and consume virtual resources.

- **Proof of Elapsed Time**- Proof of Elapsed Time (PoET) consensus mechanism fairly decides which miner will get to create the next block. The decision depends upon the waiting period of the validator. Random waiting time is assigned to every node and the one whose wait period completes first gets to create the next block. The algorithm has additional checks to stop nodes from consistently winning the election. In PoET, every miner gets a fair chance to create their block. In the next chapters, consensus mechanisms will be discussed in detail.

## 1.5   How Blockchain Works?

In the previous sections, we have discussed the basics of blockchain and its key technologies. Then, we have mentioned the different consensus mechanisms that can be adapted in various applications of blockchain [38]. Now, we can understand the basic working of blockchain in the following steps to add validated transactions to the blockchain, illustrated in Fig. 1.11.

- Step 1- We can understand the working of blockchain, considering an example of a transaction in which user A wants to transfer some money into user B's account.
- Step 2- Before sending the money to the receiver, the transaction needs to be authenticated using digital signature and encryption techniques using the sender's private key.
- Step 3- Then, the authenticated transaction with the transaction fee is broadcast to all the other nodes of the P2P network.
- Step 4- After that, these transactions should be verified by all the nodes in the network, which is known as the mining mechanism.
- Step 5- Miners in the network compete to solve a complex mathematical puzzle. The miner who solves that puzzle first can validate the transaction, and in return, they get the transaction fee as a reward.

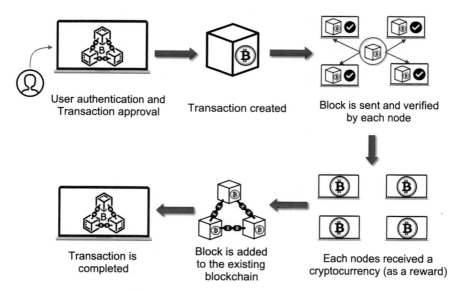

**Fig. 1.11** Working of blockchain

- Step 6- Once a transaction is verified, all the participants of the network have to come to an agreement regarding which block to append to the network. This is where the consensus algorithm can be utilized. A consensus algorithm ensures that all the users in the network are on the same stage, meaning that they all agree on which block to pick up next as the miners create many blocks.
- Step 7- When the blockchain accepts the block added to the network, it confirms that the block is verified, and user B receives the money sent by user A securely.

## 1.6 Features of Blockchain

Blockchain evolved with several advantages for industries that need to be discussed in detail.

### 1.6.1 Peer-to-Peer Network

P2P networks consist of a group of connected computers with equal permissions and responsibilities for processing the data. There is no centralized server that the whole system relies upon. All the peers/nodes are equally privileged to participate in the network. The main usage of the P2P system is file sharing. If we use a traditional client–server system, it becomes prolonged to download a file, and it depends on a central server. P2P does not depend upon the centralized authority, i.e., if one node fails, then the whole system will not crash, but in a client–server system, fault in one node becomes a failure of the complete system. The P2P feature of blockchain is used in a consensus mechanism where all nodes have the same privileges for getting cryptocurrencies as there is no single governing body in the network [39].

### 1.6.2 Distributed Ledger

It is a public ledger shared among all the nodes of the blockchain, which resides in different geographical locations of the world. The distributed ledger contains many transactions or contracts maintained in a decentralized form. All the information in a ledger is cryptographically secure using cryptographic signatures and hash keys. Every node of the blockchain maintains a copy of the ledger. Every local copy of the ledger is the same for all the nodes, ensuring the consistency of information and maintaining trust between the nodes. All local copies are always updated based on global information. Once the information is stored in the ledger, it becomes an immutable database [35]. This feature of the ledger makes the blocks hard to attack by attackers. If an attacker wants to modify the transaction, it must simultaneously attack all the distributed copies. If one wants to make changes in the transaction, it needs to generate one more transaction for change but cannot modify the previous one. The ledger includes all the historical information which may be used for future computation. Blockchain works as a distributed ledger technology, while bitcoin works by utilizing blockchain technology. Let's consider a scenario. Consider a blockchain network involving five people, namely A, B, C, D, and E. If there is an occurrence of a transaction between person A and person B, then this transaction reflects on the ledger of all the five persons of the network, whether they are involved in the transaction or not. All the persons have the knowledge of every transaction on the network; it maintains the transparency of transactions between all the nodes. Distributed ledger systems are used in various industries such as finance, music, entertainment, artwork, supply chain of multiple commodities, and many more [40].

### 1.6.3 Cryptographically Secure

Each transaction in a block is cryptographically secured by generating a digital signature using cryptographic hash algorithms. The transaction is transparent to all the participating nodes in the network, but they do not have any knowledge about the user's identity and confidential information. The transaction is secured using a digital signature, which combines private and public cryptographic keys. Blockchain mainly uses the SHA-256 algorithm to convert data transactions into hash values making it difficult for an attacker to find the original data from the generated hash value. So, a combination of hash algorithms and digital signatures ensures the security and integrity of data transactions in the network [41].

### 1.6.4  Append-only

The blockchain uses an append-only data structure. It means that only validated blocks can be added to the order based on their timestamp order. The addition of blocks should be performed based on the sequential order in which the blockchain has verified them. Also, each block in the network should refer to the previous block's hash. Append-only feature of the blockchain ensures that once the block is added to the network, it can't be altered. If some malicious attacker tries to access or modify the data blocks, then the hash value of that corresponding block will be changed, interfering with the basic working of a chain of blocks in blockchain network [42].

## 1.7  Growth of Blockchain Technology

Blockchain technology has evolved tremendously since the invention of bitcoin by Satoshi Nakamoto in 2008. Before the advancements in Blockchain technology, many researchers discussed about the previous generation of blockchain and their limitations, which lead to the evolution of the latest era of Blockchain 5.0. The newest era of blockchain is supposed to have a huge impact on every industry, which includes the financial sector, media, government, law, etc. in terms of security, reliability, and transparency. The evolution of blockchain can be explained by considering the following generations of blockchain.

### 1.7.1  Blockchain 1.0: Bitcoin

Blockchain 1.0 as a first-generation technology evolved from the term Digital Ledger Technology (DLT), which provides all the participants with a distributed network to resolve the issue of double spending. Double-spending problem can cause disruption in the network if some users trade with the same currency more than once. Thus, in early 2008, Satoshi Nakamoto released Bitcoin, which utilizes Blockchain technology to store and share the data transactions among multiple users with security, verifiability, and efficiency. Blockchain technology uses various cryptography and consensus mechanisms to discard the centralized network and facilitate users in consensus to validate the transactions which further keeps data records safe and protected. In Blockchain 1.0, there is a mining mechanism involved to mine or validate the data transactions to resolve the security issues of centralized systems [43].

### 1.7.2   Blockchain 2.0: Smart Contracts

Blockchain 2.0 as a second-generation technology emerges, combining several technologies such as smart contracts with PoW consensus mechanism to resolve the issues of Blockchain 1.0. A smart contract can be defined as a self-executable code running on the blockchain based on predetermined conditions or rules. One of the main advantages of a smart contract is that it cannot be tampered with or altered, i.e., it is impossible for a malicious attacker to manipulate a smart contract. This reduces the cost of verifying intervention and prevents the fraudulent activity of attackers. An open-source platform, i.e., Ethereum blockchain, is used to deploy smart contracts using a Solidity programming language. In the second era of blockchain, miners compete to validate the transactions, but they get the reward in the form of Ether, i.e., the currency of Ethereum [43].

### 1.7.3   Blockchain 3.0: DApps

The new era of Blockchain 3.0 also includes smart contracts but with some additional technologies such as sharding and Decentralized Apps (DApps). DApps mainly work in the backend of Ethereum to provide users and applications connectivity with the smart contract in a distributed manner, thus, eliminating the need for a centralized server in the blockchain network. Furthermore, the third-generation blockchain uses several consensus mechanisms such as PoW, Proof-of-Stake (PoS), and Proof-of-Authority (PoA), which can be considered beneficial while implementing the smart contract in the blockchain. Furthermore, Blockchain 3.0 does not involve any miner and their transaction fee for validating the transactions as they use a built-in mechanism, which further decreases the transaction costs for the users [43].

### 1.7.4   Blockchain 4.0: Industry 4.0

The main requirement of Industry 4.0 is to acquire cybersecurity and innovative technologies such as supply chain management, the Internet of Things (IoT), and AI. This is the main reason to introduce a new age of Blockchain 4.0 with the advancements in technologies to serve businesses with better security, privacy, transparency, and data integrity. Therefore, InterValue is being used to build a platform for Blockchain 4.0. The main aim of InterValue is to develop an enhanced version of DAG with improved scalability, increased usability, and reliability with Blockchain 4.0. Some industries that can benefit from using Blockchain 4.0 are supply chain management, health management, approval workflows, IoT data collection, financial sectors, and conditional payments. This version makes blockchain 3.0 disposable in real-life scenarios [44].

### *1.7.5  Blockchain 5.0*

All the previous generation of blockchain has tried to make its implementation secure, transparent, reliable, and scalable for the network. But, still, multiple industries in different fields could not utilize it in their applications to that extent. This is the reason for the evolution of Blockchain 5.0 so that various organizations can utilize it in their application with enhanced scalability, economy, high security, transparency, and confidentiality. Relictum Pro is the first-known implementation of Blockchain 5.0. It is implemented on a network in which multiple smart contracts can be executed for transactions, which proves to be more secure than any other previous era of blockchain. Many organizations are simultaneously working to improve Blockchain 5.0 and its features with Relictum Pro, which has several benefits as compared to previous generations [45]. We have seen how the advancement in technologies can improve the utilization of blockchain for industries in multiple sectors. We further want to highlight that growth in blockchain technology has been discussed in detail in Chap. 2.

## 1.8  The Need of Blockchain Technology

In this section, we will discuss what are the limitations of a traditional transaction system that lead to the invention of the blockchain technology. Traditional systems use client–server architecture to exchange transactions between users. All activities of the individuals involved in the network are monitored by the third authority such as banks or central authorities. But, it is not guaranteed that transactions between two parties through central authority will be safe or not. Thus, in 2008, Satoshi Nakamoto first discussed about the bitcoin cryptocurrency, which utilizes blockchain technology. The main aim was to overcome the privacy and trust issues of fiat money. They have introduced the blockchain technology so that bitcoin can be transferred between users with security, transparency, and verifiability. We have already discussed the limitations of previous generations of blockchain as blockchain evolved over the decade so that industries can utilize it with its features of cross-chain function, immutability, cryptographical security, and used technologies such as smart contracts, cryptographic hash functions, and consensus mechanism. Cross-chain function can be defined as multiple smart contracts on multiple blockchains that can communicate with each other leading to high security, trust, and privacy in the network.

## 1.9  Types of Blockchain

### 1.9.1  Public Blockchain

As the name suggests, public blockchains are open source, i.e., open to all the willing participants of the network. Anyone is allowed to connect with the network and become a part of core activities. As no one grants authority, they are also known as permissionless blockchains. Public blockchains are secure even though they are open and public. The information which is openly available is the transaction information like wallet number, the amount, and the date. Self-governance and a higher level of security are present in the public blockchain as people universally can proactively read and write the code of the blockchain. The uncontrollability is the main advantage here as nobody will be able to control the whole network. A decentralized network is present in the public blockchain protocol like Bitcoin. These blockchains are known to be fully distributed because the authority on the blockchain is equally divided among the nodes [46]. The principal reason for public blockchains to be secure is each node can load the ledger containing all the transaction information. And due to this, it is challenging to hack as the target is not just one node but hundreds of them. These blockchains are fixed as they cannot be changed without altering the whole blockchain record of all transactions. It is not impossible to hack, but it would be extensively resource-intensive and time-consuming. The P2P nature of public blockchain networks is one of its best features. There is complete financial freedom for a user who wants to perform a transaction from anywhere at any time with another user at a fast speed. A public blockchain has a unique feature, that is, its network-wide consensus mechanism. For achieving consensus, each node is given the right to contribute to the decision. For example, PoW and Dash's Proof of Stake (PoS). The best example of an open, public blockchain is Bitcoin. Other examples are Ethereum, Stellar, and Dash [47].

### 1.9.2  Private (permissioned) Blockchain

In a private blockchain identified as a permissioned blockchain, access and participation in the transaction are restricted. Private blockchains are mostly used by private organizations where only pre-chosen entities can join the network. The network is centralized, and the central authority is responsible for giving permissions for writing transactions and who can read the particular transaction. The central authority also determines mining rights, which can be overridden or modified. The administrator can give or revoke the permissions granted to a user. These blockchains may or may not have a token, depending on the blockchain proprietor. Some samples of activities in a private blockchain are access, visibility, storage, and execution [52]. In this type of blockchain, the members of the blockchain have knowledge about each other's identity, but the transaction details remain private. They are faster and provide more efficiency, but security is not strong like public blockchains. Here, the consensus is reached through a single party or selected entities, and hence, it can lead to manipulation even though there is cryptographic security up to some level. One of the private blockchains is IBM's Hyperledger Fabric which can be deployed in a private network. Users can participate in the network once they are invited to join the network in a private blockchain. To utilize this blockchain for tracking food from the source to the shelves, Walmart partnered with IBM. In this, every, entity from farmers to distributors to retailers, can have permission access to information regarding the source and the current location [48].

### 1.9.3  Consortium Blockchain

It can be said that the consortium blockchain is like a hybrid of public and private blockchain. It is also identified as a federated blockchain. It is partially public because it is shared and partially private because access to the blockchain is restricted for the nodes. Some nodes are allowed to participate in the transactions, while some nodes control the consensus process. The network is centralized like the private blockchain, with a single point of failure. The control is not in the hands of a single authority but a few authenticated users. The control is not entirely centralized; it is a blend of centralization and decentralization. Some nodes need to sign off each transaction, while some need pre-approval from the network. Consortium blockchains imitate the benefits of private blockchains by providing enhanced efficiency and transaction privacy. On account of JPMorgan's cryptographic money, they intend to join their JPM coin along with numerous different banks on their Quorum Blockchain. While some may say it is a private blockchain, it is open to the public (other member banks) up to some extent. The JPM coin on Quorum Blockchain will be first utilized by institutional payment customers of JPMorgan.They can use this consortium blockchain for faster international or local transactions at any time with less cost [49].

## 1.10  Implementation Platforms of Blockchain

In the previous sections, we have discussed the structure of blockchain technologies used, which involves the deployment of smart contracts on the blockchain platform to perform any transaction between two parties with security, transparency, and verifiability. In addition, cryptographic hash functions and digital signatures are used to ensure the authentication of message exchange between the users from the sender's end and the recipient's end. Now, we need to discuss how the transaction between users can be implemented using these technologies in the blockchain. For that, we will consider the following open-source blockchain tools to deploy any application within any organization such as health care, supply chain, or any industry. These tools can be discussed as follows:

### 1.10.1   Node.js

The blockchain application can be deployed by installing Node.js and Visual Studio (VS) Code. The installation can be explained in the following steps starting with the installation of VS and Node.js on your computer. These are the prerequisites required for installation according to your operating system, which can be discussed in the following steps:

Visual Studio Code: https://code.visualstudio.com/download

Node.js: https://nodejs.org/en/download/

- After installing VS code and Node.js, create a folder at desired location in your workstation as shown in Figs. 1.12 and 1.13.
- Open VS Code, click on 'File' in menu bar, click on 'Add folder to workspace', choose the folde that you created in Step 1, and click on 'Add' button.
- You will see your folder in Explorer part of VS Code.
- Just to check whether VS code and Node.js are working or not, we will run a JS file for printing hello world. For that, right click on the folder that you have created, click on 'New File', and give the name of the file with .js extension. For example, 'first.js' shown in Fig. 1.14.

**Fig. 1.12**  Add created folder to workspace

**Fig. 1.13** Folder in explorer of VS code

**Fig. 1.14** Creation of file first.js

- Write the following code in this js file that you created. var message ='hello world'; console.log(message).
- To run this code, click on 'Terminal' in the menu bar, click on 'New Terminal', and you will see the small terminal window below the code. In the terminal, write the following command to run the file. node first.js as shown in Fig. 1.15.
- If you see the output 'hello world', then the installation is successful and your code is working fine.

**Fig. 1.15**  Run file first.js on terminal

**Fig. 1.16**  Install packages for blockchain deployment

– Since, we are working with blockchain, we need to add a package for performing
cryptographic operations such as SHA-256. To install the package, in the terminal,
write the following command. npm install –save crypto-js as shown in Fig. 1.16.
– After the package gets successfully installed, you will see a folder 'node_modules'
under the folder name that you added to workspace. If you click on 'node_modules'
folder, you will see several files for encryption and decryption, out of which one
will be 'crypto-js.js', which will be used in the subsequent codes.
– Restart your VS Code once after performing the above tasks.

## *1.10.2  Remix IDE*

We have already discussed Ethereum in Sect. 1.2; it provides a secure and transparent open-source platform to deploy blockchain applications so that transactions between users can be performed with security and transparency, and less expensively due to the elimination of third-party systems. Ethereum utilizes an online blockchain platform, which is Remix Integrated Development Environment (IDE), to deploy smart contracts so that transactions can be executed through the use of Ether, which is a cryptocurrency used by the Ethereum in the form of currency between the users involved in the blockchain [33, 50]. So, Remix IDE can be used to deploy the smart contracts in the Ethereum blockchain platform, which can be explained in the following steps with its prerequisites [31, 32]:

- Firstly, open the Remix IDE with the help of url https://remix.ethereum.org/.
- After opening the Remix IDE in your browser, click on 'Create New File' in the File Explorers section and name the smart contract file as 'first.sol', illustrated in Fig. 1.17.
- Then, you can see some featured plugins in which we have to choose the 'Solidity' to deploy the smart contract using Solidity programming language, which is the reason to store smart contract with .sol extension.
- Then, write the smart contract in the 'code section' of Remix IDE.
- We have considered the example in which there is communication between patients and doctors to monitor patients remotely. The code for the smart contract is written accordingly.

**Fig. 1.17**  Create a smart contract first.sol

**Fig. 1.18** Compile smart contract first.sol

- After that, in 'Solidity compiler' section in 'Home', click on 'Compile first.sol' to compile the smart contract code written in the code section as shown in Fig. 1.18.
- After compiling the smart contract, click on 'Deploy' in 'Deploy and run transactions' in 'Home' section to run and deploy the smart contract first.sol as shown in Fig. 1.20.
- In 'DEPLOY AND RUN TRANSACTIONS', you can see various types of environments such as JavaScript Virtual Machine (JVM), Injected Web3, and Web3 Provider to deploy and run the transactions as conveyed in Fig. 1.19.
- JVM provides the platform to run blockchain on your browser only without the use of any external tool.
- While, using Injected Web3 and Web3 Provider, you need to use some external tool, i.e., Metamask can be used as an external tool for Injected Web3 and Ganache for Web3 Provider that will be discussed in the next section as shown in Fig. 1.21.
- When you select JavaScript VM as environment, it shows the account with '100 Ether'. Then, click on 'Deploy', and it shows the output successfully in the terminal corresponding to the deployed contract.
- When we click on the 'Deployed contracts', it shows the functions used in the written smart contract.
- If you want to deploy your smart contract using some external tool, then firstly add the extension of Metamask to your browser and set up an account to execute the transaction with the ether.
- Now, you can click on 'Deploy' to get the output in the terminal successfully.

**Fig. 1.19** Select environment as JavaScript VM to deploy first.sol

**Fig. 1.20** Successfully deployed smart contract with its functions

**Fig. 1.21** Deploy smart contract with Injected Web3 environment

### *1.10.3  Ganache*

In the previous blockchain tool to deploy smart contracts, we have explained the use of Remix IDE, considering the environment as JavaScript VM as an internal tool and Injected Web3 as an external tool using the Metamask. But, one more environment needs to be considered, i.e., Web3 Provider. Web3 Provider environment can be utilized to deploy smart contracts to execute the transactions between users. But, for that, we need to understand the working and installation of Ganache to be implemented with Remix IDE to deploy the smart contract, which can be explained and installed in the following steps [34, 36]:

- You can download the Ganache as provided by Truffle Suite with the help of url https://www.trufflesuite.com/ganache as displayed in Fig. 1.22.
- Now, consider the smart contract 'first.sol' created in the previous tool.
- Click on 'Compile first.sol' to compile the smart contract.
- Now, click on 'Deploy and run transactions' in 'Home' section. You can see the different available environments as we have already considered the JavaScript VM and Injected Web3. So we have to consider the Web3 Provider.
- To consider the Web3 Provider, open Ganache downloaded from the Truffle Suite.
- When you open the Ganache, click on 'QUICKSTART Ethereum' to proceed further.
- After that, at the top extreme left side of Ganache, You can see the 'Accounts', and when we click on it, it gives 10 accounts by default as displayed in Fig. 1.23.
- You can change the number of accounts by clicking on 'settings' at the top extreme right side of Ganache. In 'settings', click on 'ACCOUNTS and KEYS' to enter the number of accounts to generate, as illustrated in Fig. 1.24.
- After completing all these settings, you can go back to the Remix IDE in your browser to select the 'Web3 Provider' as an environment as displayed in Fig. 1.25.
- When you click on 'Web3 Provider', it will show a screen in which enter the RPC server, i.e., HTTP://127.0.0.1:7545 (as shown in the Ganache) in the 'Web3 Provider Endpoint' as shown in Fig. 1.26.
- After that your Web3 Provider is working and you can deploy the contract by clicking on 'deploy' and can see the functions associated with the smart contract, as demonstrated in Fig. 1.27.
- We have considered the example of smart contract similar to the previous one.

### *1.10.4  Hardhat*

We have discussed different blockchain tools to deploy smart contracts, which mainly involve Remix IDE using different environments and also with the help of Node.js and VS Code. There is another blockchain tool, i.e., Hardhat, a local Ethereum network specially designed to deploy smart contracts in VS code using Solidity. You

**Fig. 1.22**  Starting of Ganache with QUICKSTART Ethereum

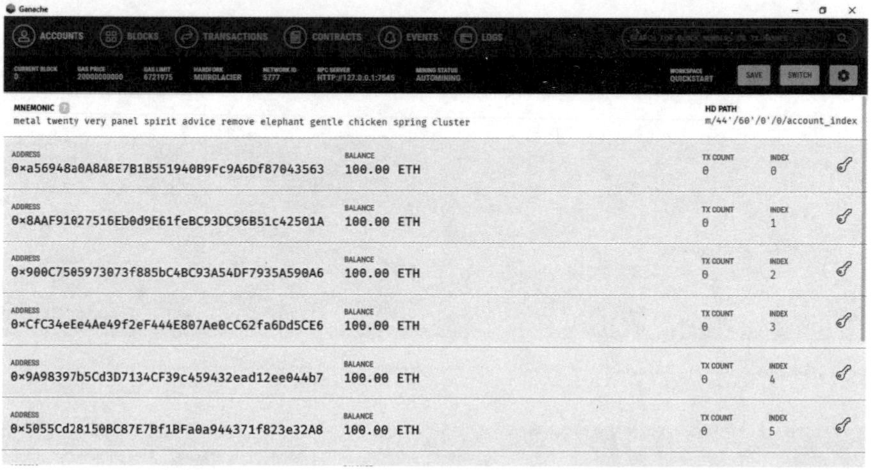

**Fig. 1.23**  Default number of accounts in Ganache

need to have Node.js downloaded to your computer, as explained in Sect. 1.10.1. So, Hardhat can be used to deploy smart contracts and its installation is explained in the following steps [51]:

– Firstly, after downloading the Node.js in your computer, create an empty folder named as 'Contract' and open 'New Terminal' to run the following command 'npm init' and act according to the instructions.
– Then, install Hardhat by running the command 'npm install –save-dev hardhat' in your terminal. It will complete the installation of Hardhat in VS code, which is also demonstrated in Fig. 1.28.

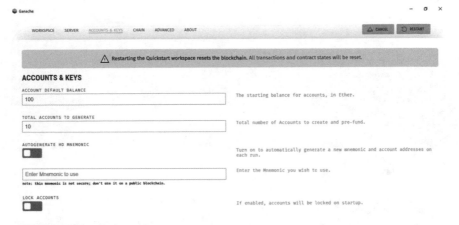

**Fig. 1.24** Number of accounts generated

**Fig. 1.25** Choose Web3 Provider as environment in Remix IDE

**Fig. 1.26** Enter RPC server from Ganache

**Fig. 1.27** Deployed smart contract with its functions

**Fig. 1.28** Installation of Hardhat

- After that, we can check whether Hardhat is installed properly or not. To check this, run the command 'npx hardhat' in your terminal. It will show you the version of Hardhat as v2.8.3 as an output in your terminal.
- After installing the Hardhat, you can follow the instructions to create a basic sample project according to your requirement as displayed in Fig. 1.29.
- But, we need to run several packages of Hardhat to run your project properly. You can run these dependencies using the command 'npm install –save-dev @nomiclabs/hardhat-waffle ethereum-waffle chai @nomiclabs/hardhat-ethers ethers' in your terminal, as illustrated in Fig. 1.30.
- You can also check the other details and version of Hardhat using command 'npx hardhat', as shown in Fig. 1.31.

**Fig. 1.29**  Creation of basic Hardhat project

**Fig. 1.30**  Installing dependencies of Hardhat

– Now, you can compile your contract, which is being created in 'contracts' folder named as 'smart.sol' in which we have considered a similar example of communication between patient and doctor for remote monitoring. For compiling the contract, you need to run a command 'npx hardhat compile' to compile the smart contract successfully, as displayed in Figs. 1.32 and 1.33.

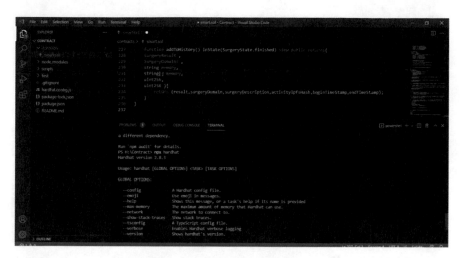

**Fig. 1.31** Display Hardhat version and configuration

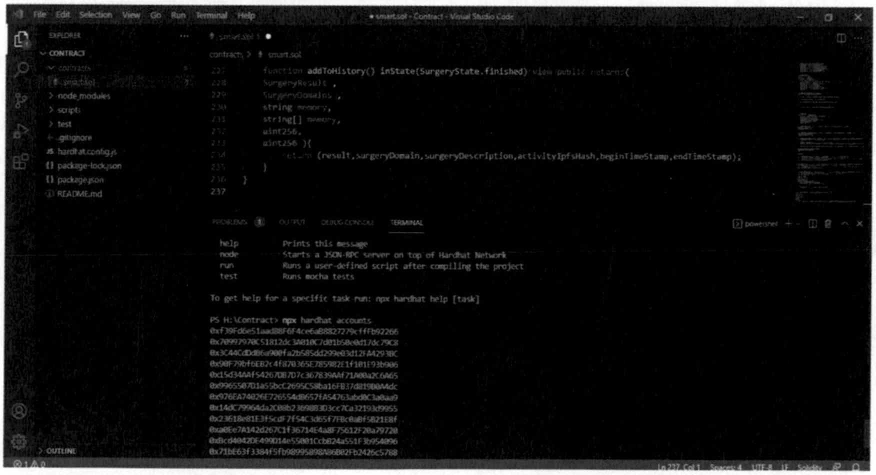

**Fig. 1.32** Display Hardhat accounts

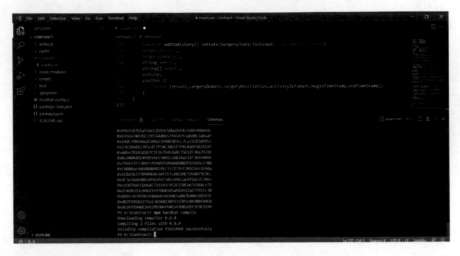

**Fig. 1.33** Smart contract compiled successfully

## 1.11  Summary

The increase in the use of the Internet and digital cryptocurrencies such as bitcoin gives rise to blockchain technology. It transforms the way of trade and interactions between people. The chapter provides a complete overview of blockchain technology, where we discussed terminologies related to blockchain, the need for blockchain, and the revolutionary perspective of blockchain. It also includes different types of blockchain and the consensus mechanism that makes blockchain fully decentralized, trustworthy, and more secure.

## 1.12  Practice Questions

### 1.12.1  Multiple Choice Questions

– What is the abbreviation of P2P system?
  - Public to Private
  - Password to Public
  - Peer-to-Peer
  - None of the above

– _____ is a part of asymmetric cryptography.

- Public key
- Private key
- Both (a) and (b)
- None of the above

– Select the characteristic of the blockchain that makes it tamper-proof.

- Cryptographic key pairs
- Immutability
- Centralization
- None of the above

– The mechanism by which bitcoins are generated _____

- Nakamutu Santoshi
- Santushi Nhiem
- Satoshi Nakamoto
- Satoshi Nguyen

– What are Orphan blocks in the blockchain network?

- Valid blocks
- Invalid blocks
- Trusted blocks
- All of the above

– What are SHA-256 and MD5 algorithms ?

- Cryptographic hash function
- Asymmetric encryption
- Symmetric encryption
- None of the above

– What is the abbreviation of MAC in the blockchain?

- Multiple access control
- Message authentication code
- Multiple authentication code
- None of the above

– _____ name of the mechanism used in Proof of Capacity (PoC) consensus.

- Sharing
- Hiring
- Authenticating
- Plotting

– What console.log() will do in .js file in visual studio code?

- Print the message
- Show the debugger
- Add a breakpoint in the code
- Run the program

– Java virtual machine platform does not need any external tool; however, Web3 provider and Injected Web3 need external took like _____

- Nodejs
- Smart contract
- Metamask
- All of the above

## 1.12.2   Fill in the Blanks

1. The record of each transaction in blockchain is _____ to all the participants of the network.
2. Blockchain works on a decentralized manner where all transaction are stored in a ledger which is _____ and _____ to all.
3. We can remove the faulty nodes from the blockchain chain by providing the system with _____, which lessens the probability of failure in the system.
4. _____ first coined the term 'Ethereum' after the release of Bitcoin to deploy blockchain-based application adopted by organizations and institutions.
5. _____ are the formal participants in the blockchain network whose main aim is to verify the new arriving transaction.
6. Staking nodes in the blockchain network uses _____ consensus mechanism to authenticate the transaction.
7. Specify the name of the node _____, which only keeps track of block headers of the blockchain.
8. A _____ is a mathematical technique used to validate the authenticity and integrity of a message, software, or digital document.
9. Specify the correct command _____ to install a package crypto.js in visual studio.
10. Name of the tool _____ by which the smart contracts can be deployed easily in the Ethereum-based blockchain platform.

## 1.12.3   Short Questions

1. Which is better between distributed and decentralized systems?
2. Why all the transactions are visible to all the nodes in the blockchain?
3. How longest chains are determined in Bitcoin?

4. Who are the miners and who selects them?
5. Who validates the transactions?
6. Why does every node need to validate the transactions?
7. How is the mining incentive provided to the miners?

### 1.12.4 Long Questions

1. How the privacy and security are maintained if all the transactions can be viewed by everyone?
2. Why do we ask all the non-malicious nodes to validate the transactions?
3. How is it possible to have a copy of the Public Ledger (which is increasing day by day) for every client?
4. How is the Merkle tree root computed for the non-even transactions?
5. What is the difference between transaction fees and block reward? Which one of them goes to the miner?

## References

1. Blockchain basics: introduction to distributed ledgers (2019). https://developer.ibm.com/tutorials/cl-blockchain-basics-intro-bluemix-trs/. Accessed 08 Jan 2022
2. Thuraisingham B (2020) Blockchain technologies and their applications in data science and cyber security. In: 2020 3rd international conference on smart blockchain (SmartBlock), pp 1–4. https://doi.org/10.1109/SmartBlock52591.2020.00008
3. Hathaliya J, Sharma P, Tanwar S, Gupta R (2019) Blockchain-based remote patient monitoring in healthcare 4.0. In: 2019 IEEE 9th international conference on advanced computing (IACC), pp 87–91. https://doi.org/10.1109/IACC48062.2019.8971593
4. Popov A (2020) How industrial sectors are using blockchain technology. https://www.forbes.com/sites/forbestechcouncil/2020/03/12/how-industrial-sectors-are-using-blockchain-technology/?sh=4a43e03a2c30. Accessed 08 Jan 2022
5. Vujicic D, Jagodic D, Randic S (2018) Blockchain technology, bitcoin, and Ethereum: a brief overview. In: 2018 17th international symposium INFOTEH-JAHORINA (INFOTEH), pp 1–6. https://doi.org/10.1109/INFOTEH.2018.8345547
6. Kumari A, Gupta R, Tanwar S, Kumar N (2020) Blockchain and AI amalgamation for energy cloud management: challenges, solutions, and future directions. J Parallel Distrib Comput 143. https://doi.org/10.1016/j.jpdc.2020.05.004
7. Gibb R (2019) What is a distributed system. https://blog.stackpath.com/distributed-system/. Accessed 08 Jan 2022
8. What does distributed system mean (2019). https://www.techopedia.com/definition/18909/distributed-system. Accessed 08 Jan 2022
9. Meador D (2018) Distributed systems. https://www.tutorialspoint.com/Distributed-Systems. Accessed 08 Jan 2022
10. Akram SV, Malik P, Singh R, Anita G, Tanwar S (2020) Adoption of blockchain technology in various realms: opportunities and challenges. Secur Priv 3:e109. https://doi.org/10.1002/spy2.109

11. Decentralized    system    (2019).    https://www.computerhope.com/jargon/d/decentral.htm.
    Accessed 08 Jan 2022
12. Giannakaris P, Panagiotis T, Zahariadis T, Gkonis P, Papadopoulos K (2019) Using smart
    contracts in smart energy grid applications, pp 597–602. https://doi.org/10.15308/Sinteza-
    2019-597-602
13. Blockchain    explained    (2022).    https://www.investopedia.com/terms/b/blockchain.asp.
    Accessed 08 Jan 2022
14. Gupta R, Tanwar S, Kumar N (2021) Blockchain and 5G integrated softwarized UAV network
    management: architecture, solutions, and challenges. Phys Commun 101355. ISSN 1874–4907.
    https://doi.org/10.1016/j.phycom.2021.101355
15. Khan SN, Loukil F, Ghedira-Guegan C et al (2021) Blockchain smart contracts: applications,
    challenges, and future trends. Peer-to-Peer Netw Appl 14:2901–2925. https://doi.org/10.1007/
    s12083-021-01127-0
16. Ekici B (2020) An overview of Ethereum and solidity. https://www.investopedia.com/terms/b/
    blockchain.asp. Accessed 08 Jan 2022. https://betterprogramming.pub/overview-of-ethereum-
    solidity-d7a0ea5fbf00
17. Wu Y, Dai H-N, Wang H (2021) Convergence of blockchain and edge computing for secure and
    scalable IIoT critical infrastructures in industry 4.0. IEEE Internet Things J 8(4):2300–2317.
    https://doi.org/10.1109/JIOT.2020.3025916
18. Blockchain: what is blockchain technology? how does it work? https://builtin.com/blockchain.
    Accessed 08 Jan 2022
19. Mining    and    consensus.    https://www.oreilly.com/library/view/mastering-bitcoin/
    9781491902639/ch08.html. Accessed 08 Jan 2022
20. Blockchain and block header. https://www.geeksforgeeks.org/blockchain-and-block-header/.
    Accessed 08 Jan 2022
21. Dutta B (2021) Types of block in a blockchain network. https://www.analyticssteps.com/blogs/
    3-types-block-blockchain-network. Accessed 08 Jan 2022
22. Geroni D (2021) Blockchain nodes: an in-depth  guide. https://101blockchains.com/
    blockchain-nodes/. Accessed 08 Jan 2022
23. Jimi S (2018) Blockchain: what are nodes and masternodes? https://medium.com/coinmonks/
    blockchain-what-is-a-node-or-masternode-and-what-does-it-do-4d9a4200938f.    Accessed
    08 Jan 2022
24. Wang Y (2020) A blockchain system with lightweight full node based on dew computing.
    Internet Things 11:100184, ISSN 2542-6605. https://doi.org/10.1016/j.iot.2020.100184
25. Blockchain hash function. https://www.javatpoint.com/blockchain-hash-function. Accessed
    10 Jan 2022
26. Cryptographic    hash    functions.    https://www.tutorialspoint.com/cryptography/
    cryptographyhashfunctions.htm. Accessed 10 Jan 2022
27. Message  authentication  and  hash  functions. https://www.vskills.in/certification/tutorial/
    message-authentication-and-hash-functions-3/. Accessed 10 Jan 2022
28. Fang W, Chen W, Zhang W et al (2020) Digital signature scheme for information non-
    repudiation in blockchain: a state of the art review. J Wireless Com Netw 2020:56. https://
    doi.org/10.1186/s13638-020-01665-w
29. Iredale G (2021) Hashing and digital signature in blockchain. https://101blockchains.com/
    hashing-and-digital-signature-in-blockchain/. Accessed 10 Jan 2022
30. Smart contracts (2021). https://www.wallstreetmojo.com/smart-contracts/. Accessed 12 Jan
    2022
31. Remix- Ethereum IDE: creating and deploying a contract (2021). https://remix-ide.
    readthedocs.io/en/latest/create_deploy.html. Accessed 12 Jan 2022
32. How  to  use  MetaMask  to  deploy  a  smart  contract  in  solidity  (blockchain)?
    (2020).    https://www.geeksforgeeks.org/how-to-use-metamask-to-deploy-a-smart-contract-
    in-solidity-blockchain/?ref=rp. Accessed 12 Jan 2022
33. Steps to execute solidity smart contract using remix IDE? (2020). https://www.geeksforgeeks.
    org/steps-to-execute-solidity-smart-contract-using-remix-ide/?ref=rp. Accessed 12 Jan 2022

34. Ethereum tutorial: ganache for blockchain (2021). https://www.tutorialspoint.com/ethereum/ethereum_ganache_for_blockchain.htm. Accessed 12 Jan 2022
35. Kakkar R, Gupta R, Tanwar S, Rodrigues JJPC. Coalition game and blockchain-based optimal data pricing scheme for ride sharing beyond 5G. IEEE Syst J. https://doi.org/10.1109/JSYST.2021.3126620
36. How to use GANACHE truffle suite to deploy a smart contract in solidity (blockchain)? (2020). https://www.geeksforgeeks.org/how-to-use-ganache-truffle-suite-to-deploy-a-smart-contract-in-solidity-blockchain/. Accessed 12 Jan 2022
37. Joshi N (2019) 8 blockchain consensus mechanisms you should know about. https://www.allerin.com/blog/8-blockchain-consensus-mechanisms-you-should-know-about. Accessed 14 Jan 2022
38. Inlea (2019) But...how does blockchain really works? https://inlea.com/but-how-does-blockchain-really-works/. Accessed 16 Jan 2022
39. IP Specialist (2019) How blockchain technology works. https://medium.com/@ipspecialist/how-blockchain-technology-works e6109c033034. Accessed 16 Jan 2022
40. Venugopal R (2021) What is blockchain: features and use case. https://www.simplilearn.com/tutorials/blockchain-tutorial/what-is-blockchain. Accessed 16 Jan 2022
41. Flair D (2022) 6 major features of blockchain|why blockchain is popular? https://data-flair.training/blogs/features-of-blockchain/. Accessed 16 Jan 2022
42. Brakeville S, Perepa B (2019) Blockchain basics: introduction to distributed ledgers. https://developer.ibm.com/tutorials/cl-blockchain-basics-intro-bluemix-trs/. Accessed 18 Jan 2022
43. GeeksforGeeks (2021) Different version of blockchain. https://www.geeksforgeeks.org/different-version-of-blockchain/. Accessed 18 Jan 2022
44. Javaid M, Haleem A, Singh RP, Khan S, Suman R (2021) Blockchain technology applications for Industry 4.0: a literature-based review. Blockchain: Res Appl 2(4):100027. ISSN 2096-7209.10.1016/j.bcra.2021.100027
45. Relictum Pro (2019) What is Blockchain 5.0, and How Relictum Pro is set to Revolutionize the Industry? https://relictumpro.medium.com/what-is-blockchain-5-0-and-how-relictum-pro-is-set-to-revolutionize-the-industry-c7aa23df7190. Accessed 20 Jan 2022
46. Kumari A, Gupta R, Tanwar S, Kumar N (2020) A taxonomy of blockchain-enabled softwarization for secure UAV network. Comput Commun 161:304–323. ISSN 0140-3664. https://doi.org/10.1016/j.comcom.2020.07.042
47. Upgrad (2019) Different types of blockchain and their uses. https://www.upgrad.com/blog/different-types-of-blockchain/. Accessed 20 Jan 2022
48. Edureka (2019) Different types of blockchain and why we need them. https://www.edureka.co/blog/types-of-blockchain/. Accessed 20 Jan 2022
49. Edureka (2019) 3 types of blockchain explained. https://hedgetrade.com/3-types-of-blockchain-explained/. Accessed 20 Jan 2022
50. Gupta R, Shukla A, Tanwar S. BATS: a blockchain and AI-empowered drone-assisted telesurgery system towards 6G. IEEE Trans Netw Sci Eng. https://doi.org/10.1109/TNSE.2020.3043262
51. Hardhat (2022) Getting started: overview. https://hardhat.org/getting-started/. Accessed 20 Jan 2022
52. Tanwar S, Parekh K, Evans R (2020) Blockchain-based electronic healthcare record system for healthcare 4.0 applications. J Inf Secur Appl 50:102407. https://doi.org/10.1016/j.jisa.2019.102407

# Chapter 2
# Blockchain Revolution from 1.0 to 5.0: Technological Perspective

**Abstract** Since the inception of bitcoin application in 2008, several trust-based global solutions have been developed to overcome the malicious intent that occurs in critical applications. Adopting blockchain technology resolves the issues related to frauds in various sectors like banks, finance, and government. It was a massive concern as fraudulent activities were carried out on a large scale to maneuver the business operation. A blockchain network's capabilities are sufficient to address these aforementioned issues. With blockchain, the third-party services were brought to a halt, and digital signature was associated with every task, thus, making blockchain a secure network. The chapter presents different forms of blockchain that evolved over the years and the methods it has adopted to transform into a system that conducts transactions in a trusted, safe, and secure environment. Further, this chapter first introduces the origin of blockchain. Then it discusses how blockchain generations varied with time as demands grew and technological innovations that keep evolving at a rapid pace. Finally, we discuss the elements of blockchain, its working, benefits, and limitations of blockchain.

**Keywords** Blockchain 1.0 · Blockchain 2.0 · Blockchain 3.0 · Blockchain 4.0 · History of blockchain

## 2.1 History of Blockchain

In the previous chapter, we have discussed the basics of blockchain with its various technologies such as smart contract, consensus mechanism, and cryptographic hash functions, so that users can utilize the blockchain platform to deploy their applications using different open-source tools, which include Nodejs, Remix IDE, Ganache, etc. Blockchain technology enables a secure and transparent storage of transactions by eliminating the centralized party systems in the network. It adapts consensus mechanisms to authenticate the request of users to add their data transactions to the blockchain network. After authentication, the transaction are added to the network, in which no one can modify or alter the data, that leads to enhanced security in the system. But, we need to know how exactly blockchain came into existence or who

© The Author(s), under exclusive license to Springer Nature Singapore Pte Ltd. 2022
S. Tanwar, *Blockchain Technology*, Studies in Autonomic, Data-driven and Industrial Computing, https://doi.org/10.1007/978-981-19-1488-1_2

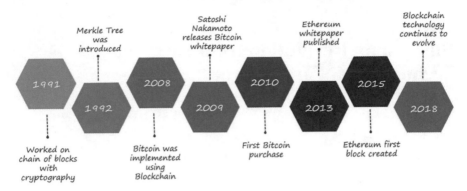

**Fig. 2.1** Evolution of blockchain

proposed the idea of blockchain technology. It first came into existence with the concept of Bitcoin launched by Satoshi Nakamoto in 2008.

Even before the evolution of Bitcoin, there were some technologies discussed based on the blockchain network. In 1991, the idea was put forward by research scientists Stuart Haber and W. Scott Stornetta. They introduced a method that dealt with tampering of documents to make the document tamper-resistant. This mechanism used a cryptographic chain of blocks, which can be used to store the time-stamped documents [1]. In 1992, Merkle Tree was introduced, as illustrated in Fig. 2.1 to tackle the enormous amount of data in a secure manner. The structure of Merkle Tree ensures the storage of digital documents in the block without compromising their security. However, this technology has not affected the market to that extent. This led to the introduction of a digital currency in 1998, i.e., bit gold proposed by Szabo, which utilizes the various cryptography mechanisms [2, 3].

Further, researcher and cryptography scientist Hal Finney discussed Reusable Proof of Work (RPoW). This system has gradually evolved cryptocurrencies in the form of digital cash, which was the early step of the Bitcoin evolution [4]. Later in 2008, motivated by the concepts of cryptography and cryptocurrencies, Satoshi Nakamoto (an individual or group of individuals) put forward a paper that proposed a decentralized peer-to-peer (P2P) cash system called Bitcoin. Blockchain was the technology behind the cryptocurrency, i.e., bitcoin [5]. The first bitcoin block was mined on the 3rd of January 2009, called 'genesis block', which had 50 bitcoins as a reward. Blockchain has a distributed ledger system that can build a trusted and secure environment and has the potential to be free from any malicious activities [6]. This is because every participant in the blockchain network holds their copy of data in the ledger. The ledger is a database that contains all the data records, which increases exponentially with the increase in transactions in the network. However, data stored in the blockchain differs from the database, as it emphasizes on the centralized system for data storage that can influence the adversaries for security threats. Contrarily, blockchain does not involve any centralized system; instead, it has a decentralized system that provides a robust solution toward any cyberthreat [7]. Also, blockchain uses different cryptographic primitives, consensus mechanisms, and smart contracts so that data can be stored securely without any interruption from malicious attackers [8].

Bitcoin served as an inspiration for other applications. However, after a few years, more platforms like Ethereum have emerged as an open-source blockchain platform that provides a decentralized, distributed smart contract-based virtual network over which developers can deploy applications. The goal of this new generation of blockchain was to build a platform to build applications called decentralized applications (DApps) and deploy contracts on the blockchain that were written using a programming language named as smart contracts. Still, Ethereum has been a popular platform for a long time for building DApps such as identity systems and financial applications. It has many benefits, such as the efficient mining time compared to Ethereum, using the Ethash algorithm for Proof of Work (PoW). There can also be drawbacks associated with the platform, such as inefficient processing of transactions (transaction per second), energy consumption, and an increase in the number of users in the network. So, to resolve these drawbacks in the second generation of blockchain, many organizations have started working on various projects to take the blockchain to the next level. For the past few years, blockchain technology has evolved worldwide quickly. The blockchain technology that organizations and industries are utilizing nowadays in their applications has evolved entirely from what it was a decade ago. Due to these changes in blockchain technology, it is getting popular worldwide across several industries such as financial, healthcare, and government sectors [9, 10]. In the following sections, we will discuss how blockchain has been evolved gradually from Blockchain 1.0 to Blockchain 5.0, adopting advanced technologies to make it beneficial for industries and organizations.

## 2.1.1 Blockchain 1.0

*Era: 2008–2013*

**Motivation**—In the early 90s, many researchers and scientists have worked on various projects to digitize the transactions for the benefit of industries and institutions. They have mainly introduced cryptocurrencies so that users from both parties can trade without any centralized system [11]. As earlier, possible mediums of transactions used to be in the form of fiat money, which increases the risk of fraudulent activities and is more expensive than digital currency. But, introducing the digital currency using the cryptography mechanisms was insufficient to resolve the double-spending issues of digital currency [12].

Thus, Blockchain 1.0 as a first-generation technology evolved from the term Digital Ledger Technology (DLT), which provides all the participants with a distributed network to resolve the issue of double spending. A double-spending problem can disrupt the network if some users trade with the same currency more than once [13]. Thus, in early 2008, Satoshi Nakamoto released Bitcoin, which utilizes decentralization technology to store and share the data transactions among multiple users, bringing security, verifiability, and efficiency in the application. Blockchain technology uses various cryptography and consensus mechanisms to discard the centralized network and facilitate users in consensus to validate the transactions, which further keeps data records safe and protected [14].

Many researchers and cryptographers have studied the implementation of bitcoin using blockchain for applications pertaining to the industries. They have found out that bitcoin majorly works on the principle that miners have to solve a complex cryptographic puzzle to validate the transactions and get a certain reward for solving the same, known as a mining mechanism. Thus, Blockchain 1.0 proves to be efficient in preserving and securing the transactions and data records compared to the conventional trading systems based on fiat money [15]. Blockchain 1.0 overcame the issues of traditional trading systems with centralized authority by providing a decentralized platform for secure access and storage of transactions [16]. Now, we will discuss the benefits of utilizing Blockchain 1.0 in industries and financial institutions.

**Benefits**—With the advancement of Blockchain 1.0, technology that can overcome the security issues related to fiat currency is getting accepted by organizations. We can consider the example of Bitcoin, which many organizations at present are using. It has been employed in different platforms and applications due to its various benefits, which can be defined as follows [17]:

- It enhances the reliability of the network by masking the identity of users or organizations as it only reveals the transactions performed by the person or organization increasing the privacy of users.
- It ensures the transparent communication between the participants in the network, as any member of the network can add themselves to the blockchain with the condition of satisfying the consensus mechanism.
- It exhibits low cost due to the use of a decentralized and distributed network as compared to the high cost of conventional systems due to the involvement of a centralized trustless party.

**Implementation**—Bitcoin is the first-known implementation of Blockchain 1.0 and can also be considered as the firstborn of the blockchain era. Bitcoin was started as a project in 2008 and published for open use in 2009. Person/group named 'Satoshi Nakamoto' published it. The currency 'Bitcoin' serves a reward for a mining process performed by miners for the work done, i.e., transaction validated by other members in the network. The miners who validate the transactions by solving the cryptographic puzzle get the deserving reward for their work. Following are the steps to better understand how bitcoin can be transferred between two users utilizing blockchain technology. Figure 2.2 shows a scenario in which a user John wants to send a bitcoin to the user Peter by initiating a transaction with the help of blockchain technology, which can be explained in the following steps [18, 19]:

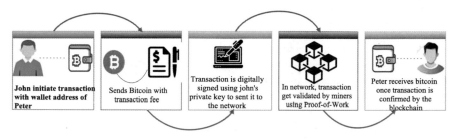

**Fig. 2.2** Transaction of bitcoin between users

- Firstly, John initiates the transaction by getting the wallet address of the Peter. Wallet address is used so that users can trade with each other through the use of wallet.
- After that, John sends the bitcoin with the transaction fee so that miner can be rewarded for validating the transaction.
- Then, there has to be a way to secure the transaction and prevent any malicious activity. The transaction is digitally signed using John's private key so that no one can manipulate the transaction for their benefit.
- The transaction preserved by the digital signature using John's private key can be sent to all the participant nodes in the network, where it gets validated by miners.
- There is competition between miners to validate the transaction using PoW mechanism. Those who solve the complex mathematical puzzle will be rewarded for verifying the transactions.
- Now, if all the nodes in the blockchain verify the transaction, then this transaction can be added to the blockchain network.
- Lastly, Peter receives the bitcoin, which is being verified by the blockchain.

**Flaws**—Although, Blockchain 1.0 overcomes the double spending, security, and transparency issues of fiat currency or physical money by eliminating the centralized authority in the network. However, blockchain 1.0 still faces some challenges that can be understood in the following steps [17]:

- In Blockchain 1.0, miners can validate transactions by applying the PoW consensus mechanism. However, they have to simultaneously solve a cryptographic puzzle to get the reward in the form of a transaction fee.
- But, solving a complex puzzle using PoW increases the computation time of transactions executed in the network and utilizes the maximum energy of the network, making the applied PoW inefficient for the network.
- It also leads to the delay in execution of transactions, which will not be favorable for the user waiting at the receiver's side to receive the transaction.
- There can still be some security issues associated with Blockchain 1.0 due to the inclusion of selfish or corrupt miners in the network. They will try to corrupt the network to get more revenue or reward while mining the transaction for the blockchain network.

- There is one more disadvantage associated with the blockchain, i.e., its inability to incorporate smart contracts and its applications. This is why Blockchain 1.0 has security and privacy issues in the network.
- These security and privacy flaws in Blockchain 1.0 led to the new era of Blockchain 2.0 that will be discussed in the next section.

**Summary**—The primary purpose of Blockchain 1.0 in this era was to idealize payment mechanisms while maintaining the anonymity of users but revealing the transaction and required information related to it. There can be various mechanisms to prevent plenty of malicious or double-spending attacks that can occur in simple digital currency transfer, but Blockchain 1.0 prevents these attacks by utilizing the PoW mechanism and digital signature, which further preserve the identity of the particular users. We have considered an example of the transaction between John and Peter to get insights into the utilization of blockchain technology while transferring the bitcoin between the users. We have specified several benefits of Blockchain 1.0, which include providing anonymity to the users and preventing double-spending attacks in the network. But, this era of Blockchain 1.0 also comes with several disadvantages, such as increased computation time leading to the utilization of maximized energy in the network. There are also security issues associated with it, which involve malicious miners trying to corrupt the network using malpractices. Thus, this evolves the need for the next era of blockchain, i.e., Blockchain 2.0, with improved security measures and aspects.

## 2.1.2  Blockchain 2.0

*Era: 2013–2015*

**Motivation**—Blockchain 1.0 and its implementation are limited to the execution of digital transactions using the PoW consensus mechanism. But, it consumes a huge amount of resources due to the increased computation time to mine transactions, which affects the network's scalability. Thus, Blockchain 2.0 as a second-generation technology emerges, which combines several technologies such as smart contracts with PoW consensus mechanism to resolve the issues of Blockchain 1.0. The smart contract can be defined as a self-executable code running on the blockchain and when the predetermined conditions or rules are satisfied. The code written for a smart contract in the blockchain is quite complex and can't be modified or altered easily to ensure the security of deployed transactions by the user in the network. It contains some predetermined conditions, which need to be obeyed by both the parties involved in the transaction preserving the integrity of the transaction [20].

Ethereum, a blockchain-based platform, can implement the transactions written in the form of smart contracts into the blockchain. Smart contracts that are designed in Ethereum use a Solidity programming language that can be defined as a high-level object-oriented programming language. There is a currency, i.e., Ether, included in the Ethereum wallet that the users can use during the transactions. In the case of Ethereum, miners validate transactions to obtain the maximum reward in the form of Ether. There is one requirement to execute smart contracts, i.e., the gas needed for transactions basically, it is sent by the users with the transaction to remunerate the energy consumption for executing the transactions. There have been many works on utilizing Ethereum with the smart contract to deploy the applications in blockchain for industrial purposes. For example, the enhanced security of a smart contract can be beneficial for several industries, such as the financial sector, government sector, and the healthcare sector, which need utmost security for their data stored in the blockchain network. There are various Ethereum-based blockchain platforms such as Remix IDE, Ganache, and Metamask to implement any application with security and privacy, which has already been discussed in Chap. 1. We can mention the various advantages of Blockchain 2.0 over Blockchain 1.0, which is why smart contracts are adopted in the blockchain [21].

**Benefits**—To overcome the issues of Blockchain 1.0, Blockchain 2.0 has evolved with the additional implementation of smart contracts in the Ethereum-based blockchain platform. It provides the network with numerous benefits in terms of security, transparency, and computation time during the execution of transactions, which can be discussed in the following steps [22]:

- Firstly, Blockchain 2.0 with the introduced Ethereum platform increased its usability for financial organizations with its enhanced security and privacy as Blockchain 1.0 was mainly focused on the cryptocurrency, i.e., Bitcoin.
- The usage of the Ethereum platform in Blockchain 2.0 with the deployed smart contract makes the communication between users more transparent. Due to this, any user can access the open-source Ethereum platform to deploy their transactions in the blockchain.
- Smart contracts also improve the data rate of the transaction, i.e., it can process more number of transactions as compared to Blockchain 1.0.

**Implementation**—Ethereum is considered as a new era implementation of Blockchain 2.0. The purpose of developing Ethereum was to provide developers with a platform where they could develop various smart contracts. A community of individuals who constantly work on developing decentralized applications would also encourage new people to join this domain and can smoothen the learning convenience. The idea of Ethereum was proposed by Russian–Canadian programmer 'Vitalik Buterin', works on the PoW consensus mechanism principle. Figure 2.3 shows how sellers and customers can exchange assets by executing the smart contract so that their transactions can be stored in the blockchain with security, transparency, and accuracy. This is because, to execute smart contracts, some set of predetermined rules or conditions are to be satisfied between two users. If the conditions are satisfied, an exchange of assets can be performed efficiently. Then, according to the code

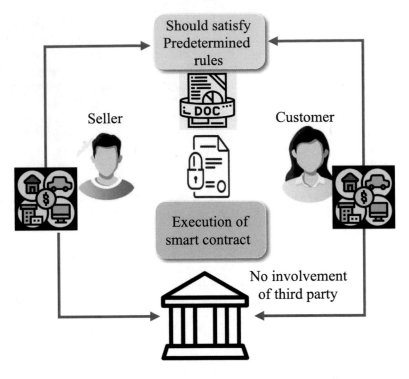

**Fig. 2.3** Execution of smart contract for transaction between users

written in the smart contract, all the participating nodes in the network should agree with the decision for validating the transaction fulfilling the criteria of the consensus mechanism. Also, deploying smart contract eradicates the involvement of third-party systems in the network, and the addition of smart contract with PoW in Blockchain 2.0 improves the data rate at which transactions can be processed in the network [23]. Chapter 1 shows the basic procedure of deploying smart contracts in Ethereum blockchain using an online Ethereum platform Remix IDE based on the working environment such as JavaScript Virtual Machine (JVM) and Injected Web3 using Metamask. We have considered one more environment in Remix IDE, i.e., Web3 Provider, which works by installing the Ganache, discussed in Chap. 1.

**Flaws**—We have discussed several advantages of Blockchain 2.0 and how it can overcome the limited usability of Blockchain 1.0 in various financial and government sectors. Also, Blockchain 2.0 proves to be beneficial in processing more transactions than Blockchain 1.0 due to the addition of smart contracts. But, there are some disadvantages of Blockchain 2.0 that can be discussed as follows [24]:

- As Blockchain 2.0 involves the execution of smart contracts in Ethereum, which can be defined as complex lines of code that need to be written very carefully.

- Even if there is any minor mistake in writing the code, then it can overwork the network as writing the whole smart code again, and then deploying it in Ethereum can be quite time-consuming.
- Also, a delay in execution of the smart contract can also give a chance to intruder to manipulate the transaction in the network.
- Thus, it is very necessary to implement smart contract accurately and correctly for timely execution of transaction using PoW consensus mechanism.
- There is one more disadvantage that using only PoW consensus mechanism is not beneficial for the network to that extent as it will not be able to resolve the issue of energy consumption of Blockchain 1.0 due to the increased computation time to solve a quite complex mathematical puzzle.
- Thus, there is a need to introduce a new era of blockchain, i.e., Blockchain 3.0 which will be discussed in the next section.

**Summary**—The main purpose of this new era, i.e., Blockchain 2.0, was to improve the processing of transactions between two parties. It majorly focused on introducing Ethereum, which implements the smart contract between users involved in the network. Users as participating nodes in the network should follow the rules set by the smart contract so that security can be enhanced in the network, preventing any suspective activity of the attacker. As we have discussed, the Ethereum platform uses Ether to exchange transactions that include gas value, i.e., that is being sent for energy consumption of transactions of users. We have also mentioned some of the advantages of Blockchain 2.0 over Blockchain 1.0, which make it more useful for the banking, government, or financial sectors. But, there are also some disadvantages in utilizing Blockchain 2.0; for example, use of PoW consensus is quite similar to Blockchain 1.0, and the complex nature of smart contracts can be tricky to write. Therefore, there is a need to advance the technologies used in Blockchain 2.0, which raises a new era of Blockchain 3.0, which is being explained in the next section.

### 2.1.3 Blockchain 3.0

*Era: 2015–2017*

**Motivation**—The previous generations of blockchain mainly suffer from scalability issues in the network, i.e., utilizing the PoW mechanism takes more time to process and verify transactions. This led to the introduction of a new era of Blockchain 3.0. The new generation of blockchain also includes smart contracts but with some additional technologies such as sharding and Decentralized Apps (DApps). DApps mainly work in the backend of Ethereum to provide users and applications connectivity with the smart contract in a distributed manner, thus, eliminating the need for a centralized server in the blockchain network [25]. The sharding technique can be used as a concrete solution to overcome the scalability issue of the network hindered in the previous generation of blockchain. This technique splits the whole network into multiple groups so that each node of the group participating in the network is responsible for managing their transactions. The third-generation blockchain uses several consensus mechanisms such as PoW, Proof of Stake (PoS), and Proof of Authority (PoA), which can be considered beneficial while implementing the smart contract in the blockchain. Applying the sharding technique with the consensus mechanisms focuses on enhancing privacy and trust in transactions. It will be difficult for an intruder to identify the transaction, which is being managed and processed by the individual nodes. Also, Blockchain 3.0 does not involve any miner and their transaction fee for validating the transactions as they use a built-in mechanism, which further decreases the users' transaction costs. So, additional consensus mechanisms and additional technologies introduced in Blockchain 3.0 prove to be beneficial in replacing the issues that occurred in the previous generations of the blockchain [26].

**Benefits**—The blockchain has evolved from the second generation to the third generation, adapting additional consensus mechanisms such as PoA, PoW, and PoS and introducing DApps to make smart contracts visible and accessible to the users. The advancement of Blockchain 3.0 with its several benefits makes the network scalable, which can be explained in the following steps [27]:

- The DApps involved in the third generation of blockchain eliminate the dependency on a centralized server as all the participant nodes work on verifying and accepting the transactions.
- Also, consensus mechanisms implemented with the smart contracts secure the exchange of transactions between users using the sharding technique, making it difficult for malicious users to access the transactions.
- There is no requirement for miners to mine the transactions, making it less costly for users as they don't have to include the transaction fee.
- Therefore, Blockchain 3.0 has improved the scalability of the previous generation and enhanced the network's security with the applied advanced techniques.

**Implementation**—Directed Acyclic Graph (DAG) can be built utilizing the properties and technologies of Blockchain 4.0. It usually consists of a chain of blocks in which each block refers to the previous hash of the block, due to which it is easy to trace blocks in the network. But, in DAG, blocks should be arranged in such a way that you can't revert to the previous block as blocks are connected in a forward manner only. Figure 2.4 shows the structure of the third generation of blockchain, i.e., DAG, in which blocks are connected in one way only that can incorporate parallel nodes while increasing the scalability of the network. We have already discussed the sharding technique that enables individual nodes to handle their transactions in the network of a connected graph. There are many other projects such as Byteball, IOTA, and IoT Chain (ITC), using the third generation of blockchain for industrial purpose [27, 28].

**Flaws**—We have discussed several advantages of Blockchain 3.0 over its previous generations. But, Blockchain 3.0 has not been explored to that extent to incorporate

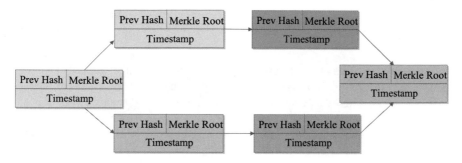

**Fig. 2.4** Structure of directed acyclic graph (DAG) using blockchain

the advancements in the blockchain to make it reliable and efficient for business purposes. There is an enormous usage and adoption of blockchain in various industries and organizations. Still, Blockchain 3.0 does not include innovative technologies such as Artificial Intelligence (AI), Human–Computer Interaction (HCI), and many more to make it accessible and adaptable for Industry 4.0. One more limitation is that consensus mechanisms used in Blockchain 3.0 are way more complex to implement with smart contracts. Thus, Blockchain 4.0 has been developed with the emerging technologies, which will be discussed in the next section [29].

**Summary**—We have explored how Blockchain 3.0 is better to deploy smart contracts involving DApps to enhance the security, scalability, and processing of transactions in the network. Also, they have used various consensus mechanisms to make the network scalable. But, Blockchain 3.0 is not utilized in Industry 4.0, which formally needs the integration of advanced technologies. That is the reason Blockchain 4.0 was initiated in 2018.

## 2.1.4 Blockchain 4.0

*Era: 2018–2019*

**Motivation**—Previous generations of blockchain have not expanded to fulfill the demand of industries involving breakthrough and advanced technologies. For instance, we need a new generation of blockchain to achieve the required industrial automation for Industry 4.0. The main requirement of Industry 4.0 is to acquire cybersecurity and innovative technologies such as supply chain management, the Internet of Things (IoT), and AI. This is the main reason to introduce a new age of Blockchain 4.0 with the advancements in technologies to serve businesses with better security, privacy, transparency, and data integrity. Therefore, InterValue is being used to build a platform for Blockchain 4.0. The main aim of InterValue is to develop an enhanced version of DAG with improved scalability, increased usability, and reliability with Blockchain 4.0. It has improvised the previous generations of blockchain, leading to a new generation with improved features so that Industry 4.0 can benefit from it by deploying their applications [30].

**Implementation**—Many works are going on for the implementation of InterValue in Blockchain 4.0. InterValue is working on releasing its latest version of the v2.0 testnet, which involves a Hashnet consensus mechanism through deployed smart contracts. This new release of InterValue Hashnet will enhance the scalability and data rate at which transactions can be processed, which will be increased up to 140000 transactions per second. However, InterValue has its previous version of the v1.0 testnet that has already been released. The earlier version of testnet consists of Hashnet and Byzantine Agreement based on Verifiable Random Function (BA-VRF) consensus to execute transactions with improved data rate and enable an environment that can incorporate multiple applications. Hashnet is based on the improved and restructured DAG to be deployed in Blockchain 4.0. InterValue group is continuously working to integrate the advanced technologies and make Blockchain 4.0 accessible for organizations such as healthcare, financial, and educational sectors in different application-based scenarios, as shown in Fig. 2.5 [30, 31].

The Internet has completely transformed the lifestyle of people. It is being greatly impacted by emerging technologies such as AI, IoT, and blockchain leading to Web 3.0. Therefore, Blockchain 4.0 and its applications can also be applied in Web 3.0 other than Industry 4.0. With the emergence of Web 3.0, we require a decentralized system so that all confidential data can be stored and accessed securely as the previous generation of Web 3.0, which is Web 2.0, used to have a centralized system that manages all the data transactions making the user's data insecure. Thus, Blockchain 4.0 and its advanced features are the primary motivation that enhances Web 3.0, which provides a distributed network for securing data transactions [32].

**Benefits and Flaws**—We have discussed the applications of Blockchain 4.0 and how we can apply it in various industrial sectors. We want to remark that it has numerous advantages of scalability and better transaction rate, and it supports interoperable blockchains, i.e., one or more blockchain platforms can communicate with each other, which is known as cross-chain transactions. But, still, there is scope to increase the scalability, security, and transparency of the network for multiple industries working in different sectors [32]. Thus, Blockchain 5.0 as a fifth-generation technology is the current evolution of blockchain that will be discussed in the next section.

**Summary**—Blockchain 4.0 is also known as the 'Cross-chain' function, as it is implemented in almost all major domains involving healthcare, sports, elections, postal services, and the government sector. Furthermore, we have discussed how the group of InterValue will release the new version of testnet implementing the Hashnet consensus mechanism, a kind of DAG structure. It also facilitates organizations with interoperable blockchain platforms to process transactions faster with the applied

**Fig. 2.5** Blockchain for
Industry 4.0

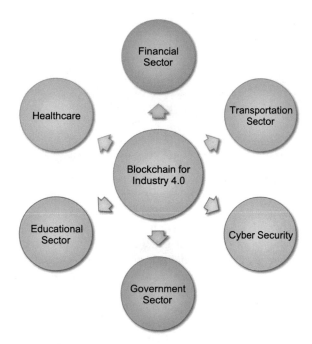

security measures. Still, Blockchain 4.0 can be evolved further to a new generation
of Blockchain 5.0, which is the latest generation of blockchain at present.

## *2.1.5 Blockchain 5.0*

*Era: 2020–Present*

**Motivation**—We have discussed all the previous generations of blockchain, which
provide several benefits by adopting innovative technologies. In the previous era of
blockchain, the InterValue group is working to make blockchain less cost-efficient
and have the fastest transactions data rate. This is the reason for the evolution of
Blockchain 5.0 so that multiple organizations can utilize it in their application with
enhanced scalability, economically reliable, highly secure, transparent, and confidentiality. Relictum Pro is the first-known implementation of Blockchain 5.0. It
is implemented on a network in which multiple smart contracts can be executed
for transactions, which proves to be more secure than any other previous era of
blockchain. Many organizations are simultaneously working to improve Blockchain
5.0 and its features with Relictum Pro, which has several benefits as compared to
previous generations [33],

**Benefits**

- Relictum Pro can be considered as a backbone to store all data transactions of users in a single system so that any confidential data related to the health care or government can be made secure with its help.
- It combines all the advanced technologies such as AI, supply chain, and HCI to implement multiple number of smart contracts in the network with high security and high reliability.
- It helps to improve the performance of industries by implementing complex projects and their solutions with Blockchain 5.0.

We have discussed all the generations of blockchain, starting from Blockchain 1.0 to Blockchain 4.0. As we forward toward the new generation, blockchain and its features are getting advanced adapting the technologies to implement smart contract with security, efficiency, reliability, and many more advancements. The main aim of blockchain's newest generation is to expand its usability for industries so that they can implement multiple applications in different fields. Now, we can show the comparison between different generations of blockchain in tabular form to highlight how the latest era of Blockchain 5.0 is better than the previous generations. Table 2.1 shows the comparative analysis of different eras of blockchain [2] based on the various parameters such as underlying technology, consensus mechanism, data rate, energy consumption, and scalability.

**Table 2.1** Comparative analysis of different generations of blockchain

| Parameters | Blockchain 1.0 | Blockchain 2.0 | Blockchain 3.0 | Blockchain 4.0 |
|---|---|---|---|---|
| Underlying technology | Distributed ledger technology (DLT) | Smart contracts | Decentralized Apps (DApps) | Blockchain with AI |
| Consensus mechanism | Proof of Work | Delegated Proof of Work | Proof of Stake, Proof of Authority | Proof of Integrity |
| Validation | By miners | Through smart contracts and miners | In-built verification mechanism via DApps | Automated verification via sharding |
| Scalability | Not scalable | Poorly scalable | Scalable | Highly scalable |
| Intercommunication | Not possible | Not possible | Possible | Possible |
| Data rate | 7 TBS | 15 TBS | 1000 s of TBS | 1 M TBS |
| Cost | Expensive | Cheaper | More cheaper | Cost effective |
| Energy consumption | Highest | Moderate | Energy efficient | Highly efficient |
| Example | Bitcoin | Ethereum | IOTA, Cardano, Anion | SEELE, Unibright |
| Application | Financial sector | Non-financial sector | Business platforms | Industry 4.0 |

## 2.2  Summary

Over the years, the level of sophistication in technology eventually led to the evolution of blockchain. We saw in detail what kind of forms blockchain has taken over the years and its methods to transform into a system that conducts transactions in a trusted, safe, and secure environment. The chapter first introduced the origin of the blockchain; we then discussed the historical perspective of the blockchain with time as demands grew and technological innovations that keep evolving at a rapid pace. Then we discussed the elements of blockchain, its working, benefits, and limitations, and finally, we discussed a case study on bitcoin.

## 2.3  Practice Questions

### 2.3.1  Multiple Choice Questions

- The year in which the Merkel tree was introduced to tackle the enormous amount of data in secure manner.

  - 1991
  - 1902
  - 1892
  - 1992

- The year in which the first digital decentralized currency 'Bit Gold' was proposed.

  - 1888
  - 1999
  - 1998
  - 1898

- Name the inventor who first proposed digital currency 'Bit Gold'.

  - Nick Manison
  - Nick Szabo
  - Satoshi Nakamoto
  - Nick Martin

- Select any one benefit of Blockchain 1.0 that benefits various industries and institutions.

  - Reduces the double spending
  - Increases latency
  - Reduces scalability
  - None of the above

- What is the first-known implementation of Blockchain 5.0.

    - Relictum Pro
    - Remix IDE
    - Hyperledger Fabric
    - None of the above

- What is the era of Blockchain 2.0?

    - 2008–2013
    - 2013–2015
    - 2010–2015
    - None of the above

- What is the main requirements of Industry 4.0?

    - Acquire cybersecurity
    - Involvement of AI
    - Internet of Things
    - All of the above

- Name the inventor who has proposed Ethereum.

    - Vitalik Buterin
    - Nikolay Lobachevsky
    - Aleksander Stoletov
    - Dmitri Mendeleev

- What is the data rate of Blockchain 3.0?

    - 1000 TBS
    - 15 TBS
    - 200 TBS
    - None of the above

- Select the tool that is introduced in the Blockchain 3.0.

    - Decentralized Apps (DApps)
    - Hyperledger Fabric
    - Hyperledger Sawtooth
    - None of the above

## 2.3.2    Fill in the Blanks

1. The first bit coin block was mined in 2009, it is called as _____.
2. The goal of new-generation blockchain was to build a platform to build an application called as _____.

3. Name the algorithm _____ which is used for Proof of Work.
4. Blockchain 1.0 overcomes the issue of centralized system by providing _____ system.
5. The currency 'Bitcoin' serves as a reward for a process called _____.
6. Blockchain 1.0 has a few flaws while including selfish or corrupt miners which get more _____ while mining the transaction.
7. Blockchain 2.0 as second-generation technology combines technology such as _____ with consensus mechanism.
8. Smart contracts are designed in the Ethereum using a _____ programming language.
9. The new-generation Blockchain 4.0 comes with new additional technologies such as _____ and _____.
10. The new release of InterValue Hashnet in Blockchain 4.0 enhances the scalability and data rates upto _____.

### 2.3.3 Short Questions

1. When was blockchain invented?
2. What are the limitations of the previous generation of blockchain?
3. Explain the advantages of newly generated Blockchain 5.0.
4. Enlist and compare different versions of Blockchain in a tabular format.
5. Explain the drawbacks of Blockchains 1.0, 2.0, and 3.0 which led to Blockchain 4.0

### 2.3.4 Long Questions

1. Explain Blockchain distributed ledger and how it is different from a traditional ledger?
2. Explain how Blockchain 4.0 is assisting in the Healthcare sector?
3. What are smart contracts and how it is helping in Blockchain 3.0 and onwards?
4. How Blockchain 2.0 is enhancing the transaction between two parties?
5. Explain Directed Acyclic Graph (DAG) in the context of Blockchain 4.0.

# References

1. Gupta R, Kumari A, Tanwar S, Kumar N (2020) Blockchain-envisioned softwarized multi-swarming UAVs to tackle COVID-19 situations. IEEE Netw. https://doi.org/10.1109/MNET.011.2000439

2. Mukherjee P, Pradhan C (2021) Blockchain 1.0 to Blockchain 4.0—the evolutionary transformation of blockchain technology. In: Panda SK, Jena AK, Swain SK, Satapathy SC (eds)
   Blockchain technology: applications and challenges. Intelligent systems reference library, vol
   203. Springer, Cham. https://doi.org/10.1007/978-3-030-69395-4_3
3. 'A timeline and history of blockchain technology' by Robert Sheldon (2021). https://whatis.
   techtarget.com/feature/A-timeline-and-history-of-blockchain-technology. Accessed 18 Jan
   2022
4. Patel MM, Tanwar S, Gupta R, Kumar N (2020) A deep learning-based cryptocurrency price
   prediction scheme for financial institutions. J Inf Secur Appl 55:102583. https://doi.org/10.
   1016/j.jisa.2020.102583. ISSN 2214–2126
5. 'Blockchain tutorial: history of blockchain' by java T point (2021). https://www.javatpoint.
   com/history-of-blockchain. Accessed 18 Jan 2022
6. Gupta R, Tanwar S, Kumar N, Tyagi S (2019) Blockchain-based security attack resilience
   schemes for autonomous vehicles in industry 4.0: a systematic review. Comput Electr Eng 86.
   https://doi.org/10.1016/j.compeleceng.2020.106717
7. Kakkar R, Gupta R, Tanwar S, Rodrigues JJPC, Coalition game and blockchain-based optimal
   data pricing scheme for ride sharing beyond 5G. IEEE Syst J. https://doi.org/10.1109/JSYST.
   2021.3126620
8. 'History of blockchain' by Binance Academy (2021). https://academy.binance.com/en/articles/
   history-of-blockchain. Accessed 18 Jan 2022
9. 'Ethereum: decentralized applications and autonomous organizations' by Zeitgeist Lab
   (2021). http://www.zeitgeistlab.ca/doc/Ethereum_Decentralized_Applications_Autonomous_
   Organizations.html. Accessed 18 Jan 2022
10. 'Ethereum' by Wikipedia (2022). https://en.wikipedia.org/wiki/Ethereum. Accessed 18 Jan
    2022
11. Akram SV, Malik P, Singh R, Gehlot A, Tanwar S (2020) Adoption of blockchain technology
    in various realms: opportunities and challenges. Secur Priv 3:e109. https://doi.org/10.1002/
    spy2.109
12. 'Blockchain evolution: from 1.0 to 4.0' by Unibright.io (2017). https://unibrightio.medium.
    com/blockchain-evolution-from-1-0-to-4-0-3fbdbccfc666. Accessed 20 Jan 2022
13. Ølnes S, Ubacht J, Janssen M (2017) Blockchain in government: benefits and implications of
    distributed ledger technology for information sharing. Gov Inf Q 34(3):355–364. https://doi.
    org/10.1016/j.giq.2017.09.007. ISSN 0740-624X
14. 'A brief history of blockchain' by Future Learn (2022). https://www.futurelearn.com/info/
    courses/introduction-to-blockchain-dlt/0/steps/250288. Accessed 20 Jan 2022
15. Kumari A, Gupta R, Tanwar S, Tyagi S, Kumar N (2020) When blockchain meets smart grid:
    secure energy trading in demand response management. IEEE Netw 34(5):299–305. https://
    doi.org/10.1109/MNET.001.1900660
16. Narayanan A, Bonneau J, Felten E, Miller A, Goldfeder S (2016) Bitcoin and cryptocurrency
    technologies: a comprehensive introduction. Princeton University Press, Princeton
17. Xu M, Chen X, Kou G (2019) A systematic review of blockchain. Financ Innov 5:27. https://
    doi.org/10.1186/s40854-019-0147-z
18. 'Blockchain transaction life-cycle' by GeeksforGeeks (2021). https://www.geeksforgeeks.org/
    blockchain-transaction-life-cycle/. Accessed 20 Jan 2022
19. Ankalkoti P, Santhosh SG (2017) A relative study on bitcoin mining. Imp J Interdiscip Res
    (IJIR) 3
20. 'Blockchain 2.0 - from bitcoin transactions to smart contract applications' by Jerome
    Kehrli (2016). https://www.niceideas.ch/roller2/badtrash/entry/blockchain-2-0-from-bitcoin.
    Accessed 20 Jan 2022
21. Bhutta MNM et al (2021) A survey on blockchain technology: evolution, architecture and
    security. IEEE Access 9:61048–61073. https://doi.org/10.1109/ACCESS.2021.3072849
22. 'Blockchain 2.0, smart contracts and challenges' by Martin von Haller Gron-
    baek (2016). https://www.twobirds.com/en/news/articles/2016/uk/blockchain-2-0--smart-
    contracts-and-challenges. Accessed 21 Jan 2022

23. '101 smart contracts and decentralized apps in ethereum' by Pablo Cibraro (2021). https://auth0.com/blog/101-smart-contracts-and-decentralized-apps-in-ethereum/. Accessed 21 Jan 2022

24. Chen T, Li X, Luo X, Zhang X (2017) Under-optimized smart contracts devour your money. In: 2017 IEEE 24th international conference on software analysis, evolution and reengineering (SANER), pp 442–446. https://doi.org/10.1109/SANER.2017.7884650

25. 'Blockchain 3.0 and the future of the decentralized internet' by Sara Technologies Inc. (2018). https://saratechnologiesinc.medium.com/blockchain-3-0-the-future-of-the-decentralized-internet-63ba199e2a5. Accessed 21 Jan 2022

26. 'Solving the blockchain scalability issue: sharding vs sidechains' by Linda Willemse (2018). https://medium.com/swlh/solving-the-blockchain-scalability-issue-sharding-vs-sidechains-1b0c6d1f6c0d. Accessed 21 Jan 2022

27. 'Explaining directed acylic graph (DAG), the real blockchain 3.0' by Sherman Lee (2018). https://www.forbes.com/sites/shermanlee/2018/01/22/explaining-directed-acylic-graph-dag-the-real-blockchain-3-0/?sh=3016383b180b. Accessed 21 Jan 2022

28. 'A complete guide to directed acyclic graph (DAG)' by Dr Ravi Chamira (2022). https://www.sofocle.com/blog/a-complete-guide-to-directed-acyclic-graph-dag/. Accessed 21 Jan 2022

29. 'Can blockchain 3.0 finally make the technology more mainstream' by Naveen Joshi (2019). https://www.allerin.com/blog/can-blockchain-3-0-finally-make-the-technology-more-mainstream. Accessed 21 Jan 2022

30. 'InterValue the world's first practical blockchain 4.0' by Aminur Rahaman (2018). https://medium.com/@aminurrahaman/intervalue-the-worlds-first-practical-blockchain-4-0-b9324878c262. Accessed 21 Jan 2022

31. 'InterValue's TestNet 2.0 release: implementation of HashNet consensus, committed to building the practical infrastructure of the Blockchain 4.0 era' by InterValue (2018). https://www.globenewswire.com/news-release/2018/06/28/1531189/0/en/InterValue-s-TestNet-2-0-release-implementation-of-HashNet-consensus-committed-to-building-the-practical-infrastructure-of-the-Blockchain-4-0-era.html. Accessed 21 Jan 2022

32. 'Blockchain 4.0' by Akash Takyar (2022). https://www.leewayhertz.com/blockchain-4-0/. Accessed 21 Jan 2022

33. 'What is Blockchain 5.0, and how Relictum Pro is set to revolutionize the industry?' by Relictum Pro (2019). https://relictumpro.medium.com/what-is-blockchain-5-0-and-how-relictum-pro-is-set-to-revolutionize-the-industry-c7aa23df7190. Accessed 21 Jan 2022

# Chapter 3
# Decentralization and Architecture of Blockchain Technology

**Abstract** In this chapter, we have described the necessity of decentralization in the current business market, which still resorts to a centralized system. It includes the methods of decentralization which include disintermediation which removes the intermediaries from the centralized system to provide complete decentralization, and competition which gives partial decentralization. A comparative analysis of centralized, distributed, and decentralization is conducted. However, it is also essential to measure the correctness of the decentralization once it is adopted by the application or platform. For that, a decentralization index has been formulated using Proof of Work (PoW) and Proof of Stake (PoS). The subsequent section discusses the benefits and challenges of decentralization in blockchain technology. Then in Notion of distributed consensus, the working of consensus algorithms with permissionless protocol is discussed. Lastly, this chapter involved a case study on blockchain-based decentralized supply chain management.

**Keywords** Decentralization · Blockchain · Disintermediation · Consensus

## 3.1 Introduction

The ever-increasing demand for data in businesses makes them data-driven to optimize their decision-making process and economic growth. As a result, data is an essential commodity that is maneuvered by every business. Additionally, it aims to quickly and accurately send and receive the data to their subordinates, giving confidence and escalating their business. Such businesses are constantly collecting enormous amounts of personal data to personalize services, better prediction, and evolve marketing strategies. However, despite reaping all the benefits from data-driven businesses, privacy concerns incur the business's economic growth. Additionally, the data is accumulated in a centralized system that the adversaries might exploit by discovering the vulnerabilities in the system. Further, the business entities are sharing the data and assets via third-party services, provoking data and transaction manipulation attacks. Recent controversial happenings such as Facebook Analytica [12], government mass surveillance [6], and frequent data breaches [17] made users and businesses think of practical ways to share their data and asset.

© The Author(s), under exclusive license to Springer Nature Singapore Pte Ltd. 2022  63
S. Tanwar, *Blockchain Technology*, Studies in Autonomic, Data-driven and Industrial
Computing, https://doi.org/10.1007/978-981-19-1488-1_3

Blockchain facilitates the urge of every business to trade their asset and data with a trackable, immutable, and decentralized ledger. It is a technology that eliminates the intervention of third-party services and gives complete control of data transactions in users' hands, wherein a user sends the data into immutable blocks of chains. Each entity of the blockchain poses an identical copy of this transaction which makes this technology transparent and efficient. In the blockchain, decentralization plays an essential role in transferring accountability and governance from a centralized system to a distributed system. It first came into existence when Satoshi Nakamoto developed the first blockchain-based application, i.e., Bitcoin [22]. Since it is distributed, anyone can add and remove the data in the blockchain and ensure data ownership, data auditability, and levigated access control.

## 3.1.1   Need for Decentralization

The traditional centralized system revolves around the central authority to manage and function the data. For example, social media platforms such as Facebook, Twitter, and many more are central authorities that govern and have complete control over the user data and different platform features like who can register the platform, resource accessibility, etc. Moreover, it involves the mediator governing bodies that assess the data between two communicating entities. This signifies that the centralized system has enviable control over the personal information shared with them [19]. Consequently, certain constraints make the centralized system obscure and not trustworthy. Trust is an important factor that acts as a mutual agreement between a centralized system and users. However, attackers can compromise organizations from time to time to acquire the user's data which implies it is easy to break this agreement [20]. Also, it is effortless for the attacker to reconnaissance a few IP addresses to find vulnerabilities in the centralized system. Later these vulnerabilities lead to severe attacks such as denial of service (DoS), SQL injection, and zero-day exploitation that deteriorate the user's privacy and security [11].

Moreover, centralization also suffers from a single point of failure that leads thwarts the business's productivity and compromises service availability [21]. It is an undesirable failure to the business that requires a constant data feed such as networking, supply chain, and real-time monitoring applications. Finally, the centralized system hinders scalability, i.e., the data is accumulated at a central place, so only a few users can access them. However, a large audience gets deprived of that data. To overcome the aforementioned issues, the organization has adopted the blockchain-based decentralization approach, which was first coined at bitcoin's release. Here, a user can send a bitcoin to another user using a decentralized network, eliminating the need for a central authority. The transaction is visible to everyone involved in the blockchain network utilized by bitcoin, making the network transparent to the users [5]. Further, they employ consensus algorithms that verify each transaction from the users. In the long run, it is beneficial for each organization to embrace decentralization due to its perpetual characteristics such as transparent transactions, full ownership, concrete security, and open development.

### 3.1.2 Comparative Analysis with Other Approaches

Traditional applications heavily depend on the centralized system for governing and handling the data. However, this centralization lures severe security threats, compromising user privacy and security. Therefore, the distributed system encouraged segregating the centralized data in multiple data centers that are controlled by the network or service provider. Such a network has a high fault tolerance against failures because of the interconnection of numerous data centers, so if one data center is down, we still have a replica of the data in the other data center [25, 27]. Nevertheless, the central authority does the management of all interconnected data centers. It can be observed that even though the data is distributed, it is still managed and governed by the central authority (Table 3.1).

Decentralization is a promising solution where data and resources are shared between members of the network, and each member has an identical copy of the database [8, 31]. From the viewpoint of the blockchain, decentralization is further bisected into two, i.e., full decentralization and semi-decentralization. There is no regulatory body in a full decentralization-based blockchain network that supervises and controls the network. It requires a considerable amount of computation to maintain the blockchain ledger for higher scalability. Every node aims to achieve consensus to verify the transaction by unraveling a resource-intensive complicated problem. The shortcoming of this approach is that if a few nodes have better computation resources, they get the consensus quickly. Every time those nodes verify the transaction, not all, raises privacy issues in the fully decentralized network. Therefore, semi-decentralization-based blockchain was introduced to confront the aforementioned issue wherein only a few authorized bodies verify the transactions. Consequently, in semi-decentralization, all nodes are not trying to get the consensus; only a few authorized participants are involved in obtaining the consensus. A comparative analysis of centralized, distributed, fully decentralized, and semi-decentralized network is presented as follows:

### 3.1.3 Methods of Decentralization

Technological advancements have transformed business and business activities to rely on participative stakeholders and bottom-up approaches instead of centralized and top-down approaches. As a result, the regulatory bodies that provide reliable confidence among business entities have been eliminated from the current businesses, adopting confidence among entities in a decentralized manner. Every sector has a primary centralized authority that manages the flow of transactions, which can be in

**Table 3.1** Comparative analysis of centralized, distributed, fully decentralized, and semi-decentralized network

|  | Centralized | Distributed | Fully decentralized | Semi-decentralized |
|---|---|---|---|---|
| Resource management | Manage by a single central authority | Managed by network provider | Managed by internal members of the network | Managed by the authorized regulatory body |
| Data ownership | Controlled by a single central authority | Controlled by a service provider | Controlled by the nodes who passed the consensus | Owned by authorized node who passed the consensus |
| Single point of failure | Yes | No | No | No |
| Fault tolerance | Low | High | High | Extremely high |
| Security | Provided by the central entity | Shared responsibility between network and service provider | Depends on the number of nodes in the blockchain | Provided by the authorized regulatory body |
| Performance | Managed and controlled by the central authority | Increases as the network and resources increase | Decreases as the number of nodes increases | High performance as a few competent nodes are managing the network |
| Reliability | Less reliable | Moderate reliable | High reliable | Extremely high reliable |

the form of bank law for the bank sector, supervision of energy consumption for the energy producers, tariffs, and safety for public transportation. However, managing and assigning responsibilities in a centralized system are cumbersome [16]. At the same time, it has been noticed that many industry sectors are adapting the decentralization approach, such as energy producers are incorporating smart grids, banking sectors are using peer-to-peer currencies (cryptocurrencies), and many more. This transformation provokes other sectors to embrace decentralization without the intervention of central authority. Recently, blockchain technology has attracted a lot of attention in providing decentralization to various industries, especially in the financial sector. There are two methods through which we can achieve blockchain-based decentralization, which are as follows:

– *Disintermediation*—It is the process of eliminating the third-party regulatory bodies from the business pipeline, and allowing the business entities to transact directly. This elimination ensures the privacy and security of the user, further it can increase the efficiency of the transaction and reduce cost [9, 28]. Let us take an analogy as shown in Fig. 3.1 to understand this method. Due to the COVID-19 pandemic, the patient received an electronic record and it is not allowed to meet anyone in person.

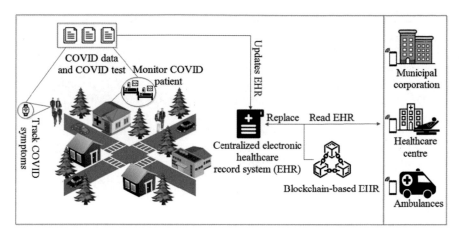

**Fig. 3.1** Blockchain-based decentralized electronic healthcare record system

The electronic record contains patient's private information such as phone number, home address, and social security number. He chooses an online platform that has a third-party service to securely share this information with the doctor. Hence, he needs to pay the fees to use this platform to maintain his privacy. However, the third party can be compromised by attackers, or it can lose the trust factor; in that case, the user's personal data is at high risk of being exploited. But if he uses a blockchain-based Electronic Health Records Monitoring System, then all he needs to do is get the blockchain id of the doctor and send the reports to him [7]. As a result, it removes the dependency on the third-party services and accomplishes the decentralization to send his report to the doctor, also he doesn't have to pay any fees to use the platform.

– *Competition*—This method involves a centralized control by the authorized service provider, but these service providers need to compete with each other to get selected for providing the service [23]. Although it does not provide complete decentralization to the users, it makes sure that there is no monopoly by a certain service provider. When it comes to blockchain technology, the consensus mechanism works as the tiebreaker between the service providers. This mechanism selects a service provider based on the quality of service it provides, reviews and reputation earned, and credits earned.

### 3.1.4 Correctness of Decentralization

Once the application adopts decentralization, it is essential to measure its reliability, whether it fully adapts to decentralization or an abstracted version of it. There are two possible ways to measure the correctness of decentralization. The existing centralized system managed and censored our data by third-party services in which the user relinquished his privacy. We can apprehend our privacy using decentralization, but the applied decentralization has to be measured for its correctness. Therefore, a decentralization index can be formulated that adheres to each application or platform. Then, a user can assess the index to determine the correctness of decentralization. The Decentralization index can be acquired using Proof of Work (PoW) and Proof of Stake (PoS) [4]. In PoW, each node that wants to participate in the blockchain has to solve the hash puzzle. To solve this puzzle, each node should have ample computation power to obtain the block generation authority. A simple rule is the higher the computation power, the higher the probability of obtaining block authority. The nodes succeed the competition of block authority assigns the hash rate. Then, we can pile up the hash rate in ascending fashion to obtain the decentralization index as shown in Eq. 3.1, which is in the range from 0 to 1.

$$D_{ind,\text{pow}} = 1 - 2 \times \left( \frac{1}{2} - \frac{1}{m} \sum_{i=1}^{m} \mathcal{H}_i \right) \tag{3.1}$$

where $D_{ind,\text{pow}}$ is the decentralization index in PoW, $m$ represents the nodes who are participating in the blockchain, and $\mathcal{H}_i$ represents the hash power of the $i$th node.

Another approach is through PoS, where the block and transaction are verified using coins. The coin owners provide their coins as security collateral to get the chance to become validators. A simple rule of PoS is the higher the coin holding, the higher the opportunity to become a validator. We can measure the decentralization as shown in Eq. 3.2 of the application by checking their coins holding.

$$D_{ind,\text{pos}} = 1 - 2 \times \left( \frac{1}{2} - \frac{1}{m} \sum_{i=1}^{m} \mathcal{S}_i \right) \tag{3.2}$$

where $D_{ind,\text{pos}}$ is the decentralization index in Pos, and $\mathcal{S}_i$ represents the share or coin's holding of the $i$th node.

The nodes which are distributed in different geographic regions can also influence the correctness of decentralization. For example, if nodes that are selected to maintain decentralization are located in the same region, then those nodes are heavily dominant in the blockchain network [4]. Therefore, the decentralization index using variance and rescaling is calculated in accordance with the geographic diversity of the node; it is represented as

$$\mathcal{D}'_g = \left( 2 - \frac{\log_{C+1} C_t - \log_{C_t+1} C_t}{\log_2 C_t - \log_{C_t+1} C_t} \right) \times \sqrt{\frac{\sum_{i=1}^{C_t} (n_i - \mu)^2}{C_t}} \tag{3.3}$$

$$\mathcal{D}_g = \frac{\left( \mathcal{D}'' - \mathcal{D}'_g \right) - \mathcal{D}^{\lceil \rangle \int}_g}{\mathcal{D}'' - \mathcal{D}^{\lceil \rangle \int}_g} \tag{3.4}$$

where $\mathcal{D}'_g$ is the decentralization index for selected application in geographic diversity, $C_t$ is the total number of countries, $\mu$ is the average value of nodes in countries, $\mathcal{D}_g$ represents the decentralization index for geographic diversity, and $\mathcal{D}''$ represents the index of all nodes located in one region.

## 3.1.5 Benefits of Decentralization

Various benefits of decentralization in blockchain technology are described as follows:

1. *Offers Trustless Environment*: A decentralized network gives a trustless environment to the blockchain network, where no member has to trust other members. Each blockchain member has an exact replica of the distributed ledger, making it impossible for an attacker to update all instances of the data. If anyone's ledger is modified, then that modification is being rejected by other blockchain members [2].
2. *Single Point of Failure*: Decentralization reduces the chances of a single point of failure in the blockchain network. It allows distributed members to keep the same data, which creates multiple breaking points which means if any member node goes down, it does not affect the overall data availability and security [26].
3. *Leadership Opportunity*: Centralized systems do not offer room for their members to display their leadership skills, as the entire management is taken care of by the centralized controller. It can overburden the centralized controller and can even lead to system performance degradation. Decentralization provides leadership opportunities to each Members/blockchain nodes to step up and utilize their leadership qualities, which benefits the blockchain network.
4. *Minimizes Data Transmission Errors*: Organizations generally do data exchanges with other companies, which need to be transformed into their standard formats. Data transformations create a space for data loss or modification, which leads to incorrect results. So, implementing decentralization (using blockchain) in an organization allows for the real-time data view.
5. *Data Transparency*: Decentralization offers transparency to the blockchain network where each member views and validates the modifications performed by the other members. It is because each member has a real-time replica of the distributed ledger.
6. *Fair Execution*: If there is an Insurance company centrally managed by some group, it may delay the payment of the insurance. But if the same business was carried out using a decentralized application, the fund gets transferred immediately as per the smart contract conditions. So, decentralization guarantees fair execution of the agreements.
7. *Immediate Decision-Making*: In centralized systems, the remote nodes have to wait for decisions to be approved by the controller. This increases the waiting time and security risks, whereas in a decentralized blockchain system, the members

have the real-time data and can take decisions independently in an efficient way [29].

### 3.1.6 Challenges of Decentralized Organizations

Various challenges involved in achieving decentralization in the blockchain network are elaborated as follows:

1. *Data Consistency*: In a blockchain network, each member has an exact copy of distributed ledger, which creates a multi-data redundancy problem. If any blockchain member updates its distributed ledger, then it is quite challenging to update its all instances (in the redundant copy of all distributed ledgers with all blockchain members).
2. *Computational Capacity*: In decentralization, each blockchain member has independence of decision-making and leadership opportunities. But it is a myth to assume that all blockchain members have the same storage and computation capabilities, due to the cost associated with them. So, validating transactions at the same pace by all blockchain members is challenging.
3. *Efficient Consensus Mechanism*: The consensus mechanism plays a key role in validating the transaction by agreeing upon common conditions. In the case of a centralized system, developing a consensus mechanism is quite simple as only one controller is involved in taking decisions for transaction validation. But decentralized systems involve many distributed blockchain members that decide upon the validation of a particular transaction, which increases the complexity of consensus mechanism design. So, designing an efficient and scalable consensus mechanism is challenging for blockchain developers.
4. *Minimizing Storage Cost*: Blockchain is a decentralized network, where each member node has an exact replica of the real-time distributed ledger. It increases the data storage cost, as it involves huge data redundancy. Removing data redundancy can increase the chance of a single point of failure. It is challenging to minimize the data storage cost in the blockchain network.

## 3.2 Investigating Decentralized Consensus Mechanisms

A consensus mechanism (also called consensus algorithm) is a set of methodologies used to achieve agreement, faith, and trust among the distributed systems (blockchain nodes or miners) working together. It also maintains the authenticity and security of blockchain transactions. It also ensures blockchain miners agree on the next block and distribute the new block information to all miners [24]. Consensus mechanisms can be either centralized or decentralized. In the centralized consensus mechanism (not preferred in blockchain network), any failure or interruption in the

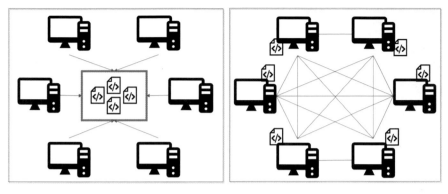

a) Centralized consensus mechanism                    b) Decentralized consensus mechanism

**Fig. 3.2**  Centralized versus decentralized consensus mechanisms [30]

communication link can affect the overall trust and security of the system [30]. Moreover, in a decentralized consensus mechanism, the methodologies are distributed across the blockchain network to achieve a perfect agreement among the blockchain nodes [18]. A decentralized consensus mechanism is effective if it achieves an inescapable agreement under adverse conditions, such as a blockchain node behaving maliciously for its benefit [1]. Figure 3.2 shows the pictorial representation of centralized and decentralized consensus mechanisms. Figure 3.2a shows that the consensus between distributed systems can be achieved with the involvement of a centralized controller or authority. All distributed systems must have to contact the centralized authority for consensus, which is not always feasible due to network channel conditions as well as a single point of failure. Figure 3.2b shows that the consensus can be achieved distributively by eliminating the intermediary. It also frees from a single point of failure and can to share information without relying on the centralized controller.

The effective design of blockchain relies on a good and reliable decentralized consensus mechanism. A transaction in a blockchain network can be stored with the collaboration of a good consensus mechanism. Table 3.2 describes the comparative analysis of various consensus mechanisms regarding permissioned, permissionless, public, private, consortium, and decentralized levels. This table signifies that the consensus mechanism (either permissioned or permissionless) can achieve what level of decentralization, i.e., low, medium, or high. If mechanisms are solely designed for private blockchains such as PoS, PoET, PoC, PaXos, and many more, then they manage to achieve low/medium decentralization levels.

**Table 3.2** Comparative analysis of various consensus mechanisms [10, 30]

| Consensus mechanism | Per | Pl | Pu | Pr | Cor | D_level |
|---|---|---|---|---|---|---|
| Proof of work (PoW) | Y | Y | Y | Y | | High |
| Proof of stake (PoS) | Y | – | – | Y | – | Low |
| Delegated PoS (DPoS) | Y | – | – | Y | – | Low |
| Proof of elapsed time (PoET) | Y | – | – | Y | Y | Medium |
| Practical Byzantine Fault Tolerance (PBFT) | Y | – | – | Y | – | Medium |
| Proof of property (PoP) | – | Y | Y | – | – | High |
| Proof of burn (PoB) | – | Y | Y | – | – | High |
| Proof of authentication (PoAh) | Y | Y | Y | – | – | High |
| Proof of capacity (PoC) | Y | – | – | Y | – | Medium |
| Proof of authority (PoA) | – | Y | – | Y | – | Low |
| Credit-based PBFT (CPBFT) | Y | – | – | Y | – | Medium |
| PaXos | Y | – | – | Y | – | Low |
| Raft | Y | – | – | Y | – | Medium |
| Ripple | – | Y | Y | – | – | Medium |
| Tendermint | Y | – | Y | Y | – | Medium |
| Simplified BFT (SBFT) | Y | – | – | Y | · | Low |
| Proof of distribution (POD) | Y | – | – | Y | – | Medium |
| Fault resilient chain (FRChain) | Y | – | – | Y | – | Low |
| Proof of QoS (PoQ) | – | Y | Y | – | – | High |
| Multi-Level BFT | Y | – | – | Y | – | Low |

Per: Permissioned; Pl: Permissionless; Pu: Public blockchain; Pr: Private blockchain; Cr: Consortium blockchain; D_level:
Decentralization level

## 3.3　Blockchain-Based Decentralized Supply Chain Management System—Case Study

Supply chain management systems are the agencies that manage the distribution of different goods. The people that are involved are the producer, the buyers, and the management team. When the supply chain management adopts the centralized network, it works in the following ways:

- The Buyer places an order of some item. The address of the buyer will be provided at this moment.
- Producer will start the process of production for the order placed.
- Once the product is ready, the producer will forward the details to the supply chain management company. The details include the pick-up address, the delivery address, weight of the package, dimensions, item description, etc.
- Then the supply management will take the package and deliver it to the destination.

**Drawbacks of the centralized supply chain management**

- As the system is centralized, there are chances of monopoly as the producer may not be aware of the transportation charges.
- The product may not be up to the mark.
- A lot of time is wasted as the entire process is manual.
- If the tracking system is not involved, then damage or stealing of goods in the middle of transportation may also take place.

### 3.3.1 Decentralized Supply Chain Management

We can convert the entire supply chain management to a decentralized system using the blockchain. This means that all the steps that are involved in the chain are recorded and all the actors that are included throughout the process are aware of the steps. With the incorporation of decentralization, the supply chain system poses the following properties:

- *Decentralized*—This means that all the data involved in the process will not be owned by some agency centrally but will be shared among all the people that are involved. So, the data that was hidden in the centralized systems will remain hidden no more.
- *Transparent*—This means that the transactions performed by the people that are involved in the entire system are visible to all the others. The privacy of the users will also be protected as the transactions will be displayed by the public names assigned to them when they join the network. But the users need to be honest as all the transactions will be recorded and visible publicly.
- *Immutable*—The transactions will be clubbed into a block and the block will be added to the chain. These blocks are cryptographically secured and are related to the previous blocks in the chain. So, the block will contain the hash of the previous blocks and a change in one block will force the change in the entire chain. This makes the system tamper-proof and immutable.

Whenever the package is handed from one actor to another, the transaction is recorded in the public ledger. The public ledger is the log data present with every user and he can read the log details. The decentralized architecture benefits the system in the following ways:

- It reduces the errors caused due to human involvement.
- It decreases the time delays at every step.
- It also cuts out the added costs.
- It helps people to gain trust in each other and become honest.
- It makes the chain transparent and involves all the actors.

### 3.3.2 Use Case: Food Supply Chain

When there are quality issues in the food industry, then the consumer's health is adversely affected. So, the trust is lost and the business also gets affected. If the entire food supply chain is made decentralized using the blockchain, then we can get the following benefits:

- The food supplied will be fresh as it is an open system.
- It ensures food safety.
- It protects the food from fraudulent members.
- The consumers are fully aware of the quality of their food because the data is available to all the users involved.
- It stops the wastage of food because every packet is recorded and visible to everyone.

### 3.3.3 Use Case—Logistic Supply Chain Management

Blockchain-based decentralization poses a distributed ledger that records immutable transactions of individuals. The distinctive features of blockchain make the technology competent in supply chain management. Recently, blockchain-based supply chain management is introduced in the logistic industry that helps the alliances to track and manage the transactions in a more secure and efficient manner. The logistic sector has many intermediaries through which the supply data is passed; this hackled the operation of the logistic industry as it has to go between various partners and services. The decentralization eliminates the intermediaries, and the industry members themselves can verify and record the transactions. It benefits the logistic sector in the following ways:

- The transaction data is trustworthy as it is verified by the blockchain regulatory bodies.
- Incorporating smart contracts can automate the transaction, which improves the efficiency of logistic operation.
- In the logistic industry, trade and commerce are solely based on documents that can be forged or manipulated by adversaries. Blockchain-based solutions can assist in verifying the documents and the transaction made by the party.

## 3.4 Blockchain-Based Decentralized Healthcare System—Case Study

The world is facing a catastrophic pandemic, that is, COVID-19 which has strained and disrupted the economic growth of nations. In such a situation, people used to

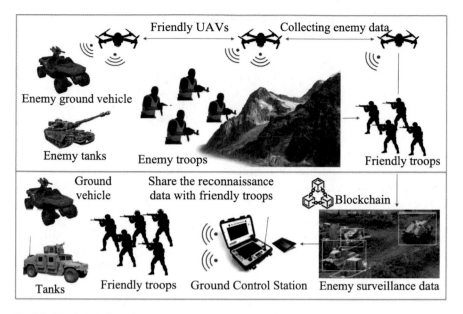

**Fig. 3.3** Blockchain-based decentralized Battlefield application

stay in their homes and were restricted by the regulations imposed by the government. Moreover, patients who were COVID-infected or had other medical disorders cannot directly reach the hospitals due to the risk of getting infected [13, 15]. Therefore, they rely on centralized healthcare system to upload their medical report and consult a suitable doctor for a diagnosis. Centralized systems are not reliable as it always keen under the eye of an attacker; also, it has many intermediate authorities by which the user's data is passed on, which questions the user's privacy. To overcome the aforementioned issues, the involvement of blockchain technology in the healthcare system is inevitable. Here, all healthcare providers have a direct connection to the blockchain network, and all medical data is verified, tracked, and stored in the existing healthcare system [3]. The medical data related to the patient is then stored securely using APIs with the assigned patient ID. As blockchain is a decentralized network, the data is shared among the blockchain member, where smart contracts are executed to convey the inward transaction. Finally, if the patient wants to consult other medical practitioners, they can share their private keys to access the patient's clinical information.

## 3.5 Blockchain-Based Decentralized Battlefield System—Case Study

In early battlefield operations, military personnel use a centralized system to communicate with each other. It could be any conventional communication system such as telegram, telephone, or message forwarding utilizing the Internet [14]. Recently, military applications have been upgraded to the latest advancements, such as communication via unmanned aerial vehicles (UAVs), incorporating artificial intelligence and the blockchain. The inclusion of blockchain in military applications ensures that communication and mission-critical data are secured by the indispensable characteristics of blockchain, such as immutability, decentralization, and transparent transaction. The decentralized blockchain stores the reconnaissance data of enemies collected by military personnel and reliably shared with friendly troopers. It stores this data in immutable blocks that are cryptographically secured. Additionally, smart contracts eliminate the requirement of third-party services to ensure data privacy in a battlefield operation. In Fig. 3.3, the UAVs collect the intelligence data such as the enemy's current location, artillery detection, mine detection, monitoring missile threats, and tracking ground vehicle movements. Then, the collected critical data is forwarded to the blockchain network where it is securely stored by public–private key pairs and only authorized troopers can view it. The ground control station (GCS) at the friendlies extracts the information from the blockchain and informs the current situation of enemies to the friendly troops. This way, the mission-critical information is not manipulated by the attackers and reliably it is passed to its authorized destination.

## 3.6 Summary

The decentralization can solve the user privacy issue, but it brings other risks with it. We need to use both centralized and decentralized architecture to get a near-perfect application. A perfect balance between both architectures gives maximum benefits to the users. An open network is beneficial for distributed consensus. The chapter elaborates on the need for decentralization and its comparative analysis with a centralized and distributed network. Later, it presents decentralized consensus mechanisms along with the parameters by which we can measure the decentralization. Lastly, this chapter present different use cases that have adopted decentralization in their regular business operation to make it free from intermediaries and bring trust and reliability into the business.

## 3.7 Practice Questions

### 3.7.1 Multiple Choice Questions

– The importance of data in business activity is

  • To improve the decision-making process.
  • Because it is required in day-to-day operations.
  • To optimize the economic growth of the business.
  • All of the above.

– What is the most common factor that reaps the economic growth of a data-driven business?

  • User privacy.
  • Distribution of data.
  • Decentralization of data.
  • All of the above.

– What is a node in the context to the blockchain?

  • A blockchain itself.
  • A computer on a blockchain network.
  • A transaction.
  • A cryptocurrency.

– The name of the Bitcoin inventor who coined the term 'decentralization' is

  • Nakamutu Santoshi.
  • Santushi Nhiem.
  • Satoshi Nakamoto.
  • Satoshi Nguyen.

– What is DAPP?

  • A cryptocurrency.
  • A consensus mechanism.
  • A decentralization application.
  • None of the above.

– What happens if one site fails in a distributed system?

  • All sites will continue operating.
  • The remaining sites will continue operating.
  • Whole system gets shut down.
  • None of the above.

– Which of the following is not an disadvantage of centralized system?

  • Easy and quick updates.

- Can lose the data.
- Single point of failure.
- Security.

– Which of the following are the real-world use cases of the blockchain?

- Healthcare system.
- Supply chain management.
- Cloud technology.
- All of the above.

– Who owns the data in distributed network.

- Users.
- Service provider.
- Central authority.
- None of the above.

– How does a centralized system affect the decision-making system?

- Increases the waiting time.
- Maximizes the efficiency.
- Ensures trust.
- All of the above.

### 3.7.2   Fill in the Blanks

1. Blockchain facilitates the urge of every business to trade their asset and data with a trackable, _____, and _____ ledger.
2. Each entity of the blockchain poses an identical copy of this transaction which makes the blockchain technology _____.
3. _____ is an important factor that acts as a mutual agreement between a centralized system and users.
4. The centralization lures severe security threats, compromising user _____ and _____.
5. The blockchain-based decentralization is further bisected into two, that is, _____ and _____.
6. _____ is the process of eliminating the third-party regulatory bodies from the decentralized business pipeline.
7. In a blockchain-based electronic healthcare system, a patient only needs a _____ of the doctor to share the medical report.
8. One can apprehend his privacy using decentralization, but the applied decentralization has to be measured to check its _____.
9. To measure the correctness of decentralization, a decentralization index can be acquired using _____ or _____.

10. Consensus mechanism plays a key role in _____ the transaction by agreeing upon common conditions.

### 3.7.3 Short Questions

1. Compare the centralized, distributed, fully decentralization, and semi-decentralization considering XYZ oil company.
2. Discuss the shortcomings of a centralized system and how decentralization has overcome them.
3. Explain the term ledger. Enlist some common types of ledgers used by the users in blockchain technology.
4. Explain the terms Proof of Work (PoW) and Proof of Stake (PoS) in the context to measure the correctness of decentralization.
5. Enlist the benefits of decentralization and explain them in a detailed manner.

### 3.7.4 Long Questions

1. There is an XYZ oil company which has various central authorities and has a central database to manage the business. Discuss how decentralization can be incorporated into the XYZ oil company to improve the business.
2. Justify the statement "Semi-decentralization is better than Full decentralization". Explain it in the context of the XYZ oil company.
3. Discuss how a blockchain-based healthcare system assists during the COVID-19 pandemic.
4. India is a democratic country, where people elect government ministers using an electoral system that is managed by the Election Commission of India. However, recently, we have seen several issues with the Indian electoral system, which raises the integrity question for the user vote. Discuss how the blockchain-based decentralized network resolves the problems of vote integrity.
5. Explain the use of smart contracts and how they can be incorporated into blockchain technology.

# References

1. Consensus protocols: how are blockchains secure? https://www.ledger.com/academy/consensus-protocols-how-are-blockchains-secure, accessed: 2021-04-01
2. What is decentralization in blockchain? https://aws.amazon.com/blockchain/decentralization-in-blockchain/, accessed: 2021

3. Blockchain-based electronic healthcare record system for healthcare 4.0 applications. J Inf Secur Appl 50:102407 (2020)
4. Dq: two approaches to measure the degree of decentralization of blockchain. ICT Express 7(3): 278–282 (2021). https://doi.org/10.1016/j.icte.2021.08.008
5. Amazon: what is decentralization in blockchain? https://aws.amazon.com/blockchain/decentralization-in-blockchain/
6. Beens RE. The state of mass surveillance. https://www.forbes.com/sites/forbestechcouncil/2020/09/25/the-state-of-mass-surveillance/?sh=1f96510bb62d
7. Bhattacharya P, Tanwar S, Bodkhe U, Tyagi S, Kumar N (2021) Bindaas: blockchain-based deep-learning as-a-service in healthcare 4.0 applications. IEEE Trans Netw Sci Eng 8(2):1242–1255. https://doi.org/10.1109/TNSE.2019.2961932
8. Blockchain. Centralized vs. decentralized: what are the core differences? https://101blockchains.com/centralized-vs-decentralized-internet-networks/
9. bobsguide: will disintermediation happen through the blockchain? https://www.bobsguide.com/articles/will-disintermediation-happen-through-the-blockchain/
10. Bodkhe U, Mehta D, Tanwar S, Bhattacharya P, Singh PK, Hong WC (2020) A survey on decentralized consensus mechanisms for cyber physical systems. IEEE Access 8:54371–54401. https://doi.org/10.1109/ACCESS.2020.2981415
11. Chen S, Wu Z, Christofides PD (2021) Cyber-security of centralized, decentralized, and distributed control-detector architectures for nonlinear processes. Chem Eng Res Des 165:25–39. https://doi.org/10.1016/j.cherd.2020.10.014. https://www.sciencedirect.com/science/article/pii/S0263876220305207
12. Confessore N. Cambridge analytica and facebook: the scandal and the fallout so far. nytimes.com/2018/04/04/us/politics/cambridge-analytica-scandal-fallout.html, accessed: 2021-12-08
13. Gupta R, Kumari A, Tanwar S, Kumar N (2021) Blockchain-envisioned softwarized multi-swarming UAVs to tackle COVID-19 situations. IEEE Netw 35(2):160–167. https://doi.org/10.1109/MNET.011.2000439
14. Gupta R, Tanwar S, Kumar N (2022) B-iomv: blockchain-based onion routing protocol for d2d communication in an iomv environment beyond 5g. Veh Commun 33:100401. https://doi.org/10.1016/j.vehcom.2021.100401. https://www.sciencedirect.com/science/article/pii/S221420962100070X
15. Gupta R, Tanwar S, Tyagi S, Kumar N, Obaidat MS, Sadoun B (2019) Habits: blockchain-based telesurgery framework for healthcare 4.0. In: 2019 international conference on computer, information and telecommunication systems (CITS), pp 1–5. https://doi.org/10.1109/CITS.2019.8862127
16. Hathaliya J, Sharma P, Tanwar S, Gupta R (2019) Blockchain-based remote patient monitoring in healthcare 4.0. In: 2019 IEEE 9th international conference on advanced computing (IACC), pp 87–91. https://doi.org/10.1109/IACC48062.2019.8971593
17. Hill M, Swinhoe D. The 15 biggest data breaches of the 21st century. https://www.csoonline.com/article/2130877/the-biggest-data-breaches-of-the-21st-century.html
18. Krawiec-Thayer MP. What's the big deal about decentralized consensus? https://blog.insightdatascience.com/whats-the-big-deal-about-decentralized-consensus-12876bb80064, accessed: 2018-11-16
19. Leetaru K. What does it mean for social media platforms to "sell" our data? https://www.forbes.com/sites/kalevleetaru/2018/12/15/what-does-it-mean-for-social-media-platforms-to-sell-our-data/?sh=646a45862d6c
20. Menon C. 'Trust' in a Centralized World & Emerging Solutions for the Protection of Human Data. https://blog.streamr.network/trust-in-a-centralized-world-emerging-solutions-for-the-protection-of-human-data/
21. N-Able. Centralized networks vs decentralized networks. https://www.n-able.com/blog/centralized-vs-decentralized-network
22. Nakamoto S (2008) Bitcoin: a peer-to-peer electronic cash system. Decentralized Bus Rev 21260

23. Quiniou M (2019) Blockchain: the advent of disintermediation. Wiley
24. Rosenberg E. What is a consensus mechanism? https://www.thebalance.com/what-is-a-consensus-mechanism-5211399, accessed: 2021-12-01
25. Seal A. Centralized vs decentralized network: which one do you need? https://www.vxchnge.com/blog/centralized-decentralized-network
26. Service N. A decentralized, secure and trustless system. https://www.netservice.eu/en/technologies/blockchain, accessed: 2021
27. Staff C. Centralized, decentralized, & distributed networks. https://www.gemini.com/cryptopedia/blockchain-network-decentralized-distributed-centralized
28. Vermaak W. How decentralized are decentralized networks? https://coinmarketcap.com/alexandria/article/how-decentralized-are-decentralized-networks
29. Wooll M. What are decentralized organizations? https://www.betterup.com/blog/decentralization-in-management, accessed: 2021-04-20
30. Yu D, Li W, Xu H, Zhang L (2021) Low reliable and low latency communications for mission critical distributed industrial internet of things. IEEE Commun Lett 25(1):313 317. https://doi.org/10.1109/LCOMM.2020.3021367
31. Zarrin J, Phang HW, Saheer LB, Zarrin B (2021) Blockchain for decentralization of internet: prospects, trends, and challenges. Cluster Comput 1–26

# Chapter 4
# Basics of Cryptographic Primitives for Blockchain Development

**Abstract** Cryptographic algorithms are the basic building blocks of a secure system and protocols. A security protocol is a set of measures to accomplish required security objectives by employing suitable security mechanisms. Security mechanisms are generally referred to as cryptographic functions, which have the fundamental property of representing the data in another form. Diverse kinds of security protocols are in practice, such as authentication, non-repudiation, and key management protocols. As one of the crypto-intensive technologies, Blockchain has become a scorching topic. Many security and privacy issues have been addressed for Blockchain supported by cryptographic primitives. Basic cryptographic primitives include hash primitives, digital signature, and encryption primitives which are incorporated in Blockchain. This chapter has discussed hash primitives such as SHA-256, SHA-512, and Ethash. Digital signatures such as Elliptic Curve Digital Signature Algorithm are the current signature scheme in Bitcoin. Schnorr signatures and BLS signatures rectify issues faced by ECDSA. All the algorithms work on symmetric and asymmetric key encryption, which are discussed in detail. This chapter also provided implementations for some of these topics, which will help in understanding the concepts better. The study of cryptographic primitives will be helpful for cryptographers to research and evaluate cryptographic solutions in Blockchain.

**Keywords** Blockchain · Cryptographic primitives · Hash functions · Digital signature · Encryption primitives

## 4.1 Introduction

Generally, the cryptographic primitives are essential in applications with an open nature, i.e., available to all for using and joining applications. These applications can easily be forged or targeted by malicious users or entities, making more robust security mechanisms. Security mechanisms are generally referred to as cryptographic functions, which have the fundamental property of representing the data in another form. While transmitting the data or even in the store, the data cannot be easily understood or stolen by malicious entities or users. Basic cryptographic primitives

S. Tanwar, *Blockchain Technology*, Studies in Autonomic, Data-driven and Industrial Computing, https://doi.org/10.1007/978-981-19-1488-1_4

include the hash functions, digital signature, implementation of hash pointers in various applications, hashchain mechanism, and many other concepts. They are required in a computer security system to construct cryptographic protocols. The concepts mentioned above are applied to a specific application to gain the maximum security from malicious users or entities. These concepts are widely used in many different fields and widely accepted as concrete mechanisms for security and thus not easily be forged by malicious users. Blockchain is an open environment. The kind of application blockchain is open, distributed, and decentralized. In this case, the data can be transmitted from any node to another node in the system. We cannot track the data transmission until some transaction has been committed. So, we need to ensure that while transmitting the data, the data must be flowing securely. Thus, we apply some security measures to ensure the integrity of the data.

We apply cryptographic hash primitives here and other properties of the blockchain that cannot easily modify the blockchain's block. This property strongly supports the hashchain mechanism, which is applied using a basic concept of hash pointers and hashing. Although it can apply these concepts in any application, the application's kind of applicability and nature modify the basic structure of the basic cryptographic primitives and use it in a way that can be helpful in that particular application. In this chapter, we have presented some of the most used basic cryptographic primitives from the perspective of blockchain. Blockchain is considered the most full-proof, secure, and unmodifiable system for digital currency and other applications. The primary concept behind these properties is the application of the basic cryptographic hash primitives.

Since the inception of blockchain in 2009, Bitcoin has been evolving ever. Table 4.1 shows the list of different types of blockchain applications compiled from various sources [1, 2]. Apart from that, corresponding hashing and digital signature algorithms are also explained in this chapter. These hashing and digital signature algorithms are nothing, but cryptographic primitives only. We can derive that the complexity and robustness of these cryptographic algorithms have also been increased over the years. Moreover, blockchain applications do not rely on a single algorithm for security purposes nowadays. Newer applications use different cryptographic algorithms according to their needs and requirements. Blockchain applications need to be more secure with the increase in computational capabilities. That is why these newer applications use more secure and robust cryptographic algorithms. Blockchain cryptographic primitives have always been leveled up in complexity and security as time passed. The most significant threat to the blockchain is preventing its security wall from breaking. In this case, cryptographic primitives support that security. The cryptographic primitives used in the blockchain keep the blockchain safe from various attacks and malicious users. With the increase in computational capabilities, researchers need to develop more secure algorithms which can be unbreakable. So far, there have been many different applications of blockchain developed and still in use. All these blockchain applications use cryptographic primitives, but one thing can be derived that with time, these blockchain applications have employed

**Table 4.1** Increase in the complexity and robustness of the cryptographic algorithms used in blockchain applications over the years

| Blockchain application | Release year | Hashing algorithms | Digital signature algorithms |
|---|---|---|---|
| Bitcoin | 2009 | SHA-256, RIPEMD-160 | ECDSA |
| Litecoin | 2011 | SHA-256, SCrypt, RIPEMD-160 | ECDSA, multisignature |
| Ripple | 2012 | SHA-256, RIPEMD-160 | ECDSA, multisignature |
| Bytecoin | 2012 | SHA-256, KECCAK, BLAKE256,RIPEMD-160 | ECDSA, ring, one-time, multisignature |
| Dash | 2014 | SHA-256, X11 | ECDSA, multisignature |
| Ethereum | 2015 | SHA-256, Ethash, RIPEMD-160, SHA-3 (KECCAK-256) | ECDSA & secp256k1 |
| Hyperledger | 2015 | SHA2-256 & SHA-3 family | ECDSA, secp384 & 521r1 |
| R3-Corda | 2016 | SPHINCS-256 & SHA-512 | ECDSA |
| Quorum | 2018 | KECCAK & SHA-3 | ECDSA with secp256k1 & ECIEC with AES |
| Stratis | 2018 | SHA-256, SCrypt, X13 | ECDSA, multisignature |

more robust, secure, and challenging-to-break cryptographic algorithms to combat malicious entities.

Figure 4.1 shows the comparative analysis of the benefits of blockchain [3]. This graph shows that the most important benefit of blockchain is transparency, immutability, and reliability, which comes under one broad umbrella, i.e., security and privacy. Blockchain achieves all properties using cryptographic primitives in the blockchain. This graph shows that organizations prefer blockchain technology more in their projects due to its benefits. The most significant benefits blockchain gives are security and reliability. The cryptographic primitives in the blockchain are essential and can be termed as one of the pillars of blockchain technology. In subsequent sections, we consider all basic cryptographic primitives used in the blockchain for our study.

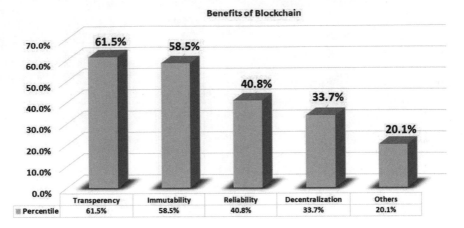

**Fig. 4.1**   Usage of cryptography in blockchain

## 4.2   Blockchain Technology Hash Primitives

### 4.2.1   Message Digest in Blockchain

Message Authentication is a mechanism or technique used to verify and validate the integrity of a message using the cryptographic hash function often termed as Message Authentication Code (MAC). With the help of message authentication, we can check the integrity of the original data of the sender that has been sent to the receiver. Basically, the message is considered as an input that will be passed through the hash function, i.e., message digest function, to get the desired output as message digest as shown in Fig. 4.2. We call it a message digest because it is a smaller representation of larger data, i.e., it can be considered a fixed-length representation of the message. There are multiple message digest algorithms such as MD, MD2, MD3, MD4, and MD5. The latest and most popularly used algorithm is MD5 and MD6. Ron Rivest developed the MD5 algorithm. It is a cryptographic hash function used as an encryption function for a file and produces the message digests of 128 bits [4]. The generated message digest using the hash function should fulfill certain conditions [5] mentioned as follows:

- Firstly, it should not be possible that someone can find the original message with the help of a message digest generated using a hash function. Otherwise, it can tamper with the original message, which will lessen the security in the network.
- Another condition is that it should not be possible to get the same message digest for two different messages.

**When can we use the MD5 algorithm?** MD5 algorithm can be used to verify the authentication of the file, i.e., if it is corrupted or modified due to the issues such as bit loss during the download or sharing of the file, corrupted due to any virus,

**Fig. 4.2** Generation of
message digest

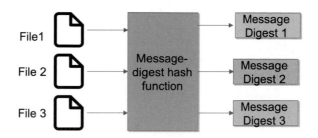

or modified by anonymous without permission of user or owner. So, the file can be verified by passing it through an MD5 algorithm, which will generate a 128-bit hash value. Then, we can check the file's authenticity by passing it through the MD5 algorithm and generating a 128-bit hash value. If comparing both hash values, i.e., previous and new hash values, results in different hash values, then it can be declared that the file has been modified or corrupted. Nowadays, many websites are providing malicious software on the Internet, which can hack your system or be more hazardous for the system. The property of the message digest algorithm can ensure security in the system by comparing the hash value of the original file and malicious file to prevent the corruption of the system.

## 4.2.2   Secure Hash Algorithms (SHA)

**SHA-256** SHA-256 is an algorithm that generates a cryptographic hash. A cryptographic hash is not encryption, but can secure the data or file by digitally signing the content. SHA-256 converts variable size input to a fixed 32 bytes or 256-bit unique output. Also, it is not possible to retrieve the original message or file from its hash value. Nowadays, it is being widely used in utilizing the core blockchain technology. The working of SHA-256 [6] can be explained in the following steps:

- **Append: Padding Bits**
  The first step of a hashing function begins with padding or appending bits to the original input message to change its length into the standard length necessarily required for the hash function to operate on it. The number of bits is calculated so that after the addition of bits, the length of the message should be 64 bits shorter than a multiple of 512. The mathematical representation of the above can be given as follows:

$$M + P + 64 = n * 512 \qquad (4.1)$$

  where M is the no. of bits in the original message and P is the no. of padding bits. The appended bits should begin with '1' followed by zeroes.
- **Append: Length Bits**
  We have appended the padding bits such that the total bits should be 64 less than a multiple of 512. To make the overall message a multiple of 512, we can add a

string of 64 length bits. Further, the new message after the addition of 64 bits can be divided into blocks of 512 bits each in which each block is known as a message block.

- **Compression Function**
  The final hash can be computed by performing computations on message blocks of 512 bits each. The message blocks are passed through 64 rounds of operation and then the obtained output will be considered input for the further blocks.

**SHA-512** SHA-512 also belongs to the family of SHA-2 algorithm, which also includes SHA-256 algorithm, which can be abbreviated as secure hash algorithms. SHA-512 converts variable size input to a fixed 64 bytes or 512-bit unique output. The working of SHA-512 can be understood with the help of the following steps [7]:

- **Append: Padding Bits**
  SHA-512 algorithm begins with padding or appending bits to the original input message to change its length into the standard length necessarily required for the hash function to operate on it. The number of bits is calculated so that after the addition of bits, the length of the message should be 128 bits shorter than a multiple of 1024.
  The appended bits should start with '1' followed by zeroes till we reach the last bit.
- **Append: Length Bits**
  After appending the bits to the original message, we can add extra 128 bits to the complete block further to split it into message blocks of 1024 bits each so that hashing operation can be applied easily on the message.
- **Compression function** The final hash can be computed by performing computations on message blocks of 1024 bits each. The message blocks are further passed through the 80 rounds of operation to get the desired output hash value of 512 bits. Also, the output can be considered as the input for the next block.

So, it can be observed that SHA-512 works somewhat similarly to the SHA-256.

### 4.2.3   RIPEMD

RACE Integrity Primitives Evaluation Message Digest can be abbreviated as RIPEMD, which further contains a group of hash functions such as RIPEMD-128, RIPEMD-160, RIPEMD-256, and RIPEMD-320. The version of RIPEMD of 128 bits does not guarantee to preserve the integrity of the message to that extent. RIPEMD with 160 bits provides the network with higher security. The release of RIPEMD with extended versions of 256 bits and 320 bits provides the same security level as the previous version. It can be implemented considering the block associated with the RIPEMD-160 hash algorithm. The integrity of the message can be preserved by splitting the message into blocks of 512 bits by utilizing the compression function.

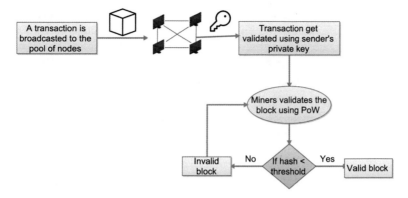

**Fig. 4.3** Working of Ethash algorithm

After that, it can go through two rounds of the considered block (sub-block) using different values of constant k associated with the block [8].

### 4.2.4 Ethash

Nowadays, industries utilize Ethereum blockchain-based platforms to deploy their applications with security and privacy. Ethash can be defined as an algorithm to perform mining on Ethereum based on the Proof-of-Work algorithm. We have already discussed that transactions appearing in the network need to be validated to append them to the network. For that, miners have to be involved in mining and verifying the block of transactions to make them valid for the blockchain network. So, Ethash is also a mining algorithm to provide authenticity in the network. In the Ethash algorithm, Fig. 4.3 input message, i.e., a transaction can be passed through the hash function using the sender's private key to generate an output hash value, which should be less than the threshold value. The threshold value depends on the difficulty, i.e., increasing the number of miners competing to validate the transaction can increase difficulty. With the increase in difficulty, it will be a challenging task for miners to mine the block leading to the loss of miners and high energy consumption [9].

### 4.2.5 SCrypt

SCrypt is also a PoW mining algorithm, first released with Tenebrix in 2011. After that, many researchers are utilizing it to deploy different blockchain-based projects. It is proposed to enhance the security maintained in the SHA-256 algorithm. It is efficient to prevent brute force attacks by securing passwords with the help of a security key retrieved from the master password. Now, when we pass an input message through the SCrypt algorithm to generate the hash value, it produces some noise, distracting the malicious attackers from securing the content of the message. We can say that the SCrypt mining algorithm protects the data from malicious activity due to its features [10, 11]:

- Firstly, it hashes the password associated with the original message so that it will be difficult for an attacker to retrieve the message.
- Due to its low energy consumption, computing power, and low cost, many cryptocurrencies, such as Litecoin, utilize SCrypt in their projects.

## 4.3  Digital Signature

Digital Signature [12] is a mathematical scheme or technique that can be used to verify the authenticity of digital messages. When the receiver verifies the digital signature, if it turns out to be valid, the receiver can believe that the message has come from an authentic and known source or sender. The receiver can also ensure that the message was not modified or altered during the transit of the message along the way. Digital signatures are the most common techniques used in cryptographic techniques for security and privacy maintenance. Most commonly, digital signatures employ asymmetric key cryptography. If the digital signature is implemented correctly, we can say that the digital signature gives the receiver a reason to be assured that the received message has come from a particular known sender. The digital signature can be used for non-repudiation also. It means that the sender cannot claim that it did not send the message because the digital signature of a particular sender is associated with the sent message and associated with the sender's private key. Hence, a digital signature preserves three main properties, i.e., authentication, integrity, and non-repudiation. The general digital signature algorithm has three main steps :

1. Key generation step where the private key is selected by the algorithm and the corresponding public key is also generated.
2. In this step, with a message and private key, the algorithm outputs the digital signature of the particular sender.
3. The receiver verifies the authenticity with the message, digital signature, and public key.

### 4.3.1  Authentication

The digital signature provides authentication [13] by using asymmetric key cryptography [14]. As shown in Fig. 4.4, we can see that the message sent by Alice is verified by Bob using the digital signature mechanism. Hence, Bob is ensured that the message has been sent by Alice only and not any other malicious entity. The

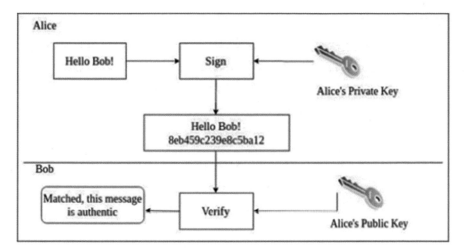

**Fig. 4.4**   Authentication using digital signature

process takes place using the sender's public key and private key. The private key of Alice is only known to herself and not to others. But, the public key of Alice is known to everyone. Alice's public key can only decrypt the message or data which is encrypted by Alice's private key. Thus, these two keys are interconnected. This property of asymmetric key cryptography provides the base for the authentication mechanism to play its role. The sender Alice uses her private key to encrypt the message and this encrypted message is sent to Bob. Anyone can see the message after decryption using Alice's public key between the transmission because Alice's public key is available to everyone. The goal is not to provide confidentiality [15] but to provide just authentication. We can combine other techniques to ensure the confidentiality of the message also. But here, we need only authentication. So, after the Bob has received the encrypted message, this message can be decrypted by Bob using Alice's public key. Thus, he can compare the original message and decrypted message to check if both messages are equal. If both the messages are equal, we can say that Alice sent the message, thus providing authentication.

### 4.3.2   Integrity

The integrity [16] of the data is essential. The data can be altered, or the integrity of the data can be breached along the way of transmission of the data. Thus, a mechanism to ensure the integrity of the data is required. The digital signature provides a widely used mechanism for ensuring the integrity of the received data. As shown in Fig. 4.5, we consider both the hash and encryption to generate the encrypted message digest. Then at the receiver end, the receiver decrypts the message digest and compares it

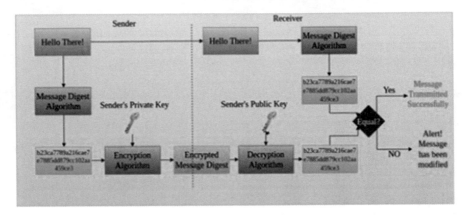

**Fig. 4.5** Integrity using digital signature

with its calculated message digest. If both are matched, then we can say that the data is not altered in between. The sender first creates the hash of the data, also called a message digest. Then it encrypts it by using the private key of its own to provide authentication. Then both the message and the encrypted message digest or hash are directed to the receiver end. At the receiver end, the receiver decrypts the encrypted message digest by using the sender's public key, and thus, it gets the message digest of the original message. Now the receiver generated message digest of the received plain text message with the same algorithm used at the sender side. Both of the message digests are compared and matched. If both the message digest are equal, we can say that the message is not altered. And if the message has been altered in between, the message digest calculated by the receiver would be different from the sender's message digest because, as per the property of the cryptographic hash functions, any minute change in the message or plain text can result in a significant difference in the message digest. This way, the integrity of the message or plain text can be preserved using a digital signature.

### 4.3.3   *Non-repudiation*

Non-repudiation [17] is a property or security measure that a digital signature provides. The sender cannot deny sending a message that he sent. If the sender has sent some message to the receiver and after some time the sender refuses to admit that he has sent that particular message, if the digital signature was applied in this situation, then the sender could not deny that they have not sent that specific message. From the concept of the digital signature, we know that if the sender sends a message by encrypting the message with its private key, then that encrypted message can be decrypted only by its public key. A similar mechanism is applied here also. If the sender's public key can successfully decrypt the message, it is evident that the message has been encrypted by using the sender's private key, and the sender itself can only possess that. So no other entity could send this encrypted message which that sender's public key can successfully decrypt. This way sender cannot deny sending the message.

There are different algorithms for digital signature, which are mentioned below and discussed later in this chapter:

1. RSA (Rivest, Shamir, Adleman) Algorithm.
2. ElGamal Encryption System.
3. DSA (Digital Signature Algorithm).
4. ECDSA (Elliptic Curve Digital Signature Algorithm).

## 4.4 Digital Signature in Blockchain

The blockchain is a distributed and decentralized environment. The participating nodes can be located at different geographical locations and those nodes can be of different configurations. The nodes participating in the blockchain are heterogeneous in nature. In this open environment, the message exchange mechanism requires security and authentication. The nodes cannot know who their neighbors are in this open environment. The nodes do not have information about their neighboring nodes. They know only the unique identification value for every node. In this situation, we need to ensure the authenticity of the data coming from different sources. If a node gets a request for a particular transaction, it must ensure that this message is coming from which node, so that it can process further checks to validate the transaction. We need to apply digital signatures to the messages transferred between different nodes in the environment to do that. In this open environment, non-repudiation is necessary because malicious nodes often attempt invalid activities. To detect that, we need to ensure the non-repudiation in this blockchain network. This can be deployed using digital signatures, as stated earlier. The digital signature can be used in the consensus mechanism Practical Byzantine Fault Tolerance (PBFT) [18], where the votes from different nodes should be considered. Thus, authentication is required in this case. The nodes can provide their private key to sign the smart contract that can be part of some operations. In cryptocurrencies such as Bitcoin [19], the exchange wallets' processes should have signatures from multiple signers, so that the digital signature can also be used. For every transaction to be approved by the miners, the authenticity of the transactions and parties need to be ensured. Whenever a transaction is validated in the blockchain, the nodes involved in the transaction digitally sign the complete set of transactions carried out, thus making it a multisigned document or information. Therefore, for verification, this information is sent to all the nodes or users in the network. Since this information has multiple signs of multiple nodes associated with that particular set of transactions, the validator nodes validate the authenticity of the data and check for the possibility of the transactions.

Several novel and promising ideas have come up for using digital signatures differently and as per the need of the situation and organization. As discussed earlier, these ideas use digital signatures differently instead of using them in their raw form. These ideas are briefly explained in the coming sections.

### 4.4.1  Elliptic Curve Digital Signature Algorithm (ECDSA)

The Elliptic Curve Digital Signature Algorithm (ECDSA) [20] is the current signature scheme in Bitcoin. It is based on elliptic curve cryptography (ECC) [21]. ECDSA depends on elliptic curves over finite fields. It also depends on the elliptic curve discrete logarithm problem's (ECDLP) [22] difficulty rather than prime factoring difficulty. Figure 4.6 shows how an elliptic curve looks like. ECDLP states, "Let $a$, $b$, and $c$ be integers such that $a^b = c$. If the value of $c$ and $a$ is provided, it is tough to find the value of $b$ if $b$ is a sufficiently large number. Now, compute $Q = nP$ by applying the equation to the elliptic curve group. Here, $n$ is an integer value, $P$ is a point on the curve, and $Q$ is the operation result, i.e., multiplying points. It is manageable to compute $Q$ given the value of $n$ and $P$ in elliptic curves, but it is challenging to find $n$ if the value of $P$ and $Q$ is provided.

The ECDSA algorithm depends on the aforementioned problem to induce signatures that are tough to forge and easy to verify. Bitcoin utilizes a curve called 'secp256k1' which is a 256-bit elliptic curve, standardized by the U.S. government agency, the National Institute of Standards and Technology (NIST). ECDSA keys and signatures are shorter for the same security level than RSA. A 256-bit ECDSA signature has the identical security strength as a 3072-bit RSA signature. Elliptic curves define a reference point, $G$, used for point multiplication on the curve for multiplying integer value with elliptic curve [23] point. It also contains an order, $n$, which defines the length of the private key and is generated by $G$.

**Fig. 4.6** An elliptic curve

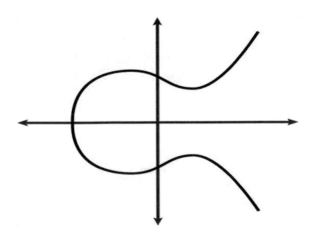

**Key Generation** The key pair of ECDSA is a combination of the private key (PRK) and public key (PUK). PRK is a random integer in the range of $[0...n - 1]$ where $n$ is the order that defines the length of the private key. PUK is a point on the elliptic curve given by scalar multiplication on the curve given by the following equation:

$$PUK = PRK * G \qquad (4.2)$$

PUK, i.e., EC 'secp256k1' point [24] $\{x, y\}$ coordinates + 1 bit (parity). For the 'secp256k1' curve, the PRK is a 256-bit integer (32 bytes), and the compressed PUK is a 257-bit integer ($\sim$33 bytes).

**Signing** A signature $\{rp, sp\}$ is created as output by taking input as a 'message' along with PRK. The algorithm of signing works as follows.

1. Apply the cryptographic hash function on the message like SHA-256 and generate message hash $mh$ as
$$mh = hash\_func(message) \qquad (4.3)$$

2. Generate a random number $k$ securely in the range $[1..n - 1]$.
3. Calculate a random point $RP = k * G$ and calculate x-coordinate as

$$rp = RP \cdot \mathbf{x} \qquad (4.4)$$

4. Calculate the signature proof as

$$sp = k^{-1} * (mh + rp * PRK)(mod\, n) \qquad (4.5)$$

5. Return the calculated output signature $\{rp, sp\}$.

The signed knows about $mh$ and PRK, and the proof $sp$ is verifiable by utilizing the corresponding PUK. ECDSA output signatures are two times longer than the private key of the signer for the curve. For example, the ECDSA output signature is 512 bits for a 256-bit elliptic curve.

**Verification of Signature** To verify the signature, the algorithm takes as an input a combination of the signed message, PUK, the signer's PRK, and the output signature $\{rp, sp\}$. A boolean value denoting a valid or invalid signature is generated as an output. The verification algorithm works as follows:

1. Using the same cryptographic hash function used while signing, calculate the message hash as
$$mh = hash\_func(message) \qquad (4.6)$$

2. Compute the modular inverse of proof as

$$sp' = sp^{-1}(mod\, n) \qquad (4.7)$$

3. Retrieve the random point calculated during signing as

$$RP' = \left(h^{\star}sp'\right) * G + \left(rp^{\star}sp'\right)^{\star} \text{PUK} \tag{4.8}$$

4. Find the x-coordinate as

$$rp' = RP' \cdot \mathbf{x} \tag{4.9}$$

5. Validate the signature by comparing whether $rp' == rp$ or not?

If $rp' == rp$, then the signature is valid, else it is valid. If $rp' \neq rp$, then either of message, signature, or PUK is incorrect. An implementation of ECDSA is shown in Sect. 4.6.1.

### 4.4.2  Schnorr Signatures

Bitcoin has always used ECDSA, but it lacks to compress and verify signatures together efficiently. Hence, there is an inclination to change to a new strategy that enhances Bitcoin's scalability, privacy, storage, and security. That will be the Schnorr signature [25]. Claus-Peter Schnorr initially proposed it in 1988. Schnorr's patent expired in 2008 and the same year Bitcoin was invented. Due to lack of popularity and testing, Schnorr was not chosen for Bitcoin. Schnorr signatures have the following characteristics that enable it to be better than ECDSA:

1. It has log discretion.
2. It is non-malleable. Using the same key and message, a third party cannot modify a valid signature to construct another valid one.
3. It provides linearity. To enforce multisignature transactions [26] where multiple parties can collaborate and integrate their public keys to construct a single aggregated key and a valid signature. It enforces key and signature aggregation.
4. It increases privacy as the list and number of participants are hidden by public-key aggregation [27] into a single and valid signature.
5. It increases the storage and capacity of a block by enabling cross-input aggregation. Each input of Bitcoin transaction requires an individual signature. This occupies a lot of space in a block. Due to the aggregation of signatures into a single one, all the inputs in transactions require only a single signature. This creates space for transaction data in the block and increases the capacity.

Like ECDSA, the Schnorr Digital Signature Scheme employs elliptic curve cryptography. Schnorr utilizes the exact private–public key pairs as already explained in ECDSA. The only distinction is in the signing and verification algorithm, which is much simpler than ECDSA.

**Signing** Steps 1 to 3 will remain the same as the ECDSA signing. Now for calculating signature proof, the Schnorr algorithm uses the equation as

$$sp = k + \text{Hash}(rp||\text{PUK}||\text{message}) \star \text{PRK} \qquad (4.10)$$

Now, return the output signature $\{rp, sp\}$.

**Verification of Signature** For verifying the signed signature, first obtain the signature $\{rp, sp\}$, the public key $(PUK)$, and the message. Now, calculate the random point $RP$ from x-coordinate $rp$.

$$sp^\star G = P + \text{Hash}(rp||\text{PUK}||\text{message})^\star \text{PUK} \qquad (4.11)$$

For a signature to be valid, $sp^\star G$ will be equal to $P + \text{Hash}(rp||\text{PUK}||\text{message})^\star \text{PUK}$. If they are equal, the signature is valid; otherwise, it is invalid.

### 4.4.3 Multisignatures

Multisignature [28] is created when multiple users create one signature. To acquire a compact signature for a group, multisignature is employed. Since individual signatures are large enough and utilize a lot of storage space, multisignature is beneficial. $N/N$ and $(N-1)/N$ are two available schemes for creating multisignatures. The private and public keys represented as PRK and PUK, respectively, are calculated the same as in ECDSA. Multisignatures use the concept of view key and spend key which is described as follows:

- Private view key is used for allowing access to view every incoming transaction for that address.
- Public spend key is used by the network to verify the signature of the key image and accept the transaction as valid.
- Private spend key is used to sign a key image when the owner wants to spend funds.

*N/N* **Scheme** This scheme works as follows:

1. The sender and recipient participant generate their key pairs, i.e., private view key and public spend key. So, at the end of this step, the keys are

   - Sender's private view key $(PRK1)$ and public spend key $(PUK1)$.
   - Similarly, recipient's private view key $(PRK2)$ and public spend key $(PUK2)$.

2. They will share their private view and public spend keys with each other.

3. Later, calculate their respective sums. To compute the wallet's private and public key, perform the summation of the participant's view keys and spend keys with each other, i.e.,

- Wallet's private key $(PRK) = PRK1 + PRK2$
- Wallet's public key $(PUK) = PUK1 + PUK2$

The process of constructing an N/N multisignature in terms of two participants, i.e., 2/2 wallet, is shown in the above-mentioned steps.

$(N - 1)/N$ **Scheme** The step-wise illustration of $(N - 1)/N$ scheme in terms of 2/3 wallet is shown as follows:

1. All three participants will generate their own key value pairs, i.e., view keys and spend keys.

---

```
           Participant 1                        Participant 2
      Private View Key (PRK1)              Private View Key (PRK2)
      Private Spend Key (PRK'1)            Private Spend Key (PRK'2)
      Public Spend Key (PUK1)              Public Spend Key (PUK2)

                          Participant 3
                     Private View Key (PRK3)
                     Private Spend Key (PRK'3)
                     Public Spend Key (PUK3)
```

---

2. Everyone sends their private view keys and public spend keys to each other as shown in Fig. 4.7.
3. Each participant multiplies their private spend key with each other's public spend keys and hashes the product.

---

```
                      Multisignatures
           Participant 1                        Participant 2
      Hash(PRK'1, PUK2)                    Hash(PRK'2, PUK1)
      Hash(PRK'1, PUK3)                    Hash(PRK'2, PUK3)

                        Participant 3
                   Hash(PRK'3, PUK1)
                   Hash(PRK'3, PUK2)
```

---

4. Each participant multiplies its multisignatures with G and shares it with each other as shown in Fig. 4.8.
5. At the end, generate a common private view key and public spend key.

---

```
   Common Private View Key (PRK) = PRK1 + PRK2 + PRK3
   Common Public Spend Key (PUK) = H(PRK'1, PUK3)*G +
                            H(PRK'3, PUK2)*G + H(PRK'2, PUK3)*G
```

---

Any two participants are enough in this scheme for sharing the multisignature key.

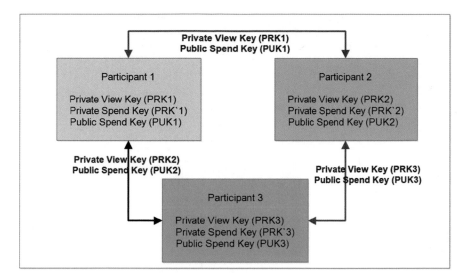

**Fig. 4.7** Exchange of private view keys and public spend keys

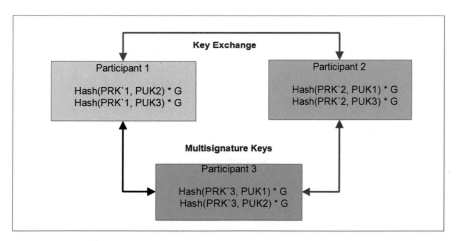

**Fig. 4.8** Exchange of multisignatures

### 4.4.4   Ring Signature

Ring signature [29] was developed in 2001 by Rivest, Adi Shamir, and Yael Tau-manby. These digital signatures authorize multiple group members to sign the message securely and anonymously. Only the original signer will know which group member actually signed the message.

**Signing** The process of creating signature is as follows:

1. The first step is to define the group of $N$ members. Each member will have their own key pair, i.e., $(PRK, PUK)$ pair. For signing a message, a member $i$ uses its own $PRK_i$ and public keys of all other members in the group. There can be a chance to review the group's validity by knowing the group members' public keys. For checking the validity of the signature, the group members' private keys should be known.
2. For signing the message, first compute the hash digest of the message as

$$MH = Hash(message) \tag{4.12}$$

3. Generate a random value $k$.
4. Use the message hash on $k$ to generate a value $m$ as $m = MH(k)$.
5. Every member of the group except the signer computes the value of $f$ as follows:

$$f = PRK_i^{PUK_i} \pmod{N_i} \tag{4.13}$$

6. Compute the value of $m'$ as
$$m' = m \oplus f \tag{4.14}$$

7. At the end, calculate $sig_i$ for the signed member $i$ as

$$sig_i = (m \oplus k)^d \pmod{N_i} \tag{4.15}$$

where $d$ is the secret key of signing party.

**Verification of Signature** Ring verification is as follows:

1. For signing the message, first compute the hash digest of the message as

$$MH = Hash(message) \tag{4.16}$$

2. Generate a random value $k$.
3. Use the message hash on $k$ to generate a value $m$ as $m = MH(k)$.
4. Calculate the secret key of each participant except the signer as

$$m = MH\left(PRK_i^{PUK_i} \oplus m\right) \tag{4.17}$$

5. The participant who wants to verify the signature calculates as

$$v = MH\left((k \oplus m)^d\right)^f \tag{4.18}$$

6. If the value of $v$ obtained from above equation is same as the value of $m$, then the signature is valid; otherwise, it is invalid.

As already mentioned, one has to get the private keys of all the members to unsign the message. Even if an attacker gets its hands on all the private keys of the group members, it is still challenging to find the original signer out of them. The probability of fining a signer is $1/N$, where $N$ is the total members in the group.

## 4.5  Encryption Primitives for Blockchain

### 4.5.1  Symmetric Encryption

In the encryption-based system, the transmitter and recipient utilize a single common key to encrypt and decrypt the message [30]. Figure 4.9 shows the symmetric encryption mechanism. However, the public-key cryptography approach uses two keys for encryption and decryption purposes, public and private keys. Though the symmetric encryption method is faster, the main concern is security, and the keys should transfer their information in a very secure way. This encryption method is also known as secret-key cryptography. At the same time, public-key cryptography is not facing this issue because it has not transmitted their private key and easily distributes their public key. In this following subsection, we will discuss various symmetric encryption algorithms.

**Data Encryption Standard (DES)** A famous symmetric encryption algorithm that was developed in 1975 is Data Encryption Standard (DES) approach [32]. The block cipher method and 56-bit key DES split the text into 64-bit blocks and encrypt that text. Generally, the DES algorithm uses the same key for the encryption and decryption so that the transmitter and recipient both use the same private key. Figure 4.10 shows the vital feature of the DES algorithm such as block cipher, numerous rounds of encryption, 64-bit key, backward compatibility, and replacement and permutation [33]. DES uses the block cipher method to mean that the key and algorithm apply to an entire data block rather than one bit at a particular time. DES encrypts the text 16 rounds in four different modes. It can encrypt the block separately and make the

**Fig. 4.9** Symmetric encryption [31]

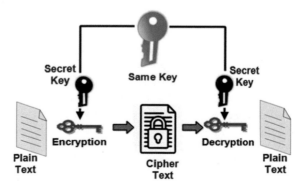

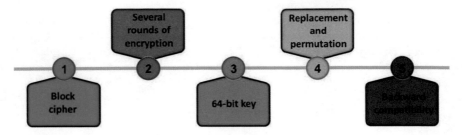

**Fig. 4.10**  Key features that affect DES algorithm

cipher block depending on its previous block. The ciphertext describes the sequence of permutations and replacements during the encryption process. However, this algorithm has benefited that it is not secure enough against brute force attacks. So, DES is replaced by a more advanced algorithm, discussed in a subsequent topic.

**Triple Data Encryption Algorithm** A mode of the DES encryption approach, which is known as triple DES or 3DES, encrypts message three times [34]. It has a key length of 192 bits and utilizes 64-bit keys. It utilizes the cartographic block method, in which the text is split into 64-bit size text blocks. Encryption is done after that. In triple DES, the first encryption key encrypts with the second encryption key. Likewise, the resultant ciphertext is encrypted using a third encryption key. That is the reason 3DES is more secure than the DES algorithm. The encryption and decryption process [35] in 3DES are as follows:

$$\textbf{Encryption:} \qquad ct = E3(D2(E1(P))) \qquad\qquad (4.19)$$

$$\textbf{Decryption:} \qquad P = D1(EH2(D3(ct))) \qquad\qquad (4.20)$$

In the above-mentioned Eqs. 4.19 and 4.20, E() represents encryption and D() shows the decryption of the DES approach. H, P, and ct denote key, plain text, and ciphertext.

**Advanced Encryption Standard (AES)** United States government standard replaced the DES algorithm in 2002 with the Advanced Encryption Standard (AES) approach. Joan Daemen and Vincent Rijmen—Belgian cryptographers—evolved a symmetric 128-bit block message encryption approach [36]. The AES algorithm can work simultaneously in multiple network environments. Initially, the National Institute of Standards and Technology (NIST) of the U.S. Department of Commerce has determined the Rijndael approach from all the given five methods. This approach is similar to the AES algorithm. The only difference between Rijndael and AES is that Rijandael can define with any key and block size that are multiples of 32 bits such as a minimum of 128 bits and a maximum of 256 bits. In contrast, AES has a fixed block size (128-bit) and 128-bit, 192-bit, and 256-bit key size. This algorithm is secure enough against the brute force attack compared to the DES and 3DES [37].

**Fig. 4.11** Asymmetric
encryption [31]

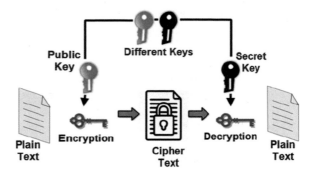

### 4.5.2  Asymmetric Encryption

As discussed in Sect. 4.5.1, the public-key cryptography has two keys. This mechanism is known as the asymmetric encryption technique. Figure 4.11 shows the plain text encrypted by two different keys, such as public key and secret or private key. The public keys are excessive to the Internet. It verifies that the unauthenticated user is not able to access the keys. Those who have a public key can easily decrypt the message. Due to this reason, this approach uses two related keys that enhance the security [31]. The data encrypted by the public key, which is freely available, is only decrypted using the private key. In the same way, the data encrypted by a private key is only decrypted using the public key. The Rivest–Shamir–Adleman (RSA) and Elliptic curve techniques (ECC) are famous asymmetric key encryption algorithms. We will discuss these algorithms in detail.

**RSA (Rivest–Shamir–Adleman) Algorithm** There is the belief that there is no efficient technique to factor a huge number. RSA algorithm reduces the key and requires more computer processing power and time. The RSA algorithm uses four steps: key generation, distribution, encryption, and decryption [38]. A general principle behind the RSA algorithm is monitoring three huge positive integers: p, q, and r.

$$(a^p)^q \equiv a (mod b) \tag{4.21}$$

In the equation mentioned above, once you have the value of a,p, and b, it is still challenging to identify the value of q.

$$(a^q)^p \equiv a (mod b) \tag{4.22}$$

For some operations, it is the same as changing the value of p and q, respectively. In general, the information encrypted by the public key is only decrypted using the private key. In the above equation mentioned, the public key is depicted by b and p, whereas the private key is represented by q.

As mentioned earlier, RSA is a four-step process discussed as follows:

- **Key generation:** The keys generated for the RSA algorithm includes the following steps:

  - Select two discrete prime numbers such as x and y.
  - Calculate that numbers like $b = xy$, where b is the private- and public-key modulus.
  - Use Carmichael's totient function $(\sigma)$ to determine $\sigma(n)$.
  - Select integer p.
  - Stimulate q using $q = p^-1$.

- **Key distribution:** Here, we consider one example where Jay wants to send a message to Raj. Jay must be aware of Raj's public key to encrypt the information using the RSA algorithm. In contrast, Raj has to utilize his private key to decrypt the information sent by Jay. Jay sends its public key $(b, p)$ to Raj using a reliable and secure way.
- **Encryption:** Once Raj gets Jay's public key, he can transmit data D to Jay.

$$ct \equiv a^p (mod b) \tag{4.23}$$

  where c is a ciphertext.
- **Decryption:** In the same way, Jay recovers a from a to ct by utilizing his private key.

$$ct^q \equiv (a^p)^q \equiv a(mod b) \tag{4.24}$$

**Elliptic Curve Cryptography (ECC)** In asymmetric cryptography, ECC is the alternate method of RSA that is used to encrypt a message so that the malicious user can ot access it. It uses elliptic curves on finite fields that make it more secure than crack. ECC's 256-bit key generated security same as RSA 3072-bit key [39]. When a longer key is used in RSA, making the process slow, ECC mitigates this issue by providing a smaller key size and security. The ECC technique is divided into two parts, such as ECDSA algorithm that is used to sign messages and the Elliptic Curve Diffie–Hellman key exchange (ECDH) that is used to share symmetric keys for the encryption [40]. Generally, blockchain uses the ECDSA technique for the signature that we had already discussed in the previous Sect. 4.4.1. As ECC follows the elliptic curve, it uses the following equations [41]:

$$Y^2 = X^3 + mX + n \tag{4.25}$$

where 'm' is the co-efficient of X and 'n' is the equation's constant. A plane curve over a finite field point satisfies the equation mentioned above, it can mirror any curve point over the X-axis, and the curve remains the same.

### 4.5.3 Key Management and Exchange

Key management refers to cryptosystem generating, managing, storing, exchanging, and replacing the cryptokeys whenever required. Cryptosystem facing issue for secure and efficient key management. The malicious user theft the details of the system by cracking the keys. They perform the malicious activity through different attacks such as replay attacks, man-in-the-middle attacks, and brute force attacks. In such cases, key management plays a crucial role in the cryptosystem. Two key management protocols, like El Gamal and Diffie Hellman, overcome this issue, discussed in detail in subsequent topics.

**Diffie–Hellman** Whitfield Diffie and Martin Hellman—American cryptographers—established the Diffie–Hellman (DH) approach in 1976. It is also known as key exchange protocol, used for secret communications, whereas the transmitter and recipient exchange information over a public network. It creates a unique session key having the property of forwarding secrecy at the time of encryption and decryption. DH has 1024 bits key size, and it is faster than RSA approach [42]. Two different parties who are unknown can share their secret key on the Internet; that is why it suffers from a Man-in-the-Middle attack. To overcome this issue, authentication certificates require the parties involved in the communication, and they are required to prove their identity using a session-to-session key agreement.

**El Gamal** Now, looking into El Gamal's key management cryptosystem, which published a public-key scheme based on discrete logarithms, similar to the Diffie–Hellman key exchange in 1984. As with Diffie–Hellman, the global elements of El Gamal are prime numbers M and N, which is a primitive of the root of M. This cryptosystem is based on the difficulty of finding discrete logarithms in a cyclic group. Each user generates a public–private key pair likely to Diffie–Hellman key exchange. The public keys are generated using the exponentiation, raising N to the power of the private key, where the secrecy of the private keys is subjected to the difficulty level of computing discrete logarithms. The basic idea with El Gamal encryption is to select a random key, and further, it uses that key to scramble the message by multiplying that message with it. For example,

- User X creates public and private keys.
- User Y encrypts data using User X's public key.
- User X decrypts the message.

The key feature of El Gamal encryption is that it is helpful in the digital signature standards. It sends a one-time key with a message, uses exponentiation, and relies on discrete log problems.

## 4.6   Implementation

### 4.6.1   Elliptic Curve Digital Signature Algorithm (ECDSA) Implementation

We have used the *pycoin* [43] Python package, which implements the ECDSA algorithm with the curve *secp256k1* used by Bitcoin. Three functions are used, namely calculateHash, computeSignature, and verifySignature for the purpose of hashing, ECDSA signing, and ECDSA verifying the signature, respectively.

```
from pycoin.ecdsa.secp256k1 import secp256k1
import hashlib, secrets

def calculateHash(message):
    hBytes = hashlib.sha3_256(msg.encode("utf8")).digest()
    print('messageHash={hBytes.hex()}')

def computeSignature(message, PRK):
    messageHash = calculateHash(message)
    signMessage = secp256k1.sign(PRK, messageHash)
    return signMessage

def verifySignature(message, signMessage, PUK):
    messageHash = calculateHash(message)
    valid = secp256k1.verify(PUK, messageHash, signMessage)
    return valid

#ECDSA Message Signing
message = "Signing Message using ECDSA"
PRK = secrets.randbelow(secp256k1.order())
signMessage = computeSignature(message, PRK)
print("Message:", message)
print("Private key:", hex(PRK))
print("Output Signature: rp=" + hex(signMessage[0]) + ", sp=" +
    hex(signMessage[1]))

#ECDSA Verify Signature
PUK = secp256k1 * PRK
valid = verifySignature(message, signMessage, PUK)
print("\nMessage:", message)
print("Public key: (" + hex(PUK[0]) + ", " + hex(PUK[1]) + ")")
print("Valid Signature?", valid)

#ECDSA Verify Signature - If Tampered
message = "Tampered message"
valid = verifySignature(message, signMessage, PUK)
print("\nMessage:", message)
print("Tampered?", valid)
```

The hashing function *calculateHash(message)* calculates and returns a SHA3-256 hash, defined as a 256-bit integer number. It will be used for signing and verification purposes. The *computeSignature(message, PRK)* function takes arguments as the message and a 256-bit private key. It computes the ECDSA signature $\{rp, sp\}$ and returns a 256-bit pair of integers. The *verifySignature(message, signMessage, PUK)* function takes arguments as the message, the signed signature $\{rp, sp\}$, and an uncompressed 2*256-bit public key. It returns whether the signature is valid or not.

```
python main.py
messageHash=6b3778a64f2675f3f76bf9f35af1fc673759ed17aed86dd56ca36c2bfd7
eb0f9
Message: Signing Message using ECDSA
Private key:
    0x1634769d9d3d37044474de843f2ee47a23e9fb4b0645206e057ed080
60a8cc23
Signature:
    rp=0xd51b3659ac5170189b09cb0ff79bd11c75ac26fc5c27ff8bc61a588
f2acdd6e2,
    sp=0x663d167d3320074ef6ce78a5d1885775d350a2af7516e956bcd37fa
38dee64cc
messageHash=6b3778a64f2675f3f76bf9f35af1fc673759ed17aed86dd56ca36c2bfd7
eb0f9

Message: Message for ECDSA signing
Public key:
    (0x35a2ce71327cab35d591caa12b617386ab86bd4b683b477a7cccbf9c6b301273,
    0xcf65fc0ea48b74565464a59cc30c59284cc0ec038c1e05c936bfe82d5dc77f58)
Valid Signature? True
messageHash=7947990ce83ee4d2ecf98279a6575df6cfe1a3b118e77ae586fe7cb69ec
3c68a

Message: Tampered message
Tampered? False
```

As per the above-displayed output, the randomly generated private key is of 256 bits. After signing, the output signature $\{rp, sp\}$ consists of a pair of 256-bit integers. The uncompressed public key, acquired by multiplying the private key by the elliptic curve's generator point, is 2 * 256 bits, i.e., two times longer than the private key. The verification holds for the produced ECDSA digital signature. The signature fails to verify if the message is tampered with.

## 4.7 Summary

This chapter studies various cryptographic primitives used in the Blockchain. The primary motivation behind using cryptographic primitives is the security and authenticity of these primitives. The cryptographic hash primitives can be used in any application requiring the security and privacy of the data flowing through the network. These primitives are widely used and proven to be successful in significant cases. With the increase in the computation capabilities in modern-day computers, these primitives also need to be more robust and secure to ensure the lead over the malicious users. This chapter also discussed various basic hash functions and their applications. The message digest is a part of hashing mechanism. The hash function and hash pointers are the most useful mechanisms used in most applications around the world. The advancements in the algorithms developed for hash mechanisms were also discussed. SHA and its variants were developed when there was a need for a more secure and robust algorithm from the security aspect. This chapter also discussed digital signatures and its applicability in the blockchain environment to ensure authenticity, integrity, and non-repudiation. The blockchain structure is derived from the hashchain structure, which was also discussed, and because of this hashchain structure, blockchain gets its property called an unmodifiable chain.

## 4.8  Practice Questions

### 4.8.1  MCQ Questions

1. Which of the following constitutes as the main components of the meta-data for a block in blockchain?

(a) Previous Block Hash
(b) Merkle Root
(c) Mining Statistics
(d) All of the above

2. A block in blockchain is pointed using

(a) Hash Pointer
(b) User ID
(c) Transaction ID
(d) Timestamp

3. A block in blockchain is pointed using

(a) Keyed Hash Function
(b) Hash Code
(c) Message Hash Function
(d) None of the Above

4. Cryptographic primitives include

(a) Hash functions
(b) Digital signature
(c) Hash chain

(d) All of the Above

5. Which used for signing the message digest in the public-key cryptosystem? system?

(a) Using the public key of the receiver
(b) Using the private key of the receiver
(c) Using the public key of the sender
(d) Using the private key of the sender

## 4.8.2  Fill in the Blanks

1. The property of consistency is preserved in blockchain by maintaining _____.
2. The current cryptographic hash algorithm used in Bitcoin is _____.
3. _____ operations are required to identify two messages with message digest in case of SHA-256.
4. DES uses _____ cipher method.
5. Schnorr Digital Signature Scheme employs _____ cryptography.

## 4.8.3  Short Questions

1. What cryptographic primitive is used in blockchain technology?
2. What is the most common attack on cryptographic primitives?
3. How many types of cryptographic primitives are there?
4. Which elliptic curve is used in Bitcoin?
5. What is Elliptic Curve Cryptography Used For?

## 4.8.4  Long Questions

1. What cryptographic primitives are used in securing the ledger?
2. What is the role of cryptographic primitives?
3. What cryptographic primitive should you implement to secure the communication channel between the devices?
4. Which cryptographic primitive is used to check data integrity?
5. What determines the Strength of a Symmetric Encryption Algorithm?

# References

1. Dasgupta D, Shrein JM, Gupta KD (2019) A survey of blockchain from security perspective. Journal of Banking and Financial Technology 3(1):1–17
2. Storublevtcev N (2019) Cryptography in blockchain. In: International conference on computational science and its applications. Springer, pp 495–508
3. Sebastian N (2022) The ultimate research on blockchain development for businesses - goodfirms survey. https://www.goodfirms.co/resources/blockchain-development-research. Accessed 20 Jan 2022
4. Corporation I (2022) Message digests and digital signatures. https://www.ibm.com/docs/en/ibm-mq/7.5?topic=concepts-message-digests. Accessed 30 Jan 2022
5. tutorialspoint (2022) Java cryptography - message digest. https://www.tutorialspoint.com/java_cryptography/java_cryptography_message_digest. Accessed 30 Jan 2022
6. Simplilearn (2022) A definitive guide to learn the SHA-256 (secure hash algorithms). https://www.simplilearn.com/tutorials/cyber-security-tutorial/sha-256-algorithm#what_is_the_sha256_algorithm. Accessed: 30 Jan 2022
7. Simplilearn (2022) Breaking down: Sha-512 algorithm. https://infosecwriteups.com/breaking-down-sha-512-algorithm-1fdb9cc9413a. Accessed 30 Jan 2022
8. GeeksforGeeks (2022) RIPEMD hash function. https://www.geeksforgeeks.org/ripemd-hash-function/. Accessed 30 Jan 2022
9. bit2meACADEMY (2022) What is the Ethash mining algorithm? https://academy.bit2me.com/en/what-is-the-algorithm-of-ethash-mining/. Accessed 30 Jan 2022
10. bit2meACADEMY (2022) What is the Scrypt hash function? https://academy.bit2me.com/en/what-is-scrypt-hash-function/. Accessed 30 Jan 2022
11. Rhodes D (2022) Scrypt: an overview of the scrypt mining algorithm. https://komodoplatform.com/en/academy/scrypt-algorithm/. Accessed 30 Jan 2022
12. Nist C (1992) The digital signature standard. Commun ACM 35(7):36–40
13. Kaur R, Kaur A (2012) Digital signature. In: 2012 international conference on computing sciences. IEEE, pp 295–301 (2012)
14. Hellman ME (2002) An overview of public key cryptography. IEEE Commun Mag 40(5):42–49
15. Pawar SB, Tandel LL, Zeple PK, Sonawane SR (2015) Survey of cryptography techniques for data security. Int J Sci, Eng Comput Technol 5(2):27
16. Nurhaida I, Ramayanti D, Riesaputra R (2017) Digital signature & encryption implementation for increasing authentication, integrity, security and data non-repudiation. Int Res J Comput Sci 4(11):4–14
17. Fang W, Chen W, Zhang W, Pei J, Gao W, Wang G (2020) Digital signature scheme for information non-repudiation in blockchain: a state of the art review. EURASIP J Wirel Commun Netw 2020(1):1–15
18. Sukhwani H, Martínez JM, Chang X, Trivedi KS, Rindos A (2017) Performance modeling of PBFT consensus process for permissioned blockchain network (hyperledger fabric). In: 2017 IEEE 36th symposium on reliable distributed systems (SRDS). IEEE, pp 253–255
19. Nakamoto S et al (2008) Bitcoin. A peer-to-peer electronic cash system
20. Johnson D, Menezes A, Vanstone S (2001) The elliptic curve digital signature algorithm (ECDSA). Int J Inf Secur 1(1):36–63
21. Lopez J, Dahab R (2000) An overview of elliptic curve cryptography
22. Galbraith SD, Gaudry P (2016) Recent progress on the elliptic curve discrete logarithm problem. Des, Codes Cryptogr 78(1):51–72
23. Koblitz N (1987) Elliptic curve cryptosystems. Math Comput 48(177):203–209
24. Bi W, Jia X, Zheng M (2018) A secure multiple elliptic curves digital signature algorithm for blockchain. arXiv:1808.02988
25. Maxwell G, Poelstra A, Seurin Y, Wuille P (2019) Simple Schnorr multi-signatures with applications to bitcoin. Des, Codes Cryptogr 87(9):2139–2164

26. El Bansarkhani R, Sturm J (2016) An efficient lattice-based multisignature scheme with applications to bitcoins. In: International conference on cryptology and network security. Springer, pp 140–155
27. Le DP, Yang G, Ghorbani A (2019) A new multisignature scheme with public key aggregation for blockchain. In: 2019 17th international conference on privacy, security and trust (PST). IEEE, pp 1–7
28. Pieprzyk J, Wang H, Xing C (2003) Multiple-time signature schemes against adaptive chosen message attacks. In: International workshop on selected areas in cryptography. Springer, pp 88–100
29. Fujisaki E, Suzuki K (2007) Traceable ring signature. In: International workshop on public key cryptography. Springer, pp 181–200
30. Beal V (2021) Symmetric-key cryptography. https://www.webopedia.com/definitions/xsymmetric-key-cryptography/, online. Accessed 30 Jan 2022
31. global SSL provider S (2022) Symmetric vs. asymmetric encryption – what are differences? https://www.ssl2buy.com/wiki/symmetric-vs-asymmetric-encryption-what-are-differences. Accessed: 31 Jan 2022
32. Staff W (2021) DES. https://www.webopedia.com/definitions/des/, online. Accessed 31 Jan 2022
33. Loshin P (2021) Data encryption standard (DES). https://www.techtarget.com/searchsecurity/definition/Data-Encryption-Standard, online. Accessed 01 Feb 2022
34. Staff W (2021) Triple DES. https://www.webopedia.com/definitions/triple-des/, online. Accessed 31 Jan 2022
35. Hyperchain (2021) Cryptography algorithm. https://hyperchain.readthedocs.io/en/latest/crypto.html, online. Accessed 01 Feb 2022
36. Staff W (2021) AES. https://https://www.webopedia.com/definitions/aes/, online. Accessed 01 Feb 2022
37. Crane C (2020) Symmetric encryption algorithms: live long & encrypt. https://securityboulevard.com/2020/11/symmetric-encryption-algorithms-live-long-encrypt/, online. Accessed 01 Feb 2022
38. Garrett P (2007) Cryptographic primitives
39. Wagner L (2020) Elliptic curve cryptography: a basic introduction. https://qvault.io/cryptography/elliptic-curve-cryptography/, online. Accessed 01 Feb 2022
40. Followmyvote (2022) Elliptic curve cryptography in online voting. https://followmyvote.com/elliptic-curve-cryptography/. Accessed 01 Feb 2022
41. Avinetworks (2022) Elliptic curve cryptography. https://avinetworks.com/glossary/elliptic-curve-cryptography/. Accessed 01 Feb 2022
42. Saha P (2021) Diffie-hellman key exchange vs. RSA. https://www.encryptionconsulting.com/diffie-hellman-key-exchange-vs-rsa/, online. Accessed 01 Feb 2022
43. Silva P, Pycoin, interface to some coin packages

# Chapter 5
# Smart Contracts for Building Decentralized Applications

**Abstract** Smart contracts are an integral part of blockchain technology that automates business operations by removing intermediaries. Leading organizations such as real estate, finance, banking, and health care have adopted it due to its intrinsic features: transparency, integrity, privacy, efficiency, and confidentiality. There are various platforms available supported by high-level programming languages to implement and deploy smart contracts. Researchers are molding modern technologies like Artificial Intelligence (AI) into smart contracts to make them more effective and reliable. However, a few vulnerabilities or risks need to be taken care of while implementing smart contracts. This chapter gives an overview of smart contracts, including the history, advantages, and disadvantages, and the need for smart contracts. Further, the different blockchain platforms related to smart contracts are discussed. Then, the implementation details of smart contracts are explained, i.e., fundamentals of Solidity programming. Finally, this chapter shows the implementation of different smart contract use cases with Solidity source codes such as oil mining, online banking, voting, telesurgery, and remote patient monitoring systems.

**Keywords** Blockchain · Ethereum · Solidity · Smart contracts and Applications

## 5.1 Introduction

### 5.1.1 What is Smart Contract?

A smart contract is an automated digital contract instead of a paper contract that shows a mutual agreement between buyer and seller. The smart contract code and the contained agreement are shared across a distributed, decentralized blockchain network [1]. Smart contracts are executed on the blockchain network, which removes the need for a third party for application-specific transactions [2]. There are many implementation and deployment platforms available where one can create his own smart contract and execute it on the blockchain network.

**Need for the smart contracts** The conventional way to make a contract is to agree on a deal, go to a third party like a bank or lawyer as per the requirement, and make a contract. The issue with this mechanism is that it is time-consuming, includes paperwork, and is not much trustworthy. With the help of smart contracts, one can exchange money, shares, property, etc., without any third party in a transparent way [3]. The concept of making contracts between multi-parties for making any agreement has been evolved through ages [4, 5]. Traditional methods follow contracts as a form of a legal written document. For the deployment of such contracts, we require agents called the lawyer and a centralized system for verification. The advent of blockchain technology eliminates the requirement of centralized governing systems and makes the system decentralized. As a result, we can remove the bottlenecks of the traditional centralized system to deploy contracts by implementing a digital smart contract [6, 7]. Table 5.1 presents a comparison of a traditional contract and smart contract in detail.

The benefits of smart contracts are described as follows:

i.   **Transparency:** It is a crucial and fundamental characteristic of the smart contract. Before agreeing on any transaction, both parties need to check the terms and conditions mentioned in the smart contract. So, it eliminates the chances of clashes.
ii.  **Time-efficient:** Conventionally, the contract takes a lot of time due to the documentation, intermediary, and other unnecessary steps. Smart contracts remove the need for third parties and eliminate the delay caused due to manual procedures. One needs to work with an online transaction and all other work procedures will be executed automatically.
iii. **Precision:** Manual works are prone to errors, whereas smart contracts are software codes, that will execute automatically upon initiating a request for a finan-

**Table 5.1** Comparative analysis of traditional contract and smart contract

| Parameter | Traditional contract | Smart contract |
| --- | --- | --- |
| Time required | 1–3 days | Minutes |
| Payment scenario | Manual remittance | Automatic remittance |
| Cost | Expensive | Not expensive |
| Signature mode | Physical | Digital |
| Escrow | Required | Not required |
| Layers requirement | Compulsory | Not compulsory |
| Reconciliation process | Slow | Fast |
| Trusted third party | Necessary | Not required |
| Dispute resolution via | Judges, arbitrators | Consensus mechanism |
| Specification | Natural language | Smart code |
| Archiving | Hard | Easy |
| Transparency | Available | Not available |
| Security | Limited | High level security |

cial transaction. Before deploying a smart contract into the blockchain network, proper testing must be performed to ensure high precision results.

iv. **Savings:** Smart contracts save money as it eliminates the roles of the third party, lawyers, intermediaries, witnesses, and papers for the documentation. So, the cost associated with these roles is eliminated in smart contracts.

v. **Trust:** As smart contracts have features like transparency and security, they are more trustworthy than conventional contracts.

vi. **Decreases Human Intervention:** Traditionally, contracts were written on a legal document, which was long enough, complicated, and arduous that required trained personnel for framing and interpreting them. The smart contract is just a piece of code that, once deployed successfully, decreases human intervention and decreases complexity. We now no more require the chains of a middleman like lawyers and consultants.

vii. **Decentralized:** Smart contracts are deployed on a highly decentralized blockchain network. Hence, it removes the bottleneck of a single point of failure of centralized system.

viii. **Storage and backup:** Blockchain technology provides the benefits of storing the records in a public ledger. Once the record is entered into the chain, it will last forever. This property of blockchain technology gives benefits for storing smart contracts on the chain eternally.

ix. **Accuracy:** The terms and conditions of the contract need to be expressed precisely without any ambiguity. The smart contract provides the benefit of recording all the legal details accurately.

x. **Security:** The blockchain technology guarantees smart contracts to be tamper proof. Hence, it achieves the highest level of security as it reduces the chances of arising dispute.

xi. **Speed and Efficiency:** Traditionally, the contracts were written in a legal document that required processing manually, which consumed more time. But the smart contract is a piece of code that runs on the network and is self-executable; hence it takes less time. The smart contract is more efficient as it is not processed manually, reducing human errors.

xii. **Clear Communication:** As the smart contract is accessible to all, the terms and conditions are cleared prior; hence, there are no chances of ambiguity. This reduces the chances of miscommunication and misinterpretations.

**History and evolution of smart contracts** Nick Szabo, cryptographer, presented the real use case of a decentralized ledger for the smart contract in 1994. He outlined that we can digitally write the contracts, which can be kept in blockchain-based network systems. These smart contracts can improve transparency and build trust among communicating parties [8]. In 1994, an American scientist and cryptographer, Nick Szabo, devised the term smart contract for the first time. He defined the term as follows: "A smart contract is a set of promises, specified in digital form, including protocols within which the parties perform on these promises" [9]. The smart contract is the piece of code that is used to achieve a common consensus among multi-parties. The smart contract is the digital contract that lays down all the protocols, terms, and conditions to reach a common consensus among the untrusted parties. It eliminates the intervention of the third party.

Traditionally, the contracts were framed as detailed legal documents and processed with the help of lawyers and consultants. Such systems were so-called centralized systems. With the advent of blockchain technology, smart contracts started deploying on the blockchain's decentralized network. Ethereum was proposed by Vitalik Buterin [10] in late 2013 and made live on July 30, 2015. It has now become one of the common platforms for smart contract implementation. The traditional system requires the intervention of the chain of middlemen like lawyers and higher authorities of the law. The smart contract just requires the blockchain network for its deployment. Smart contracts have been strengthened over time. The evolution of smart contracts with its detailed description is described as follows:

- **Smart Contract 1.0—Bitcoin script** Initially, the smart contracts were written in the form of scripts like pseudo-code that was stack-based with no looping statements. But due to the language limitations, the development and the progress of smart contracts were limited. It has many limitations as a lack of looping statements made the programming power.

- **Smart Contract 2.0—Ethereum smart contract virtual machine** With the evolution of Ethereum, much progress was seen in the designing of smart contracts. Along with Ethereum, the Ethereum virtual machine (EVM) played an integral role in the successful deployment of the smart contracts in the network. It ensures the integrity of the contract along with the easy execution of the contract.

- **Smart Contract 3.0**—This generation of smart contracts allowed users to solve real-life problems through blockchain technology. This happened due to the invention of Cypherium. It is the only decentralized technology that meets the real-world requirements and has its own Cypherium virtual machine that is JAVA-based and allows us to perform hierarchical calculations, supports run time and compile time security check, supports transparent billing mechanism, and also has enhanced security features.

### 5.1.2  Disadvantages of the Smart Contracts

   i. **Immutable:** Smart contracts are immutable in nature. Once if even the smallest error is encountered, then it is difficult to rectify that error and it will remain there forever.
  ii. **Need permission environment:** A smart contract works well only for the permission environment as it needs to identify the parties between which the contract is been devised.

iii. **Non-elimination of the third party:** It is not possible to completely remove the third party. Lawyers are not required for framing the contract. But we require a lawyer to understand the basic terms and conditions, to understand the terminology of law.

iv. **Uncertain legal status:** As no government has control over any of the contracts made, there arises the absence of a standardized and firm rule book accepted and implemented by all the users.

v. **The need for a technical person:** As smart contracts can only be implemented using programming languages, hence, we need a technical person who has good coding skills and who can frame all the terms and conditions specified exactly in a program.

vi. **Human errors:** As the code is written by the human, there are chances of getting errors. If the smart contract is deployed, then it is almost impossible to rectify those errors.

## 5.2 Blockchain Platforms Using Smart Contracts

There are many smart contract platforms introduced to date like Ethereum, Rootstock, Chainspace, Zether, SmartDEMAP, EOS, Codius, Counterparty, DAML, Dogeparty, List, Monax, Symbiont, Stellar, Tezos, etc. However, all of them are not much popular but some of them have contributed a lot to the world of smart contracts in the form of privacy, confidentiality, speed of execution, reducing gas cost, etc. Detailing of the smart contract platforms that we have studied is as follows:

- **Ethereum:** Ethereum [11] is the first and the most popular smart contract platform. It was introduced with the intent of merging together and improving upon the concepts of scripting, altcoins, and on-chain meta-protocols. Ethereum has its own cryptocurrency called 'ETHER'. Ethereum provides a built-in Turing-complete programming language for the ease of writing smart contracts and decentralized applications where one can create its own arbitrary rules for ownership, transaction formats, and state transition function. For the execution of the code, Ethereum has the functionality of 'Ethereum Virtual Machine code' or 'EVM code' to convert the high-level language into a stack-based bytecode. It also provides the facility of wallets to the users. It implements the 'Greedy Heaviest Observed Subtree' (GHOST) protocol to reduce the confirmation time of a transaction.
- **Rootstock:** The ROOTSTOCK (RSK) [12] is a bitcoin-powered smart contract platform introduced in the year 2015. It incorporates a Turing-complete Rootstock Virtual Machine (RVM) to bitcoin. The currency of RSK is called the 'Rootcoins' (RTC). It implements the DECOR+ and GHOST protocols. It also integrates hardware wallets like Ethereum. The average confirmation time of a transaction is very less (10 s) compared to other platforms like bitcoin (10 min).
- **Chainspace:** Chainspace [13] was introduced in the year 2017 as a 'Sharded' smart contract platform. It implements a distributed atomic commit protocol named

Sharded Byzantine Atomic Commit (S-BAC) combining two protocols 'Byzantine agreement' and 'Atomic commit' which aims to provide security against the byzantine nodes. The Chainspace platform can handle up to 350 transactions per second for 15 shards.

- **Zether:** Zether [14] is the modified version of the Ethereum platform. It was basically introduced as a fully decentralized, confidential payment method to increase privacy in the smart contract world. It has its cryptocurrency named Zether tokens (ZTH) [15]. It uses Proof-of-Stack (PoS) consensus mechanism with some modifications instead of Proof of Work (PoW) and supports high-level languages like Solidity for writing the smart contracts. Zether uses ZK-Proofs to maintain confidentiality. It uses the -protocol to reduce the cost of a transaction. The average block execution time is about 15 s for the smart contracts written on the Zether platform.
- **Bitcoins:** Bitcoin uses script language to process bitcoin transactions. This language has limited capabilities for processing documents.
- **Hyperledger Fabric:** The Hyperledger Fabric platform is used for private blockchain where Chaincode is coded programmatically on the network. It is executed and validated by Chaincode validators during the consensus process.
- **NXT:** The NXT platform used for public blockchain includes a limited number of templates for writing the smart contracts. It is used when the user cannot write the code on their own.
- **Side chains:** The blockchain platform enhances the privacy protection of smart contracts. It also increases the performance of the blockchain by adding capabilities such as secure handles and real-world property registry. From foreknown blockchain platforms, Ethereum is the most popular platform used for writing smart contracts. It is a global and decentralized platform for new types of applications. It can accept a code in any programming language.

## 5.3   Deploying Smart Contracts on a Blockchain

Vitalik Buterin, a Canadian-Russian programmer, invented Ethereum in 2015. This invention led to the new concept of implementing the smart contract on the blockchain network. Nick Szabo coined the term 'smart contracts' in 1996. His work inspired so many upcoming researchers like Vitalik Buterin, who came up with the new concept of Ethereum. Some important terminologies to be noted are as follows [16]:

- **Ethereum Virtual Machine:** As the name suggests, it is a virtual machine and not a physical entity. It provides the run time environment for the execution of smart contracts. They work at the global level by executing smart contracts globally.
- **Gas Value:** It is the unit used in Ethereum for calculating the fees for successfully carrying out a single operation in the Ethereum blockchain network. The operations that can be carried out are transactions and execution of the smart contract. So if anyone wishes to carry out such operations in the blockchain network, they require

some amount of gas. In return, miners get fees back in the form of Ether for carrying out any of the operations. The price of 1 Ether is 20,4437.57 Indian Rupee as of February 1, 2022.

- **Decentralized Applications (DApps)** These applications are similar to web applications, where web applications use the API to interact with the webserver and the database. However, DApps use smart contracts as an interface to interact with the blockchain network. The DApp runs client-side code usually written in JavaScript in the browser. To manipulate the DApps and smart contracts, we make payments in cryptocurrencies like Ether, Bitcoin, etc. [17].

- **Ethereum Ecosystem:** It provides a decentralized platform to execute smart contracts. It also provides the developers with a blockchain with a built-in Turing-complete programming language that follows a peer-to-peer network model. Here, the blockchain database is managed and maintained at each node of the network.

- **Web3—Ethereum JavaScript API:** The collection of libraries allows us to interact with the network's local or remote Ethereum nodes. We can interact by making HTTP requests with inter-process communication. It also enables the users to interact with the smart contracts deployed on the network from web applications or browsers. For any platform like Web3, it first searches the Web3 provider, the node on the blockchain network that has the inbuilt capabilities of handling the communication among the contracts and manages the transactions made on the Ethereum blockchain.

## 5.4 Applications of Smart Contract

Smart contract has applications in various fields like gaming, insurance, tax records, banking sectors, etc. [18, 19]. The following are some applications of the smart contracts:

- **Digital Identity:** Today, we all might have many identity credentials for the different web services. It is a tedious process to keep a record of all such identity references and validate them. A smart contract provides benefits by allowing users to keep all the data in one place securely and can be validated easily. Once registered to the block, the blockchain will keep our identity holistic.

- **Supply chain management:** Smart contracts in the supply chain can eliminate mediators' need to monitor the entire process at each granular level. We can write all the terms, conditions, and protocols required to be considered and implemented at each level of the supply chain management. It cuts down the cost of hiring the third-party service and decreases human errors and fraudulent activities, making the system more transparent.

- **Real Estates:** Real estate marketing with cross borders is a tedious task as it requires ample formalities to be completed before having the actual transaction. Applying smart contracts can eliminate the inclusion of the government in this entire process. It also speeds up the process of buying and selling real estate commodities.

- **Insurance:** For claiming the insurance, you need to convince the respective authority for the refund; unfairness happens to the client if the agent denies any insurance. We can frame a smart contract between the insurance company and the users and embed the IoT device, which monitors insurance entities and eliminate human intervention. If the case is genuine, then the IoT device triggers the contact, and the money gets transferred automatically to the client's account.
- **Tax Records:** Pending taxes lead to corruption and black money that hinders the nation's economic growth. It can be easily handled with a smart contract, where it automatically triggers the user's account if the terms and conditions are met. It also provides transparency and hence eliminates the chances of corrupt money.
- **Gaming Industry:** We can apply smart contracts for online games where the money is automatically regulated between players as soon as they win the game by triggering the smart contract. This is helpful in gambling and casinos as all the transactions are recorded and kept safely, and the amount will also be regulated automatically.
- **Banking Sector:** Cryptocurrencies are a boon, especially for cross-border transactions, as they can smoothly carry out such transactions without the intervention of government agents. The majority of the world's transactions are carried out at various banks across the globe, which makes the system dependent on the central authority for executing any transaction. Smart contracts speed up this process and accurately execute transactions across the world, which speeds up business operations.
- **AI and Smart Homes:** We can make our home smart by molding the IoT devices with the smart contract. We can write specific conditions and the timings in the contract that triggers the IoT devices to operate automatically. We can also trace our daily activities by logging the records.
- **Healthcare sector:** Nowadays, digital watches are embedded with IoT devices that can track our health attributes like blood pressure, heart rate, temperature, and many more. We can frame the smart contract for patient monitoring and if any abnormalities are encountered, then it triggers the watch and updates the users [20].

## 5.5   Implementation Concepts

### 5.5.1   Introduction to Solidity Programming

The blockchain network replaces the arbitrators with smart contracts that automate the execution workflow when some pre-conditions are met. To create smart contracts in various blockchain platforms, we need to incorporate Solidity programming, which is a high-level programming language specifically designed to implement smart con-

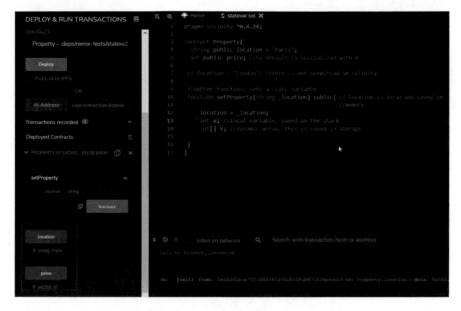

**Fig. 5.1** State variables of Solidity programming

tracts. It was first introduced and implemented by Christian Reitwiessner et al. [21] in an Ethereum-based blockchain. It supports object-oriented programming languages such as C++ and Python and executes in the run time environment, i.e., Ethereum virtual machine (EVM). In addition, the run time environment provides a robust solution to security threats such as denial-of-service and data manipulation attacks. This section gives a detailed description of Solidity programming so that a learner can understand the fundamentals and eventually grasp it to create application-specific smart contracts. The implementation steps consist of Solidity code written in Remix IDE as shown in Fig. 5.1.

**Structure of a Solidity Program**

- The Solidity code is saved as an extension of *.sol*. To understand this, below is the 'hello world' sample code of Solidity programming. The program is saved as *'helloworld.sol'*; this indicates that it is a Solidity source code file.
- Next, the *pragma solidity ^0.8.10;* defines the specific version of Solidity that is used by the developer. It allows resolving any compatibility issues in the source code file.
- The *contract helloworld* is the keyword that specifies the compiler that the program is meant to create a smart contract, with the name 'helloworld'. The contract has a scope shown in curly brackets, within which we can declare variables and functions similar to other programming languages.

- The *welcome* is a Solidity variable that is used to store information of real-world objects of different data types such as string, character, Boolean, float, and integer. The operating system assigns memory registers by looking at the data types of the variables. Here, the string 'Hello World!' is stored inside the Solidity variable 'welcome'.

- The *addition* is a Solidity function where we have multiple lines of statements aiming to perform a specific task. The necessity of the functions in any programming language is its unique characteristic to reuse the code instead of writing the entire source code. We can use the function() keyword followed by the unique function name in Solidity. Further, the function has to specify its scope of accessibility; it can be done using public, private, internal, and external keywords. Here, we have addition() as a function name, comprising three local variables x, y, and addition.

```
pragma solidity ^0.8.10;
contract HelloWorld
{
    string public welcome = "Hello World!";
    function addition() public view returns(uint)
    {
        uint x = 10;
        uint y = 20;
        uint addition = x+y;
        return addition
    }
}
```

### 5.5.2  Basics of Solidity Programming

The section represents the fundamentals of Solidity programming by providing a detailed explanation with Solidity source codes executed in Remix IDE.

**Solidity state variables:** The state variables are the variables whose value is permanently stored inside the smart contract. For example, in the below source code the variable *location* is the state variable, whose value is 'Paris', and it is permanently stored inside the smart contract. Figure 5.1 shows the output of the state variable program.

```
pragma solidity ^0.4.24;
contract Property
{
    string public location = "Paris";
    int public price; //by default is initialized with 0
    function setProperty(string _location) public
```

```
{ //_location is local and saved in memory
   location = _location;
   int a; //local variable, saved on the stack
   int[] x; //dynamic array, this is saved in storage
}
}
```

**Setter and getter:** The setter() and getter() functions of Solidity are used to set and get the values of any specific property. Here, we have two state variables 'value' and 'location', to which we can set the value using 'value' keyword and return the value using 'return' keyword. The output shows the getLocation and setLocation to return the value of location, i.e., London and then set the value, i.e., 25.

```
pragma solidity ^0.5.2;
contract Property
{
    //state variables
    uint public value = 10;
    string public location = "London";
    function setLocation(string memory _location) public
    {
        location = _location;
    }
    function setValue(uint _value) public
    {
        value = _value;
        //setter to set the value = 10
    }
    function getLocation()
    public view returns(string memory)
    {
        return location;
        //getter, returns a value of a location
    }
}
```

**Output**

```
getLocation= London
setLocation= NewYork
getLocation= NewYork

getValue= 10
setValue= 25
getValue= 25
```

Arrays in Solidity When we have large data values of the same data type, we have to rely on a data structure, i.e., arrays. It is a collection of data of the same type. Solidity supports two types of arrays, that is, fixed size and dynamic size arrays. A fixed-size array is an array whose size can be known at compilation time. On the contrary, the dynamic arrays are those whose length is not fixed; they can be changed at the run time. The below programs show the two fixed-size arrays *x* and *y*, then later using functions we have manipulated the arrays such as fetching array length and returning the element of an array.

```solidity
pragma solidity ^0.4.24;
contract BytesFixedArrays
{
    bytes2 public x; //fixed-size array of 2 bytes
    bytes3 public y; //fixed-size array of 3 bytes
    function setX_Y() public
    {
        x = 'ab';
        //the elements of x are the hexadecimal ASCII
        // codes of 'a' and 'b'
        y = 3;
        // 3 is converted to hexadecimal (0x03)
    }
    //returns an element of the array
    function getX(uint i) public view returns(bytes1)
    {
        return x[i];
    }
    //returns the fixed length of the array
    function get_y_length() public view returns(uint)
    {
        return y.length;
    }
}
```

**Output**

```
get_y_length() = uint256: 3
x = bytes2: 0x6162
y = bytes3: 0x000003
```

The below program shows the working of a dynamic array in Solidity programming. In this program, we have considered price and location as two state variables; the price variable is a dynamic array of whose elements can be traversed and deleted using get_element and delete_element functions.

```solidity
pragma solidity ^0.4.24;
contract BasicdynamicArray
{
    string public location = 'London';
    uint[] public prices;
    function add_price(uint _price) public
    {
        prices.push(_price);
    }
    function get_length() view public returns(uint)
```

```
    {
        return prices.length;
    }
    function get_element(uint index) view
    public returns(uint)
    {
        if (index < prices.length)
        {
            return prices[index];
        }
    }
    function delete_element(uint index)
    public returns(bool)
    {
        if(index >= prices.length)
            return false;
        for (uint i = index; i< prices.length - 1; i++)
        {
            prices[i] = prices[i+1];
        }
    prices.length--;
    return true;
    }
    function f() public{
        uint[] storage myArray = prices; //dynamic array
    }
}
```

**Output**

```
add_price: 30 31 32
get_length= unit256:3
get_element:1 = unit256:31
prices:2= unit256:32
```

Another manipulation of arrays can be done by resizing the array size. To do that, we have used a function *optimized_delete* which optimally deletes the element from the array and resizes its size. The normal delete function, *delete_from_array*, deletes the element from the index *i* but now the index *i* is empty, therefore all elements have to move left to fill the empty index slot.

```
pragma solidity ^0.4.18;
contract arrayresize
{
    //dynamic array
    uint[] public myArray = [1, 2, 3, 4];
    function delete_from_array(uint i) public{
        delete myArray[i];
    }
    //add an element to the dynamic array
    function add(uint item) public{
```

```
        myArray.push(item);
    }
    function optimized_delete(uint index) public {
        if (index >= myArray.length) return;

        for(uint i=index;i<myArray.length-1;i++){
            myArray[i] = myArray[i+1];
        }
        myArray.length--;
    }
}
```

## Output

```
myArray: 1 2 3 4
add: 5 6 7
myArray: 1 2 3 4 5 6 7
optimized_delete: 2
myArray: 1 2 4 5 6 7
```

**Structures in Solidity:** Solidity enables developers to create user-defined data types, i.e., structure. It consists of multiple elements of different data types.

```
pragma solidity ^0.5.0;
contract basicstructure
{
   struct student
   {
      string studentname;
      string studentaddr;
      uint id;
   }
   student std;
   student std1
     = student("Nilesh",
            "India",
             2);
   function set_book_detail() public
   {
      std = student("Rajesh",
                  "India",
                   1);
   }
   function std_info()
   public view returns(
     string memory, string memory, uint)
     {
        return(std1.studentname, std1.studentaddr,
             std1.id);
     }
   function get_details()
   public view returns(string memory, uint) {
      return (std.studentname, std.id);
   }
```

```
}
```

## Output

```
get_details= string:Rajesh, uint256:1

std_info= string:Nilesh, string:India, uint256:2
```

**Enumeration in Solidity:** Enumeration is also used to create user-defined data types. It is specifically used to create a list of records that is easy to access and maintain. In the below code, we have an enumeration of months, where we have taken two functions that set the user's default month, and the other fetches the month of the user's choice.

```
pragma solidity ^0.5.0;
contract Types {
    enum month
    {
    January,
    Feburary,
    March,
    April,
    May,
    June,
    July,
    August,
    September,
    October,
    November,
    December
    }
    month mth;
    month choice;

    month constant default_value
    = month.Feburary;

    function set_value() public {
    choice = month.December;
    }
    function get_choice(
    ) public view returns (month) {
    return choice;
    }
    function getdefaultvalue(
    ) public pure returns(month) {
        return default_value;
    }
}
```

**Output**

```
get_choice= uint8:11
getdefaultvalue= unit8:1
```

## 5.6  Smart Contracts Case Studies

In this section, we use the basic concepts of Solidity language as discussed in the previous section to develop a smart contract for specific application purposes. This section is divided into five case studies, such as telesurgery, oil mining, banking, voting, and remote patient monitoring. The detailed description of each smart contract along with the step-by-step explanation of the Solidity code is described as follows.

### 5.6.1  Telesurgery

In this case study, we present the working procedure (smart contract implementation code in Solidity language) of telesurgery or remote surgery. This contract is used to ensure synchronization, trust, security, and information tracking. It comprises various entities such as surgeon entity, patient entity, authority entity, and hospital entity [22]. Here, each entity is associated with each other while executing the telesurgery. We will discuss the step-by-step guide for the development of the telesurgery smart contract, which is as follows.

1. Define the Solidity version, entities, variables, and mapping related to the telesurgery operation. Here, *Telesurgery* is the name of the contract. *pragma solidity* $\geq 0.4.22 < 0.6.0$ specifies the version that the Solidity supports.

```
pragma experimental ABIEncoderV2;
pragma solidity >=0.4.22 <0.6.0;
contract TeleSurgery{
Entity patient;
address authorizingCommitee;
mapping(address=>Entity) entities;
mapping(address=>bool) surgeons;
mapping(address=>bool) careTakers;
SurgeryState state;
SurgeryResult result;
SurgeryDomains surgeryDomain;
string surgeryDescription;
string[] activityIpfsHash;
uint256 beginTimeStamp;
uint256 endTimeStamp;
```

2. Define the role of each entity (patient, surgeon, caretaker), telesurgery domain (heart, kidney, liver, etc.), type of service (nurse, machine operator, wardboy), state of surgery (created, active, finished), and final outcomes (successful, failed). This step also defines the recommendation about the surgeons, i.e., strongly recommended, normal recommended, and not recommended.

```
enum Role{NA,Surgeon ,Patient ,CareTaker}
enum SurgeryDomains{heart,kidney,liver,stomach,
brain}
enum ServiceProvided{nurse,machineOperator,wardBoy}
enum SurgeryState{created,active,finished}
enum SurgeryResult{successful,failed}
enum ReccommendType{strongly,normal,not}
```

3. Define the entity structure 'Entity' with *struct* keyword. Each entity must have address, name, location, and role. But surgeons have some other parameters such as specialty, recommendation type, and number of successful/unsuccessful surgeries. For patient entity, date of birth is necessary to record and for caretaker, designation and service type are must to specify.

```
//Structures
struct Entity{
    address addr;
    string name;
    string location;
    Role role;
    //Valid only if role is Surgeon
    SurgeryDomains speciality;
    ReccommendType surgeonReccommended;
    bool isCertified;
    uint32 surgeriesSuccessful;
    uint32 surgeriesUnsuccessful;
    uint32 performanceRate;
    uint32 totalReviewCount;
    //Valid only if role is Patient
    string oldTransactionIpfsHash;
    string dob;
    //Valid only if role is CareTaker
    string designation;
    ServiceProvided service;
}
```

4. Specify the modifiers for each entity, i.e., surgeon, patient, and caretaker, as well as service type.

```
//MODIFIERS
modifier onlySurgeon{
    require(entities[msg.sender].addr
    != address(0x0));
    require(entities[msg.sender].role
    ==Role.Surgeon);
```

```
    _;
}
modifier onlyPatient{
    require(entities[msg.sender].addr
    != address(0x0));
    require(entities[msg.sender].role
    ==Role.Patient);
    _;
}
modifier onlyCareTaker{
    require(entities[msg.sender].addr
    != address(0x0));
    require(entities[msg.sender].role
    ==Role.CareTaker);
    _;
}
modifier inState(SurgeryState _state){
    require(state==_state);
    _;
}
constructor ()public{
    authorizingCommitee=msg.sender;
}
```

5. Define the common functions of the smart contract, such as *addPatient*, *addSurgeon*, *addCareTaker*, and *updateRecommendation*.

```
    //Functions
// Common to all
function addPatient (
    string memory _name,
    string memory _location,
    string memory _dob,
    string memory _oldTransactionIpfsHash) public{
        require(entities[msg.sender].role!=
        Role.Patient,"Already enrolled");
        entities[msg.sender].addr=msg.sender;
        entities[msg.sender].role=Role.Patient;
        entities[msg.sender].name=_name;
        entities[msg.sender].location=_location;
        entities[msg.sender].oldTransactionIpfsHash
        = _oldTransactionIpfsHash;
        entities[msg.sender].dob=_dob;
    }
function addSurgeon (
    string memory _name,
    string memory _location,
    SurgeryDomains _speciality) public{
        require(entities[msg.sender].role
        !=Role.Surgeon,"Already enrolled");
        entities[msg.sender].addr=msg.sender;
        entities[msg.sender].role=Role.Surgeon;
        entities[msg.sender].name=_name;
```

```
            entities[msg.sender].location=_location;
            entities[msg.sender].isCertified=false;
            entities[msg.sender].speciality=_speciality;
            entities[msg.sender].surgeriesSuccessful=0;
            entities[msg.sender].surgeriesUnsuccessful
            =0;
            entities[msg.sender].performanceRate=9;
            entities[msg.sender].totalReviewCount=10;
        }
    function addCareTaker(
        string memory _name,
        string memory _location,
        string memory _designation,
        ServiceProvided _service) public{
            require(entities[msg.sender].role
            !=Role.CareTaker, "Already enrolled");
            entities[msg.sender].addr=msg.sender;
            entities[msg.sender].role=Role.CareTaker;
            entities[msg.sender].name=_name;
            entities[msg.sender].location=_location;
            entities[msg.sender].designation
            =_designation;
            entities[msg.sender].service=_service;
        }
    function updateReccommend(address _address)
    internal{
        if(entities[_address].performanceRate>9)
        {
            entities[_address].surgeonReccommended
            =ReccommendType.strongly;
        }
        else if(entities[_address].performanceRate>8)
        {
            entities[_address].surgeonReccommended
            =ReccommendType.normal;
        }
        else
        {
            entities[_address].surgeonReccommended
            =ReccommendType.not;
        }
    }
```

6. Define functions related to the patient entity, such as *addSurgery*, *addSurgery-Surgeon*, *addSurgeryCareTaker*, and *addSurgeonFeedback*.

```
//BY PATIENTS
function addSurgery (
    SurgeryDomains _surgeryDomain,
    string memory _surgeryDescription) onlyPatient
    public{
        patient=entities[msg.sender];
        surgeryDomain=_surgeryDomain;
```

```
        surgeryDescription=_surgeryDescription;
        state=SurgeryState.created;
        beginTimeStamp=now;
    }
function addSurgerySurgeon(address _surgeon)
onlyPatient inState(SurgeryState.created) public {
    surgeons[_surgeon]=true;
    state=SurgeryState.active;
}
function addSurgeryCareTaker(address _careTaker)
onlyPatient inState(SurgeryState.created) public {
    careTakers[_careTaker]=true;
    state=SurgeryState.active;
}
function addSurgeonFeedback(address _address,uint32
totalPositiveFeedback) onlyPatient
inState(SurgeryState.finished) public {
    require(entities[_address].role==Role.Surgeon,
    "Can't add review if not surgeon !");
    if( totalPositiveFeedback > 2 ) {
        // if else statement
        entities[_address].performanceRate
        = ( entities[_address].performanceRate
        *entities[_address].totalReviewCount + 1)/
        ( entities[_address].totalReviewCount + 1);
    }else {
        entities[_address].performanceRate=
        entities[_address].performanceRate/
        ( entities[_address].totalReviewCount + 1);
    }
    entities[_address].totalReviewCount++;
    updateReccommend(_address);
}
```

7. Define functions related to the surgeon entity, such as *viewPatientData*, *addActivity*, *finishSurgery*, and *addToHistory*.

```
//BY SURGEONS
function viewPatientData(address _address)
onlySurgeon view public returns(string memory){
    require(entities[_address].role==Role.Patient,
    "Can't view data if not patient !");
    require(entities[msg.sender].speciality ==
    surgeryDomain, "Can't access data of patients
    of other Surgery Domain");
    return entities[_address].oldTransactionIpfsHash;
}
function addActivity(string memory _activityIpfsHash,
address _address) onlySurgeon inState
(SurgeryState.active)
public {
    require(entities[_address].role==Role.Patient,
    "Can't add activity if not patient !");
```

```
        require(surgeons[msg.sender]==true,
        "Can't add activity of patients if not selected
        by the patient");
        activityIpfsHash.push(_activityIpfsHash);
    }
    function finishSurgery(address _address,
    SurgeryResult _result)
    onlySurgeon inState(SurgeryState.active) public{
        require(entities[_address].role==Role.Patient,
        "Can't finish if not patient !");
        require(surgeons[msg.sender]==true,
        "Can't add activity of patients if not selected
        by the patient");
        //set if successful or not
        result=_result;
        //set ending timestamp
        endTimeStamp=now;
        state=SurgeryState.finished;
        if(_result==SurgeryResult.successful){
            entities[msg.sender].surgeriesSuccessful
            +=1;
        }
        else{
            entities[msg.sender].surgeriesUnsuccessful
            +=1;
        }
    }
    function addToHistory()
    inState(SurgeryState.finished)
    view public returns(
    SurgeryResult,
    SurgeryDomains,
    string memory,
    string[] memory,
    uint256,
    uint256 ){
        return (result,surgeryDomain,surgeryDescription,
        activityIpfsHash,beginTimeStamp,endTimeStamp);
    }
}
```

Upon executing the aforementioned steps in the Remix IDE, the *Telesurgery* contract produces the output as mentioned in Figs. 5.2 and 5.3. Figure 5.2 shows the Remix interface for writing the Solidity code and Fig. 5.3 shows the final generated output. The output comprises status, transaction hash, gas value, transaction cost, execution cost, and many more.

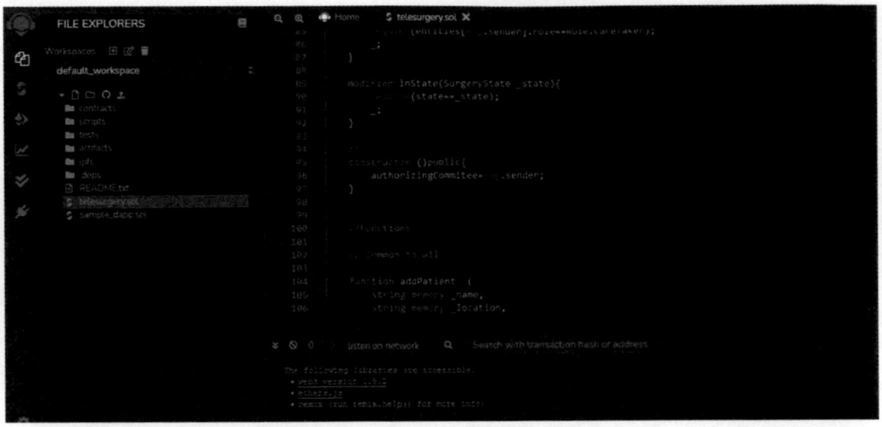

**Fig. 5.2**  Telesurgery smart contract Remix IDE interface

**Fig. 5.3**  Telesurgery smart contract output with cost, gas value, transaction hash, etc.

## 5.6.2   Oil Mining

Oil mining is now noticing the capabilities of blockchain technology to track and maintain its resources efficiently. In this case study, we have created a smart contract to efficiently manage the mining of oil and gas commodities as displayed in Fig. 5.4. We have assigned suitable roles for the different entities of the mining operation using Solidity functions, events, and structure. The detailed workflow of assigning roles is described as follows:

1. pragma solidity ^0.6.0; shows that the Solidity 0.6.0 is used for the oil mining smart contract.

**Fig. 5.1** Oil mining smart contract Remix IDE interface

2. The proposed smart contract conveys role-based access control; therefore, Open-Zeppelin roles are used as they have *AccessControl* that can be used to assign any role to the members of the smart contracts. The AccessControl.sol is a helper file that provides necessary access control function that can be used in the smart contract.
3. The smart contract is given a name, i.e., OilMining; the *is AccessControl* is the member of OpenZeppelin.

```solidity
pragma solidity ^0.6.0;
//import OpenZepplin Roles
import "github.com/OpenZeppelin/
openzeppelin-contracts/
contracts/access/AccessControl.sol";
contract oilMining is AccessControl
{
    bytes32 public constant LAND_OWNER_ROLE =
    keccak256("LAND_OWNER_ROLE");
    bytes32 public constant MINING_COMPANY_ROLE =
    keccak256("MINING_COMPANY_ROLE");
    uint PortRate;
    mapping(address=>LData) LO;
    struct LData
    {
        string hash;
        bool request;
    }
    mapping(address=>CData) MC;
    struct CData
    {
        string hash;
        bool request;
    }
    mapping(address => mapping
    (address => bool))validOffer;
```

```
mapping(address => mapping
(address => deal))deals;
struct deal
{
    uint256 initialBonus;
    uint256 thresholdProductionRate;
    bool request;
    bool paidIB;
}
```

4. Next, two role identifiers are defined, i.e., LAND_OWNER_ROLE and MIN-
   ING_COMPANY_ROLE. The identifier information is hashed using keccak256()
   hash function.
5. Data structure *LData* and *CData* for both the identifiers are created which has
   members such as *hash* and *request*. Then a mapping has been formulated to
   manage the offer and relations between the stakeholders.
6. Next, different events are generated between land owner (LO) and mining company (MC) such as generation of offers between LO and MC, status of royalties,
   and inter-bank payment. The constructor assigns the authority and it grants the
   contract deployer the default admin role, which makes him to grant and revoke
   any roles.

```
event Admin_set(address newAdmin);
event Offer_generated(address MC,
address LO,uint256 initialBonus,
uint256 thresholdProductionRate);
event Offer_accepted(address MC,address LO);
event Royalty_paid_successfully(address MC,
address LO,uint256 date,uint256 Amount);
event Royalty_payment_failed(address MC,
address LO,uint256 FaultyAmount);
event IB_paid_successfully(address MC,
address LO,uint256 date,uint256 Amount);
event IB_payment_failed(address MC,
address LO,uint256 FaultyAmount);
constructor() public
{
    _setupRole(DEFAULT_ADMIN_ROLE, msg.sender);
}
```

7. According to the events, the registration request of LO and MC is requested and
   granted by the admin. *requestToBecomeLandOwner* and *requestToBecomeMi-
   ningCompany* indicate the registration request to become LO and MC, where
   both parties have to submit the hash and request created in the structure. Based
   on that, the request is granted or revoked by the admin; the related functions for
   the same are *grantLandOwner* and *grantMiningCompany*.

```
function requestToBecomeLandOwner
(string memory hashKey) public
{
```

```
    require(!hasRole(LAND_OWNER_ROLE,
    msg.sender), "Caller is already a Land Owner.");
    LO[msg.sender].request = true;
    LO[msg.sender].hash = hashKey;
}
function grantLandOwner(address _LOToBe)public
{
    require(LO[_LOToBe].request == true,
    "No Land Owner request found
    for this account!");
    grantRole(LAND_OWNER_ROLE,_LOToBe);
    LO[_LOToBe].request = false;
}
function requestToBecomeMiningCompany
(string memory hashKey) public
{
    require(!hasRole(MINING_COMPANY_ROLE,
    msg.sender),
    "Caller is already a Mining Company.");
    MC[msg.sender].request = true;
    MC[msg.sender].hash = hashKey;
}
function grantMiningCompany(address _MCToBe)public
{
    require(MC[_MCToBe].request == true,
    "No Mining Company request found
    for this account!");
    grantRole(MINING_COMPANY_ROLE,_MCToBe);
    MC[_MCToBe].request = false;
}
```

8. The below code shows the offer that is provided to the LO and MC shown in *makeOffer* and *acceptOffer*.

```
function makeOffer(address _LO,
uint256_initialBonus,uint256
_thresholdProductionRate)public
{
    require(hasRole(MINING_COMPANY_ROLE,
    msg.sender), "Caller is not a Mining Company.");
    require(hasRole(LAND_OWNER_ROLE, _LO),
    "The stakeholder to offer is not Land Owner");
    deals[msg.sender][_LO].initialBonus=
    _initialBonus;
    deals[msg.sender][_LO].thresholdProductionRate=
    _thresholdProductionRate;
    deals[msg.sender][_LO].request=true;
    emit Offer_generated(msg.sender,
    _LO,_initialBonus,_thresholdProductionRate);
}
function acceptOffer(address _MC) public
{
    require(hasRole(LAND_OWNER_ROLE,
```

```
        msg.sender),"Caller is not a Land Owner.");
        require(hasRole(MINING_COMPANY_ROLE, _MC),
        "The stakeholder to accept offer is
        not a Mining Company.");
        require(deals[_MC][msg.sender].request==true,
        "No offer found to be accepted.");
        require(validOffer[_MC][msg.sender]==false,
        "Deal already accepted.");
        validOffer[_MC][msg.sender]=true;
        emit Offer_accepted(_MC,msg.sender);
    }
```

9. Finally, the payment options are enable using *payIB* and *paywithProduction* to manage the daily royalties between LO and MC.

```
    function payIB(address payable _LO) public payable
    {
        require(hasRole(MINING_COMPANY_ROLE,
        msg.sender), "Caller is not a Mining Company.");
        require(hasRole(LAND_OWNER_ROLE, _LO),
        "The stakeholder to offer is not a Land Owner.");
        require(validOffer[msg.sender][_LO]==
        true,"Invalid deal.");
        require(deals[msg.sender][_LO].paidIB==
        false,"Initial bonus already paid
        to the Land Owner.");
        if(msg.value>=deals[msg.sender]
        [_LO].initialBonus)
        {
            _LO.transfer(msg.value);
            deals[msg.sender][_LO].paidIB=true;
            emit IB_paid_successfully(msg.sender,
            _LO,now,msg.value);
        }
        else
        {
            msg.sender.transfer(msg.value);
            emit IB_payment_failed(msg.sender,
            _LO,msg.value);
        }
    }
    function payWithProduction(uint256 PR,
    address payable _LO) public payable
    {
        require(hasRole(MINING_COMPANY_ROLE,
        msg.sender), "Caller is not a Mining Company.");
        require(hasRole(LAND_OWNER_ROLE, _LO),
        "The stakeholder to offer is not a Land Owner.");
        require(validOffer[msg.sender][_LO]==
        true,"Invalid deal.");
        require(deals[msg.sender][_LO].paidIB==
        true,"First pay the Initial bonus.");
        if(msg.value>=PR/100*12)
```

```
        {
            _LO.transfer(msg.value);
            emit Royalty_paid_successfully
            (msg.sender,_LO,now,msg.value);
        }
        else
        {
            msg.sender.transfer(msg.value);
            emit Royalty_payment_failed
            (msg.sender,_LO,msg.value);
        }
    }
}
```

## 5.6.3  Banking Smart Contract

Banking is another sector where privacy needs to be preserved; therefore, integration of smart contracts is required between the bank account holder and its deposits. The below-mentioned smart contract specifies the log events about a deposit being made by the account holder with the prescribed amount. Figure 5.5 shows the interface of the online banking smart contract in Remix IDE. The step-by-step process of the banking-based smart contract is summarized as follows:

**Fig. 5.5** Online banking smart contract Remix IDE interface

1. First, pragma solidity ^0.5.8; defines the version of Solidity code, i.e., 0.5.8 for compatibility checking.
2. Contract name *'SimpleBank'* is defined, where different parameters have been taken such as clientcount, balance, and owner.
3. A suitable event *LogDepositMade* is generated to log the details about the deposit the account holder is going to perform. Furthermore, a constructor named *payable* is generated so that the client can receive the initial funding of 30 ethers.

```
pragma solidity ^0.5.8;
contract SimpleBank
{
uint8 private clientCount;
mapping (address => uint)
private balances;
address public owner;
event LogDepositMade(address indexed
accountAddress, uint amount);
constructor() public payable
{
    require(msg.value == 30 ether,
    "30 ether initial funding required");
    owner = msg.sender;
    clientCount = 0;
}
```

4. Next, the function *enroll()* enrolls a customer with the bank by providing the initial funding of 30 ethers, and returns the balance of the user after enrolling. The *deposit()* deposits ether into the bank and returns the balance of the client after the successful deposit. *withdraw()* withdraws the ether from the bank and returns the remaining balance of the client.

```
function enroll() public returns (uint)
{
    if (clientCount < 3)
    {
        clientCount++;
        balances[msg.sender] = 10 ether;
    }
    return balances[msg.sender];
}
function deposit() public payable returns(uint)
{
    balances[msg.sender] += msg.value;
    emit LogDepositMade(msg.sender, msg.value);
    return balances[msg.sender];
}
function withdraw(uint withdrawAmount)
public returns (uint remainingBal)
{
    if (withdrawAmount <= balances[msg.sender])
    {
```

```
            balances[msg.sender] -=
            withdrawAmount;
            msg.sender.transfer
            (withdrawAmount);
        }
        return balances[msg.sender];
    }
    function balance()
    public view returns (uint)
    {
        return balances[msg.sender];
    }
    function depositsBalance()
    public view returns (uint)
    {
        return address(this).balance;
    }
}
```

### 5.6.4 Voting Smart Contract

The Current online voting system is not reliable and suffers from security threats such as denial of service, vote manipulation, and malware attack on the voting system. Utilizing blockchain makes the online voting system robust, reliable, and protected from malicious adversaries. The user votes are stored inside an immutable block with a particular transactionID, timestamp, and hash. This transaction is broadcasted to the blockchain network, where each blockchain member verifies it, and then it is added to the distributed ledger of the blockchain. Figure 5.6 illustrates the smart contract of an online voting system in Remix IDE. A detailed summarization of the voting system smart contract is described as follows:

1. First, we have created a contract name *Ballot* which has a different structure with its members such as voter weight, voter has voted or not, address of the delegate, and vote count. The constructor named *_numProposals* has been used to create new ballots for different proposals.

```
pragma solidity $\ge$0.4.22 $<$0.6.0;
contract Ballot
{
struct Voter
{
    uint weight;
    bool voted;
    uint8 vote;
    address delegate;
}
struct Proposal
```

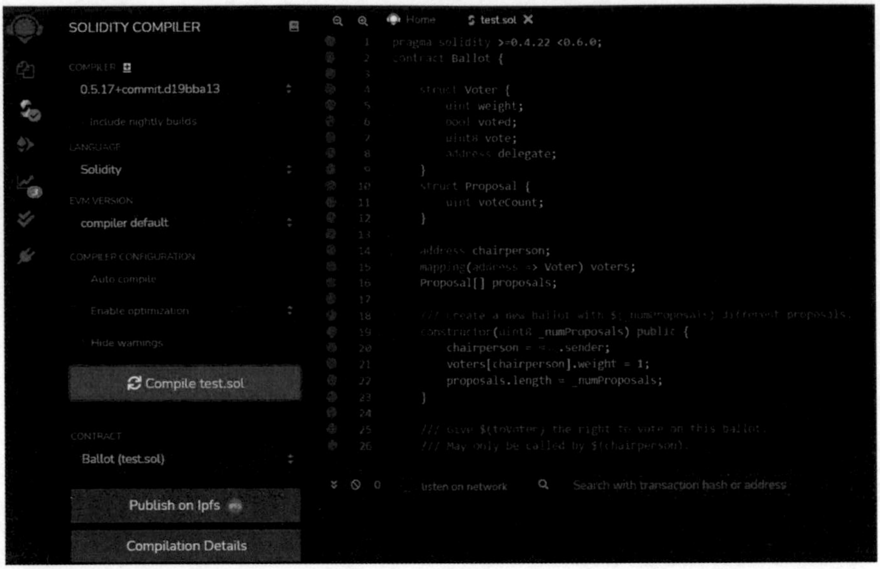

**Fig. 5.6** Online voting system smart contract Remix IDE interface

```
{
    uint voteCount;
}
address chairperson;
mapping(address $\ge$$ Voter) voters;
Proposal[] proposals;
constructor(uint8 _numProposals)public
{
    chairperson = msg.sender;
    voters[chairperson].weight = 1;
    proposals.length = _numProposals;
}
```

2. The contract has four function such as *giveRightToVote()* which specifies the voter
   the right to vote on this ballot. The *delegate()* adds a delegate the user vote to the
   voter. The *vote()* gives a single vote to the proposal.

```
function giveRightToVote(address toVoter)public
{
    if (msg.sender!= chairperson||
    voters[toVoter].voted) return;
    voters[toVoter].weight = 1;
}
function delegate(address to)public
{
    Voter storage sender = voters[msg.sender];
    if (sender.voted) return;
```

```
    while (voters[to].delegate!= address(0)&&
    voters[to].delegate != msg.sender)
        to = voters[to].delegate;
    if (to == msg.sender) return;
    sender.voted = true;
    sender.delegate = to;
    Voter storage delegateTo = voters[to];
    if (delegateTo.voted)
        proposals[delegateTo.vote].voteCount +=
        sender.weight;
    else
        delegateTo.weight += sender.weight;
}
function vote(uint8 toProposal) public
{
    Voter storage sender = voters[msg.sender];
    if (sender.voted ||
    toProposal >= proposals.length)
    return;
    sender.voted = true;
    sender.vote = toProposal;
    proposals[toProposal].voteCount +=
    sender.weight;
}
function winningProposal() public view
returns (uint8 _winningProposal)
{
    uint256 winningVoteCount = 0;
    for (uint8 prop = 0; prop <
    proposals.length; prop++)
        if (proposals[prop].voteCount >
        winningVoteCount)
        {
            winningVoteCount = proposals[prop].
            voteCount;
            _winningProposal = prop;
        }
    }
}
```

### 5.6.5 Remote Patient Monitoring

Remote patient monitoring is now noticing the capabilities of blockchain technology to track, secure, and manage its resources efficiently. In this case study, we make use of wearable devices for remote patient monitoring [23, 24]. This smart contract is bifurcated into four different functions *aUser*, *healthData*, *prescription*, and *viewPatientData*. The step-by-step detailed workflow of the Solidity smart contract is as follows.

1. *pragma solidity 0.8.7* depicts the version of Solidity compiler used in the execution of the smart contract.

```
pragma solidity 0.8.7;
```

2. This snippet of code defines the contract name 'Patient' and variables used in the smart contract, such as *peopleCount, ownerID, uniqueUserID, userType*, and many more.

```
contract Patient{
uint256 peopleCount = 0;
uint256 ownerID;
uint256 uniqueUserID;
string userType;
string medicine = " ";
mapping(uint256 => Person) people;
```

3. This step specifies the modifier for patient and doctor, such as *onlyPatient()* and *onlyDoctor()*. In each modifier, smart contract compares the hash values using *keccak256*.

```
modifier onlyPatient() {
    require(keccak256(abi.encodePacked(userType))
    == keccak256(abi.encodePacked("P")));
    _;
}
modifier onlyDoctor() {
    require(keccak256(abi.encodePacked(userType))
    == keccak256(abi.encodePacked("D")));
    _;
}
```

4. Here, the contract specifies the structure of a person (or patient) with id, temperature, spo2, and heart rate variables.

```
struct Person{
    uint256 id;
    uint256 temperature;
    uint256 spo2;
    uint256 heartRate;
}
```

5. This step defines various function definitions, such as *aUser, healthData, prescription*, and *viewPatientData*. The detailed functionalities of each function with Solidity code is written as follows.

```
function aUser(string memory _userType) public {
    userType = _userType;
}
function healthData(uint256 _temperature,
```

```
         uint256 _spo2, uint256 _heartRate)
         public onlyPatient {
             peopleCount += 1;
             people[peopleCount] = Person(peopleCount,
             _temperature, _spo2, _heartRate);
             ownerID = peopleCount;
             uniqueUserID = peopleCount;
         }
         function prescription(string memory _medicine)
         public onlyDoctor {
             medicine = _medicine;
         }
         function viewPatientData(uint256 patientID)
         public view returns(uint256 patientTemperature,
         uint256 patientSpO2, uint256 patientHeartRate,
         string memory patientMedicine) {
             if(patientID == uniqueUserID) {
                 return (people[patientID].temperature,
                 people[patientID].spo2,
                 people[patientID].heartRate, medicine);
             }
             else{
                 revert("Access Denied!");
             }
         }
     }
}
```

Upon executing the aforementioned steps in the Remix IDE, the *Patient* contract produces the output as mentioned in Figs. 5.7 and 5.8. Figure 5.7 shows the Remix interface for writing the Solidity code and Fig. 5.8 shows the final generated output. Output comprises of status, transaction hash, gas value, transaction cost, execution cost, and many more.

## 5.7  Security Threats on Smart Contracts

### 5.7.1  Known Possible Attacks on the Smart Contracts

1. **Reentrancy:** The Reentrancy attack can happen by calling the external contracts or functions. They can lead to the DAO's collapse. There are mainly two types of Reentrancy attacks, Reentrancy on a single function and Cross Function Reentrancy. It is recommended by some experts to complete internal work first and then call the external contracts in order to avoid vulnerabilities due to Reentrancy attacks [25].

2. **Front-Running:** Due to the participants can see the transactions before it commits in the memory pool, a malicious user can change the order of execution for a particular transaction. There can be three types of front-running attacks, i.e.,

**Fig. 5.7**   Patient smart contract Remix IDE interface

**Fig. 5.8**   Patient smart contract output with cost, gas value, transaction hash, etc.

displacement, insertion, and suppression. There are three ways to remove the front-running from an application. The first one is to use Batch auctions, the second is to use a pre-commit scheme, and the third one is to use the 'mitigation'. Mitigations are used to limit the price slippage.

3. **Integer overflow and underflow:** If all the users can change the state of a variable, it might be vulnerable to attack. For example, a variable with data type uint. If a user makes the value of the variable less than zero, it will cause an underflow situation and the variable will get set to its maximum value. If the value is set to some large number, then it causes an overflow situation. The solution to this problem is to be careful with smaller data types like uint8, uint16, uint24, etc.

4. **DoS with unexpected revert:** These types of attacks are performed on smart contract-based auctions. A malicious bidder can become the leader while making sure that any refunds to their address will always fail.
5. **DoS with block gas limit:** There are two types of DoS attacks on gas limit: Gas limit DoS on a contract via unbounded operation and Gas limit DoS on a network via blocks stuffing.
6. **Insufficient gas griefing:** When a contract makes a sub-call to another contract, the gas forwarded is limited to 63/64 of the remaining gas by the EVM. The attacker can take benefit of this and cause the transactions to fail by sending them with a low amount of gas. One solution to address this problem is to set a minimum gas limit along with other data. Another solution is to permit only trusted accounts to relay the transaction.
7. **Forcibly sending an Ether to a contract:** An attacker can forcibly send Ether to a contract without triggering its fallback function. This is also of a serious issue in the smart contract platforms.

## 5.7.2 Cause of the Attacks on Smart Contracts

**Language-Level Vulnerabilities**

1. **Call to the Unknown:** Sometimes, calling a function from the other smart contract may result in the call of a fallback function. This situation arrives when the callee smart contract has no such function [26].
2. **Gasless Send:** When a user sends the ether to some other contract, the gas unit available to the callee is bounded by 2300 units. The send function will invoke the fallback function of the callee as it has no signature. However, if the fallback function has any state-updating or such instruction that cannot be executed in 2300 gas units, then the call will end up in an out-of-gas exception. Such vulnerability is known as a gasless call.
3. **Exception Disorders:** Due to the irregular execution of exceptions, the security of a smart contract may be vulnerable. This is called an exception disorder.
4. **Type Casts:** Smart contracts can detect some of the type errors. However, type errors in the call to another interface are something that cannot be handled. The compiler only checks for the existence of a function in callee's interface. It does not throw an exception if there is any type of cast error. So, the caller may not be aware of this kind of error and that may result in some unwanted output.
5. **Reentrancy:** Reentrancy, as the name suggests, simply means re-entering a function. We all know that a non-recursive function cannot be re-entered until its execution is over. However, this may not always be the case in smart contracts. If a function is making the fallback function be called in between and then the fallback function is calling that function, then it may result in a never-ending loop. This may result in an out-of-gas state or loss of a huge amount of ether.

6. **Keeping Secrets:** We can set a variable public or private as per our needs. In applications of smart contracts like multiplayer gaming, some variables need to be private in order to keep the moves of a player secret. But, to set the value of a private variable, we need to make a transaction published on the blockchain. Now, the blockchain transactions can be seen by all the players. So, the value of that variable can be seen by other players too. This may result in gaming attacks.

**Compiler-Level Vulnerabilities**

1. **Immutable bugs:** A transaction or a contract published on a blockchain becomes immutable. Similarly, if a contract contains any bug, it also becomes immutable. There is no way to alter the contract. So, programmers need to find ways to alter the contract, but as this is against the nature of the blockchain, many experts do not support it. So, there is no direct way to patch a bug. This vulnerability can be an entry point for some attackers.
2. **Ether lost in transfer:** If some user wants to send some cryptocurrency (Ether) to an address, they must specify that 160-bit address. But, if the address belongs to an orphan node, there is no way to recover that Ether, i.e., the Ether is lost forever [27].
3. **Stack size limit:** When a call is made to any contract, the call stack is increased by one frame. The stack size is limited to 1024 frames. Due to vulnerabilities like re-entrancy, the calls will throw an exception after the stack limit is reached. However, the issue was addressed by the hard-fork of the Ethereum blockchain in 2016, and the stack size of a call has been limited to 63/64 of its gas limit; the maximum reachable depth of a stack is always less than 1024.

1. **Unpredictable state:** A user cannot predict the state of a smart contract, i.e., when a user sends some transaction to a contract, they cannot be sure that the state of the smart contract will remain unchanged until the transaction is addressed. Also, when two users mine a block simultaneously, we cannot guarantee that a particular state would be helpful to add that transaction. Also, this vulnerability can be a reason for issues like updating a smart contract which is totally against the blockchain rules.
2. **Generating randomness:** For the applications like lotteries, multiplayer games, the non-deterministic nature of a smart contract is needed. To meet this requirement, some smart contracts generate pseudo-random numbers. These numbers are generated with the help of future timestamps or hash of some blocks in most cases. However, this strategy can be used by a malicious user to bias the outcome of the pseudo-random generator.
3. **Time constraints:** According to the time constraints, some actions are permitted in a particular state. These time constraints are generally implemented by the timestamps of blocks. It helps maintain coherence with the state of a contract. But it may also lead to a vulnerability as the miner can choose the timestamp

arbitrarily. A malicious miner can choose a suitable timestamp for its block to benefit from this vulnerability.

## 5.8 Summary

This chapter explores more about the smart contract. Nick Szabo invented a smart contract in 1994. Initially, it didn't receive much attention as it required a proper blockchain platform for its successful implementation. After Ethereum was developed, smart contract gain became popular and emerged as one of the trending topics for researchers. We require a blockchain network to deploy any of the smart contracts. The smart contract is a code that helps achieve a common consensus among the multi-parties who don't trust each other. This chapter discussed how the smart contract is beneficial over the traditional contracts. The traditional system required the intervention of a long chain of middlemen, which smart contracts removed. Different platforms and tools are explored, which are used to deploy the smart contract. This chapter implemented smart contracts on the topic "Smart Grid for energy trading" on the remix. The code is written in the Solidity language and deployed on the remix. With the advent of blockchain technology, the smart contract has been applied in many domains. This chapter also discusses the case study of industries which applies smart contract. Various platforms have been evolved and hence the comparison has been shown on various platforms where we can deploy our smart contracts. As smart contracts gain more advantages than traditional contract systems, the application has been blooming in many business sectors. These applications have been discussed briefly in this chapter and the future possibilities. This chapter conveys a brief overview of this upcoming new trend of the smart contract, which has proved to be a boon and beneficial in many business sectors and different domains. It will surely outlay the existing traditional system within a short span.

## 5.9 Practice Questions

### 5.9.1 MCQ Questions

1. What are the advantages of using blockchain for GST collection?

(a) Automatic tax refund for the final consumer.
(b) Automatic tax payment to the tax authority.
(c) Tamper-proof invoice generation based on the taxes already levied.
(d) Transactions are done in real time.

2. What data should be part of the blockchain ledger for property transactions?

(a) Property records (all clearance certificates, etc.) along with proof of ownership.

(b) Bank account details of the buyer and seller.
(c) Sale Deed of the property.
(d) None of the above.

3. How will blockchain play provide an advantage in implementing it with GST payments?

(a) Blockchain smart contracts can calculate the invoice based on the tax amount that is already levied during the production process.
(b) Smart contract directly transfers the tax amount to the tax authority (CGST or SGST).
(c) It ensures that the return filing process is not required.
(d) The administrative burden for accounting services is drastically reduced.

4. What are the advantages of using blockchain for land registries?

(a) It will allow the government to privately store and modify land record details efficiently and in a tamper-proof manner.
(b) It will provide a mechanism to provide tamper-proof land ownership records which will help in preventing land dispute cases.
(c) Smart contracts can help automate processes such as land distribution through wills of individuals.
(d) It will allow landowners to exchange land multiple times which is currently not possible.

5. Why is a blockchain-based solution suited for managing government data at different levels?

(a) Provide different role-based access policies across different levels of the organization hierarchy in a multi-authoritative setup.
(b) Provide classification of data based on importance even in a decentralized environment.
(c) Provide centralized data management at the national level.
(d) Provide a way to access the data only via the blockchain.

## 5.9.2 Short Questions

1. How is blockchain used for land records?
2. How can blockchain be used in tax?
3. What is Transparent and incorruptible in blockchain?
4. Can blockchain change the nature of land registries in developing countries?

### 5.9.3 Long Questions

1. What area do you think would benefit from blockchain improvements to its current tax compliance methods?
2. What are the Blockchain Use Cases in Government and the Public Sector?
3. How will blockchain streamline the validation of educational and professional qualifications?
4. How will blockchain technology impact the collection of tax?
5. Explain the application of blockchain to land registries.

## References

1. Frankenfield J (2020) Smart contracts. https://www.investopedia.com/terms/s/smart-contracts. asp. Accessed 06 Feb 2020
2. Bodkhe U, Tanwar S, Parekh K, Khanpara P, Tyagi S, Kumar N, Alazab M (2020) Blockchain for industry 4.0: a comprehensive review. IEEE Access 8:79764–79800. https://doi.org/10. 1109/ACCESS.2020.2988579
3. Rosic A (2020) Blockgeeks, smart contracts: the blockchain technology that will replace lawyers. https://blockgeeks.com/guides/smart-contracts/. Accessed 08 May 2020
4. Education C, Pros and cons of smart contracts
5. Mack OV (2020). https://abovethelaw.com/2018/06/smart-contracts-taking-over-pros-cons-and-how-to-stay-on-top-of-it-all/. Accessed 06 May 2020
6. Bhardwaj C, What are smart contracts: advantages, limitations, and use cases
7. ChainTrade: 10 advantages of using smart contracts
8. Gupta R, Tanwar S, Al-Turjman F, Italiya P, Nauman A, Kim SW (2020) Smart contract privacy protection using AI in cyber-physical systems: tools, techniques and challenges. IEEE Access 8:24746–24772. https://doi.org/10.1109/ACCESS.2020.2970576
9. Contributors W (2020) Smart contracts. https://en.wikipedia.org/w/index.php?title=Smartcontract&oldid=953054365. Accessed 08 Apr 2020
10. Contributors W (2020) Vitalik Buterin. https://en.wikipedia.org/w/index.php?title=VitalikButerin&oldid=947883877. Accessed 08 Apr 2020
11. Buterin V, Ethereum white paper
12. Lerner SD (2020) Rootstock platform, bitcoin powered smart contracts. http://cryptochainuni. com/wp-content/uploads/Rootstock-WhitePaper-v9-Overview.pdf. Accessed 06 May 2020
13. Al-Bassam M, Sonnino A, Bano S, Hrycyszyn D, Danezis G (2017) Chainspace: a sharded smart contracts platform. CoRR. arXiv:1708.03778
14. Bünz B, Agrawal S, Zamani M, Boneh D (2020) Zether: towards privacy in a smart contract world. In: International conference on financial cryptography and data security. Springer, Berlin, pp 423–443
15. Tanwar S, Patel NP, Patel SN, Patel JR, Sharma G, Davidson IE (2021) Deep learning-based cryptocurrency price prediction scheme with inter-dependent relations. IEEE Access 9:138633–138646. https://doi.org/10.1109/ACCESS.2021.3117848
16. LeewayHertz: steps to create, test and deploy Ethereum smart contract (2020). https://www. leewayhertz.com/ethereum-smart-contract-tutorial/. Accessed 06 June 2020
17. Patel MM, Tanwar S, Gupta R, Kumar N (2020) A deep learning-based cryptocurrency price prediction scheme for financial institutions. J Inf Secur Appl 55:102583. https:// doi.org/10.1016/j.jisa.2020.102583, https://www.sciencedirect.com/science/article/pii/ S2214212620307535

18. Ambisafe: smart contracts: 10 use cases for business (2020). https://ambisafe.com/blog/smart-contracts-10-use-cases-business/. Accessed 17 Apr 2020
19. Shaik VA, Malik P, Singh R, Gehlot A, Tanwar S (2020) Adoption of blockchain technology in various realms: opportunities and challenges. Secur Priv 3:e109. https://doi.org/10.1002/spy2.109
20. Tanwar S, Parekh K, Evans R (2020) Blockchain-based electronic healthcare record system for healthcare 4.0 applications. J Inf Secur Appl 50:102407. https://doi.org/10.1016/j.jisa.2019.102407, https://www.sciencedirect.com/science/article/pii/S2214212619306155
21. Solidity: introduction to solidity programming (2021). https://docs.soliditylang.org/en/v0.8.11/. Accessed 01 Dec 2021
22. Gupta R, Tanwar S, Tyagi S, Kumar N, Obaidat MS, Sadoun B (2019) Habits: blockchain-based telesurgery framework for healthcare 4.0. In: 2019 international conference on computer, information and telecommunication systems (CITS), pp 1–5. https://doi.org/10.1109/CITS.2019.8862127
23. Hathaliya J, Sharma P, Tanwar S, Gupta R (2019) Blockchain-based remote patient monitoring in healthcare 4.0. In: 2019 IEEE 9th international conference on advanced computing (IACC), pp 87–91. https://doi.org/10.1109/IACC48062.2019.8971593
24. Vora J, Nayyar A, Tanwar S, Tyagi S, Kumar N, Obaidat MS, Rodrigues JJPC (2018) BHEEM: a blockchain-based framework for securing electronic health records. In: 2018 IEEE globecom workshops (GC Wkshps), pp 1–6. https://doi.org/10.1109/GLOCOMW.2018.8644088
25. Ethereum smart contract best practices, 'known attacks'
26. Nicola Atzei MB, Cimoli T, A survey of attacks on Ethereum smart contracts
27. Kabra N, Bhattacharya P, Tanwar S, Tyagi S (2020) MudraChain: blockchain-based framework for automated cheque clearance in financial institutions. Future Gener Comput Syst 102:574–587. https://doi.org/10.1016/j.future.2019.08.035, https://www.sciencedirect.com/science/article/pii/S0167739X19311896

# Chapter 6
# Distributed Consensus for Permissionless Environment

**Abstract** Day-by-day, both data and network size are growing at a rapid rate. It is essential to keep private data secure and also prevent malicious activities. In a permissionless blockchain, nodes do not take permission for participation. One can directly mine a block by performing an open task. Security can be a significant issue here. Also, there is no third-party involvement in blockchain, so keeping trust among peers is an essential feature. The distributed public ledger stores history of old transactions to maintain trust between peers. To prevent malicious activities, consensus algorithms are used, which are defined as a complex task that a miner must perform to mine new blocks into the blockchain. In this chapter, various consensus mechanisms are mentioned with merits and demerits. With high computation power and digital currencies, nodes can quickly get into the blockchain and perform malicious activities. For that, various consensus algorithms are used like Proof of Work (PoW), Proof of Stake (PoS), Proof of Burn (PoB), Proof of Capacity (PoC), etc. Every consensus is developed to solve issues of previously developed consensus and provide more efficiency concerning resource allocation, scalability, security against attacks, power consumption, etc. Bitcoin is one of the use cases of blockchain, which is developed upon the PoW consensus method. Various companies have developed cryptocurrencies that are based on consensus algorithms. Consensus can be implemented on smart contracts to govern specific rules in the blockchain. While working with extensive transactions and a large chain of blocks, scalability, efficiency, and malicious attacks are significant issues. We have done a comparative analysis of all the consensus algorithms based on such issues.

**Keywords** Blockchain · Consensus · Permissionless · Proof-based · Voting-based

## 6.1 Introduction

Blockchain (BC) technology is so trending in this digital era that knowing it becomes essential. The traditional system focuses on scientific computations, while Industry 4.0 is the era of high computation power and storage of Big data [1]. BC is a decentralized and information-sharing platform. BC is formed to store records of

© The Author(s), under exclusive license to Springer Nature Singapore Pte Ltd. 2022    153
S. Tanwar, *Blockchain Technology*, Studies in Autonomic, Data-driven and Industrial
Computing, https://doi.org/10.1007/978-981-19-1488-1_6

transferred money in the form of digital currencies. The blocks of the data are bound to each other using cryptographic principles. The three main pillars of BC are transparency, immutability, and decentralization. BC technology has become a growing trend because of bitcoin. The terms BC and Bitcoin seem similar often. Still, they are different as Bitcoin is a digital currency also called cryptocurrency. In contrast, BC can be said a distributed ledger-based technology which records all the performed transactions [2].

Bitcoin is the first-ever application of BC. The goal of Bitcoin was to implement a P2P network of digital currencies to allow secured transactions online without the interference of a financial institution. As a result, there would be no transaction fee or third-party fee. In Bitcoin's implementation, every block contains a timestamp to perform transactions within the network, and activities are hashed to prevent unauthorized access [3]. In BC, a block is structured with block index, the previous block hashed id, the timestamp, target difficulty, set of old transactions, etc. [4]. Block stores data of own activities and activities performed within the chain. Security is a significant issue while working with an open network. Third-party involvement is somehow costly and time-consuming to come up with a specific task. It may lead to security and privacy issue. Miners can add new blocks by solving problems with difficulty levels high. Replica of a newly inserted block is broadcasted to the peers involved in the chain. For insertion of a new block, it must be validated and agreed upon by the majority of nodes of the current chain. If most of the chain rejects the node, the block will be discarded.

BC provides access based on authentication by performing healthy activities and contributing to the network. A Smart Contract (SC) [5] brings a network at a high-security level and distributed manner. SC governs protocols, roles, and semantics in the BC network with authenticated entities. To prevent malicious access and maintain trust, consensus algorithms are involved. A consensus algorithm allows peers to mine a new block by performing a specific complex task. This chapter describes different types of distributed consensus mechanisms in a permissionless environment. Some companies working based on SC to implement consensus are also discussed in this chapter. After that, each algorithm is mentioned with a broader and detailed explanation followed by a comparison table. BC technology's significant issues are scalability, cost reduction, storage optimization, and efficiency. Implementation of consensus is shown at the end of this chapter.

## 6.1.1  Notion of Consensus

In the introduction chapter, it was discussed that BC is immutable and is tamper-proof [6]. A block in a BC contains two parts, i.e., the header and the data. The header of the block connects the transactions. If an attacker tries to tamper with a block, he needs to change the values of all subsequent blocks [7]. So, any change in any transaction will result in a change in the block header and the headers of the subsequent blocks. They are connected in a chain in a way that if you want to make any change in any block, you need to update the entire chain. Every peer in a BC network maintains a local copy of the BC. Now, the next interesting part of BC is how you will manage the replica. The idea of this BC is that multiple nodes in the network are interconnected, and every node contains a replica of the BC. Every node is maintaining a replica of the global BC. Hence, two requirements emerge here. Firstly, all the replicas need to be updated with the last mined block. Secondly, all the replicas need to be consistent, i.e., the copies of the BC at different peers need to be exactly similar. So, here the notion of consensus comes into practice. The idea of distributed consensus [8] was explored in the literature from the early 1990s, where people ensured that different network nodes see the same data at nearly the same point in time. In other words, all the nodes in the network need to agree or consent regularly that the data stored by them; are similar they are exactly identical. Therefore, this is achieved through what we call a consensus algorithm. The consensus algorithm ensures no single point of failure because your entire data is decentralized. If one node fails, you still have the data in multiple other nodes and thus, the system can provide you service even in the presence of failures until and unless the network gets disconnected. The basic philosophy is based on message passing [9] like when you inform your current data to other nodes, everyone that way gets the data from all other nodes, and they validate their local data. That way, one can verify whether the data that one has is the most recent or whether their data matches with the data of one's peer. This philosophy requires that the participants in the consensus algorithm know each other because you need to check or find out with which node you can validate your data.

## 6.1.2 Distributed Consensus

The notion of consensus is a stimulating topic in a distributed or decentralized [10] network. Consensus is an approach to reach a common agreement between multiple parties in a distributed or decentralized environment. Let us take an example to understand the concept of consensus mechanism. Assume that there are four generals in an army and making a few decisions. The generals have their policies to act on whether to attack or retreat when making decisions in a decentralized environment. Each general expresses their viewpoint by using a designated choice function and the decision can be toward a majority side. The system finally decides what to do next using this choice function. Figure 6.1 shows a scenario where three generals choose to attack and one chooses to retreat. The system determines to arrive at a consensus to attack as the majority is toward that side. This consensus algorithm is necessary for a message-passing environment in a decentralized or distributed system.

Let us discuss why we need consensus. We can ensure reliability and fault tolerance in a traditional distributed environment by applying consensus. Reliability and fault tolerance mean that when you have multiple parties with decision capability, some nodes may start working maliciously or work as a faulty node. It is crucial to come to a common decision or viewpoint in those particular cases. Having a common viewpoint in an environment where people can behave maliciously or crash or work in a faulty way is a difficult task. Hence, under this kind of distributed environment, our objective is to ensure reliability; that means to ensure correct operations in the presence of faulty individuals.

Considering multiple transactions taking place from numerous ATMs and banking sectors, you want to transact some money from one branch to another bank branch to another account in another bank branch. During that time, you need to agree that all the bank branches need to decide that the transactions are valid. After determining that this transaction is valid, they should commit the transaction. Another example of consensus is state machine replication. The state machine replication [11] is an essential aspect of any distributed system. When you want to run some distributed protocol over a network, every node executes the current version of the protocol, making the protocol state in various state machines. The entire execution aspect of the protocol can be portrayed as a state machine. These state machines need to be replicated into multiple nodes so that every individual node can reach a common point or common output of that protocol. Yet another example of consensus is clock synchronization [12]. Say you have multiple clocks in your network and every individual node tries to find out which is the most updated clock or the most current clock. They should make a consensus among themselves and agree to a single clock. They can do further operations by applying this kind of clock synchronous, asynchronous clock architecture across the network. These are the typical examples of consensus in a traditional distributed system environment.

Let us look into why achieving consensus can be difficult in certain scenarios or typically in a message-passing system. We are considering the scenario shown in Fig. 6.1 of multiple generals. These generals utilize a message-passing environment to communicate their viewpoints to others. Considering every individual general is making a telephone call to other generals and then communicating their viewpoint to others. A general is sending its information to its neighboring general. Out of four generals sending their decision as 'Attack' to neighboring generals, one malicious general calls and sends information as 'Retreat'. In a distributed environment, this creates confusion as messages are passed on to further neighboring generals. From one neighboring node, 'Attack' decision is informed, and the other 'Retreat' decision is circulated. That way, in a decentralized or distributed platform, achieving consensus over a message-passing system can be difficult when you have this kind of malicious node or nodes that start working maliciously. We have a technical term for this kind of node called Byzantine time nodes, and this kind of failure is termed Byzantine time failures. We will discuss Byzantine failure [13] in detail in Chap. 8. In a nutshell, this kind of Byzantine failure can cause the system to behave maliciously, or it may be challenging to achieve a consensus in a distributed environment.

**Fig. 6.1** Example to illustrate distributed consensus

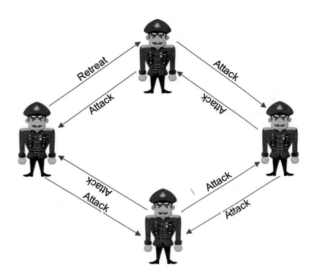

### 6.1.3 Faults in Distributed Systems

However, consensus achieving consensus can be significant in a distributed environment due to multiple types of failures. Generally, in a distributed system, we assess three different types of failures described as follows:

1. *Crash Fault*: A crash fault [14] occurs when a node unexpectedly crashes or the nodes become unreachable in between an ongoing communication. Anyone will not receive any message from the crashed node. This is one type of typical fault that can be a kind of hardware fault or a software fault due to which the node or the process, which is communicating with another one in that particular process, fails.
2. *Network or Partition Fault*: A network fault [15] occurs when a network link fails. A network failure may result in a partition in the network. Let us understand through an example. Assume that there are multiple nodes in the network and these nodes are interconnected to each other. If a particular node fails in the interconnected network, the corresponding link fails, resulting in the entire network getting partitioned into two different parts. Due to this, a node will not receive messages from any node of one partition to any node of another partition or vice versa. This kind of network failure can hamper reaching the consensus. Network fault also comes from either hardware or software failure.
3. *Byzantine Fault*: Byzantine fault [16] is more challenging to handle in a distributed environment where a node starts acting maliciously. Whenever a node starts acting maliciously, you do not know what the node's action would be. The node can send a positive vote or sometimes send a negative vote. In the case of the crash fault or a partition fault, you can find out what the fault's effect will be, but the impact of a Byzantine fault is challenging because it entirely depends on how maliciously the

node is acting and what the node is doing. Sometimes, the node can vote against a consensus, or sometimes, the node can vote for the consensus. Therefore, handling Byzantine nodes becomes difficult in a typical distributed system.

### 6.1.4   Properties of Distributed Consensus

Distributed consensus protocols need to satisfy specific properties [17] of the protocols, which are described as follows:

1. *Termination*: It states that every correct individual decides a correct value at the end of the consensus protocol. It means whoever is the correct or non-faulty node in the network must terminate the protocol and decide on one value, and that value should be the correct value.
2. *Validity*: It states that if all individuals propose the same value, all correct individuals decide on that value. For example, if all the individuals present a value '10', every valid node should reach a consensus with the value '10'. They should not deviate from that particular value.
3. *Integrity*: It states that every correct individual decides at most one value. Some individuals must propose the decided value. So, the integrity property ensures that the consensus value should not deviate from the individuals in the network proposed. For example, you should not get a value of '20' in the consensus if none of the nodes in the network proposed a value of '20'.
4. *Agreement*: It means that every correct individual must agree on the same value. Whenever all the nodes agree on the same value after termination, we call the system to reach a consensus.

### 6.1.5   Correctness of a Distributed Consensus Protocol

The following two properties can characterize the correctness [18] of a distributed consensus algorithm:

1. *Safety*: The safety property states that the correct individual must not agree on an incorrect value. It means nothing terrible will happen. The safety property ensures that one will never convert to an incorrect value, or the valid individuals in the network will never convert to an incorrect value.
2. *Liveliness (Liveness)*: The liveliness or liveness property states that every correct value must be accepted eventually, which means something good will eventually happen. If you have proposed some good values, that good value will ultimately be committed, although there can be some time lag or delay in reaching the consensus. After the consensus protocol terminates, you will expect to get a consensus value out of that.

So, these are the two correctness properties for a distributed consensus that we need to ensure whenever we design a distributed consensus algorithm.

### 6.1.6 Types of Consensus Mechanisms

All the consensus mechanisms are designed to handle the faults in a distributed system and qualify them to reach a definitive agreement. There are two general types of consensus mechanisms mentioned as follows:

1. *Leader election-based*: This mechanism demands nodes to compete in a leader election [19] lottery, and the final value is proposed by the node that wins. For example, Bitcoin's PoW belongs to this category.
2. *Byzantine Fault Tolerance (BFT)-based*: This mechanism depends on the scheme where nodes are publisher-signed messages [20]. An agreement is reached when a particular amount of messages are received. The mechanism does not require to compute-intensive operations.

### 6.1.7 Consensus in Blockchain

In a centralized network, there is a central authority and all the nodes trust the authority that it will behave honestly and share the correct information with all the nodes. Since only the central authority can modify the data, it is straightforward that it will achieve consensus. But how does BC, a decentralized peer-to-peer system with no authoritative figure, make decisions? All the nodes need to agree on the same term for making decisions, or we can say that all the nodes need to be agreed upon a consensus in BC. Let us understand consensus with the help of an example. We have some mining nodes that have the same copy of the BC ledger, and all these copies will be in the same state. The BC processes some transactions and allows peers to get into the mining process to mine blocks. But the main question is which mining node will add the block? If we agree that the most recent node will invoke the new block to BC and then be replicated in all the other nodes, there remains a problem if the recent node is malicious. So, we need to ensure that all nodes reach a consensus and have the same copy of the BC ledger.

For this, we require consensus algorithms in which all nodes must agree on what the truth is. There are various consensus mechanisms, but all serve the same purpose to ensure that records are trustworthy and honest. The difference is the way they reach the consensus. Consensus algorithms are responsible for maintaining the privacy and efficiency of BC transactions. The use of proper consensus is essential to increase the performance of the BC network [21]. There cannot be any decentralized network without a consensus algorithm. If we use a certain consensus, it does not matter whether the nodes trust each other. They have to follow specific rules for a collective agreement. Consensus algorithms do not agree with the majority votes only but agree to the one that benefits everyone. So, it is always a win for the network. There are two ways of achieving consensus in BC described as follows:

1. *Permissionless Blockchain*: It allows any user to participate in the BC without any permission [22]. It happens in a public or open environment where users do not know each other, and there is no predefined trust. The only need is to mine a block and be interactive with the BC. By performing consensus, miners can add their block into the BC network [23]. After getting included, a node can interact with all other nodes within the BC network. Permissionless BC has some characteristics as follows:

   - It must be decentralized or distributed to protect BC from malicious activities. Distributed BC is tamper-proof, having an extensive history of transactions.
   - Transparency [24] is provided as there is no third-party involvement. All peers verify and validate the occurred transaction. The transaction is inserted into the distributed ledger after validation. Peers are rewarded with cryptocurrency or incentives for participating and performing the task.
   - Permissionless BC provides reliability and scalability to the BC network.
   - Tokens are used to compensate nodes of the BC network. The utility of BC directly affects the value of a token.

2. *Permissioned Blockchain*: It is a closed environment, as the name suggests [25]. In a permissioned BC, users need to take permission to access it. While working with an organization or consortium, permissioned BC can be used. It allows authenticated users to join and interact with the network [26]. Key features of permissioned BC are as follows:

   - The decentralization of a system is dependent on the structure defined by the consortium or organization. Also, this system is managed by the organization's authority, and roles can be determined using SC. How many nodes have participated, how the structure is, which consensus mechanisms are used, and how many faulty nodes affect the decentralized BC network.
   - Transparency is measured by work done by each node, how the business relations are set up, and how the cost and time are minimized.
   - As only authenticated people can access the BC, privacy is highly prioritized.
   - As it is from a business perspective, there is no token-based system. Yet, some consortiums are trying to utilize tokens to increase interest in work.

Consensus algorithms pertaining to the permissioned BC will be discussed in Chap. 8. This chapter discusses consensus algorithms pertaining to only permissionless BC.

## 6.2 Taxonomy of Consensus Algorithms for Permissionless Environment

Figure 6.2 shows the taxonomy of consensus algorithms for a permissionless environment. The taxonomy is divided into resource-based, economy-based, capability-based, and hybrid consensus algorithms. They are explained as follows:

1. *Resource-based consensus algorithms*: Algorithms falling under the resource-based category use computing resources such as high computation power, space, and memory for mining blocks in the network.
2. *Economy-based consensus algorithms*: Algorithms falling under the economy-based category require high usage of wealth to get a chance for mining blocks in the network.
3. *Capability-based consensus algorithms*: Algorithms falling under the capability-based category are created for handling use-case scenarios such as asset handling and ownership validation, under a Trusted Execution Environment (TEE).
4. *Hybrid-based consensus algorithms*: Algorithms falling under the hybrid category combine the characteristics of resource-, economy-, and capability-based algorithms.

## 6.3 Various Consensus Algorithms

### 6.3.1 Proof of Work (PoW)

BC provides privacy to the data and a trustworthy environment, so if BC works for a long time, the data will remain private and available. To maintain security, data must

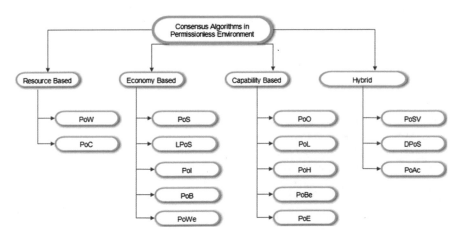

**Fig. 6.2** Taxonomy of consensus algorithms for permissionless environment

**Fig. 6.3** PoW mechanism

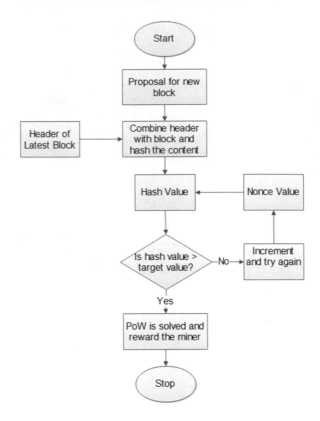

be immutable and cannot be removed with distributed architecture where each node is peer-connected. Thus, fault tolerance can be handled. Bitcoin is one part of BC technology that has not been tampered with yet. A hash function is a reason behind it. A hash function is a mathematical function, i.e., a chain of bits used to encrypt the data uniquely. PoW [27] is the most recognized and oldest consensus protocol applied on Bitcoin, introduced in 1992 by Dwork and Naor, to control e-mail spamming. It is a compute-intensive method based on the SHA-256 hash function. PoW is competitive as every miner tries to be the foremost to solve the problem issues on the network for being capable of inserting new blocks on BC. The miner uses the hash function [28] to a new block and a random variable called a nonce. The resulting hash function should meet specific requirements (for example, the hash should begin with the value 0000). The more complex the condition is, the more difficult it will be to tamper with the block data. The mining process continues with periodic attempts until a nonce that meets the requirements is found. To mine a new block, the miner needs to find nonce's value with some mathematical problem. PoW contains mathematical problem puzzles with high to low complexity to prevent malicious activities. Nodes within the BC solve this puzzle individually to get a reward after mining the block. When one node finds the value of nonce, other nodes stop finding value and verify the resultant value. If the solution is correct, the respected node gets rewarded. Miners [29] update distributed public ledger, a database of transactions held within BC. As the number of miners increases, the block creation time reduces, demanding the difficulty condition to be more stringent. The average time for block creation is 10 minutes, and it works on a sample of the last 2016 blocks.

$$D_{new} = D_{current} \times \frac{20160}{T_{2016}} \tag{6.1}$$

In the above equation, $D_{new}$ denotes a new mining difficulty, actual difficulty is denoted by $D_{current}$, and the duration of mining the last 2016 blocks is denoted by $T_{2016}$ represented in minutes. Figure 6.3 shows steps to perform PoW and is discussed as follows:

1. *Mining difficulty*: Bitcoin increases difficulty level after producing 2016 blocks each time.
2. *Collect transactions*: To find the Merkle root, collect all the pending transactions after adding the last block. After finding Merkle root, BC version, the hashed value of the current and previous block, timestamp, nonce value, etc.
3. *Calculating*: To mine a new node, a double SHA-256 hash value is calculated and nonce values are found. If the value is near the result value, the proposed block can be added based on verification.
4. *Restarting*: If the node completes the task, it will start from step-1. Else, it will begin with step-2.

**Applications:** Most cryptocurrencies are based on the PoW, such as Bitcoin, Ethereum, Litecoin, Monero, Zcash, Dash, and Hashcash.

## Advantages

1. *Prevents double-spending attack*: The double-spending [30] problem is to use the same currency for two or more transactions simultaneously. Physical currency is more secure as it is tough to create duplicates, while digital currency can be copied for rebroadcasting. Technical issues can lead to a double-spending problem since a faulty node can easily copy currency. Double spending can be prevented using consensus mechanisms like PoW in Bitcoin, which uses BC-based cryptocurrencies. BC is a decentralized network where miners need to perform some tasks to include new blocks and prevent malicious BC activities. So, miners have distributed public ledgers with a history of old transactions to avoid malicious activities like double spending.
2. *Extremely resistant to abuse*: Due to the requirement of PoW to have compute-intensive resources, it is highly resistant to malicious activities as adversaries would need superior computing power.

3. *Reaching consensus quickly*: It is difficult to find the solution to complex mathematical problems imposed on the network, but it is very easy to verify the same. Hence, consensus can be reached quickly once the hash is created and verified.

**Disadvantages**

1. *Requirement of huge computation power*: PoW demands exceptional computing power and a massive amount of electricity is strained during the process as all mining nodes try to solve the complex problem given to the network, but only one can mine a block.
2. *Not suitable for small networks*: Since PoW requires a large number of miners competing to mine a block in BC, it is best suited for large networks. For small networks, the possibility of adversaries accessing the network's computation power and thus mining a faulty block increases.
3. *Suffers from Sybil attack*: Sybil attack [31] attempts to fill the entire network with nodes/clients under his control. If the attacker can fill the network with clients under his control, then the attacker can control or get a monopoly over the network. Different kinds of actions can be performed based on the instruction from the attacker, like refusing to relay the valid blocks. They can only relay the blocks that the attackers generate. Thus, blocks can lead to double spending.
4. *Suffers from Denial-of-Service (DoS) attack*: In DoS [32] attack, attackers send a massive amount of data to a particular node. If you are sending excess data to a specific node, the node will not process the normal Bitcoin transactions as it will be busy handling the incoming data.
5. *Monopoly problem*: Where the miners have more resources or more probability of completing the work [33], there is a statistical theory called the tragedy of the common. So, from an economic perspective, this tragedy of the commons theory says that such a monopoly can increase over time. How to limit the total number of Bitcoins generated out of a mining procedure? The total number of Bitcoins is gradually going to get saturated.

## 6.3.2 Proof of Stake (PoS)

To reduce the problem of handling monopoly and the power conjunction in a PoW-based system, different other consensus mechanisms came into practice. The second most popular consensus mechanism is the PoS [34] mechanism. This PoS mechanism was proposed in 2011 by Vitalik Buterin. PoS is an updated version of PoW where users must have a stake to participate in the mining process. This consensus aims to create a secured system where peers can validate the transactions considering integrity. Miner, also known as a forger, is elected in a semi-random process. There are mainly two things to consider for the procedure. First is the user's stake. It is based on depositing a certain token in the system as a virtual account. A peer having more stakes has more chances to get selected to mine block. A degree of chances should be added to the process for avoiding biases. The second thing is to add randomly to the semi-random process. For this, there are two methods: randomized block selection and coin age selection [35]. The forger with the lowest hash and the highest stake is selected in randomized block selection. The validator is selected based on how long its token has been staked in coin age selection. As long as the node holds the coins, the node gets more network rights. The holders get the reward based on the coin age as follows:

$$coin\ age\ =\ staked\ coins\ \times\ duration\ of\ stake\ (in\ days) \qquad (6.2)$$

$$proofofhash\ <\ coin\ age\ \times\ target \qquad (6.3)$$

Here, the proofofhash is the hash value of weight factor, unspent output value, and current time's fuzzy sum. The mining difficulty is inversely proportional to the coin age. When any node is being selected to mine the next block, the node is responsible for checking whether the transactions are valid or not. If they are valid, it verifies the block and adds it to the BC. As a reward, the node receives the transaction fees associated with the block. If a node wants to stop being a validator, the stake and the reward of the node get released after a certain period. The broad difference between a PoW-based system and a PoS-based system is that in the case of PoW, mining a block depends on the work done by the miner. If the miner has vast computing resources, getting a new block gets high. On the other hand, in the case of PoS, the amount of bitcoin that the miners hold instructs which miner can generate the next block. If a minor holds one percent of the total bitcoin, the miner can mine one percent of the proof-of-stake block. In PoW, the cryptocurrency is created as a reward for miners. In PoS, the cryptocurrency is made at the launch. Figure 6.4 shows the steps to perform PoS.

**Applications:** Crytocurrencies such as Peercoin, Nxt, Ethereum 2.0, Qora, Black-Coin, and ShadowCash are based on the PoS mechanism.

### Advantages

1. *Reduced electricity consumption*: PoS does not impose any mathematical problem on the network users need to solve. Hence, it reduces energy consumption.
2. *Secure against 51% attack*: To carry out 51% attack [36], 51% of the cryptocurrency needs to be secured by the adversary, which is very costly, even if the miner accumulates the same, it will not be advantageous as a decrease in the value of cryptocurrency will prove to be a loss to the miner as it affects its own assets. Hence, PoS models are significantly less prone to 51% attack.
3. *Easy staking*: It is due to mass participation and less stress on network participants. The rate of participation increases as stakers do not have to worry about hardware utilization making it more decentralized.

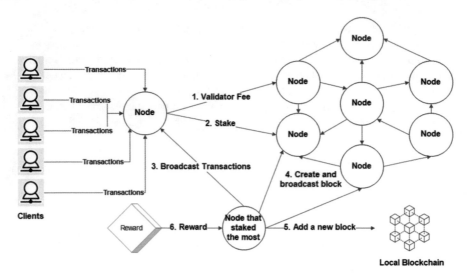

**Fig. 6.4**  PoS mechanism

4. *Decentralization*: As power consumption dependency is significantly less in PoS, it somewhat solves the centralization problem of PoW Proof of Stake solves the centralization. Moreover, it becomes more accessible and environmentally friendly.

**Disadvantages**

1. *Favors the rich*: The network can be influenced by the nodes holding more wealth, thus obtaining the chance to process transactions, charge a commission, and become richer.
2. *Suffers from nothing-at-stake attacks*: The concept is to maximize profits by putting nothing at stake [37]. Generally, when forks are created in BC, the longest chain is considered and other chains are considered orphaned. In this kind of attack, all the chains are followed so that participant gets rewards either way, and one chain out of them will be picked as a winner.
3. *Suffers from grinding attacks*: For favoring an adversarial stakeholder, malicious parties will use computational resources for biasing leader election, thus creating grinding vulnerability [38].

**Variations of PoS:** There are many variations of PoS, and each solution is an improvement on the original Proof-of-Stake solution, each providing resource efficiency and effectiveness.

- *Delegated PoS (DPoS)*: It was developed in 2014 by Daniel Larimer, founder of BitShares, Steemit, and EOS. The more stake a node owns, the more powerful voting power it has to assign the witness. In DPoS [39], stakeholders can either vote directly or give their voting power to another stakeholder to vote on their behalf. These users are called 'witnesses'. They are chosen using an election system to verify blocks. If witnesses sign and verify all transactions in the block, a reward is given to them, shared with the stakeholder who voted for them. If they fail to verify all the transactions, then no reward is provided to them, and the reward is added to the next witness verifying the transactions. There is another set of users called 'delegates' who govern the BC for any change in block size or amount to be paid to the witness. DPoS provides decentralization and better reward distribution. They do not require any high computation power and are more scalable. One of the disadvantages of DPoS is the possibility of ruling the network by creating cartels of witnesses. Applications like BitShares, Steem and Steemit, and EOSIO are based on the DPoS mechanism.

- **Leased PoS (LPoS)**: LPoS [40], launched in 2017 as a part of Waves project, is an improvement on PoS where nodes with a low number of coins can participate by taking currency on lease from other nodes with high stakes. The coins are in total control of the account holder. When the nodes with low stakes validate the block, the reward is shared with the wealthy holders who gave the lease. While in DPoS, an election-based system was used, the stakeholders can borrow and lend tokens directly to participate in the system themselves. Such a system is more scalable with high throughput, fast, and energy-efficient.

- **Proof of Importance (PoI)**: PoI [41] was introduced in 2015 and is used by the cryptocurrency New Economic Movement (NEM). The reduced transaction flow issue of PoS is addressed by PoI. Every node in PoI is assigned an importance score. It not only rewards the nodes with high stakes, but it also rewards users based on more number of transactions. A node performing transactions with a node of high importance score value will get a chance to mine the next block. PoI is highly scalable, fast, energy-efficient, and does not require any special hardware.

- **PoS Velocity (PoSV)**: PoSV [42], used by Redcoin cryptocurrency, is designed to encourage social interactions in the digital age. It is an improvement over PoW and PoS. Two main functions of Redcoin, i.e., storage of value and medium of exchange, correspond to ownership, i.e., stake and activity, i.e., velocity. The higher the value of velocity, the better is the economy. It does not use a linear coin age function but instead uses an exponential growing function to encourage node activity.

### 6.3.3 Proof of Burn (PoB)

PoB, designed by Iain Stewart in 2012, is a good alternative to the PoW consensus algorithm. It reduces energy consumption. In PoB [43] shown in Fig. 6.5, powerful

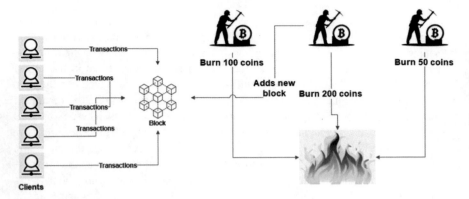

**Fig. 6.5** PoB mechanism

hardware and energy are not required; instead, cryptocurrency is burned to invest some resources in the network. Miners invest in virtual mining power. By this method, miners can show their commitment toward the network that they are not malicious. The more coins the user burns, the more mining power he gets. The concept of burning coins is to send them to an unidentified address where they cannot be accessed. These addresses are generated randomly, and there is no private key. Similar to PoW, the miners get block rewards, and within a certain time, these rewards will cover the initial investment. The similarity between PoS and PoB is that miners need to invest in the network. In PoS, if forgers decide to leave the network, they can take the coins back and sell them to the market. This will not create market scarcity. However, in PoB, validators need to destroy their coins forever, creating a market scarcity.

**Applications:** Cryptocurrencies such as Slimcoin, Counterparty, Triggers, and Redcoin use the concept of PoB.

### Advantages

1. *Sustainable*: PoB is highly sustainable because it reduces power consumption. It encourages the long-term commitment of the miners and also it reduces market scarcity.
2. *No hardware needed*: There is no need for mining hardware as coin burns are basically virtual mining rigs.

### Disadvantages

1. *Centralization*: In PoB, the rich get richer is a big problem. The users who have more coins can burn more coins and get more mining power. This makes the BC centralized as PoW.

2. *Lack of transparency*: The burning of coins is not transparent, so a random user cannot verify it.
3. *Not eco-friendly*: The coins are generated using PoW mining only, so it requires resources and hence is not eco-friendly and is time-consuming.

### 6.3.4 Proof of Capacity (PoC)

PoC was created in 2015 by Dziembowski and Ateniese. It is also known by various names, such as Proof of Space and Proof of Storage. PoC [44] overcomes the high energy consumption drawn by PoW and favors the rich issue drawn by PoS and PoB. Instead, PoC leverages available hard disk space to allow the mining devices to participate in the network. The more the available hard disk space is, the more chances are to be selected for the next block creation. PoC works in two phases, namely plotting and mining, as follows:

1. *Plotting*: Plotting is done using a 256-bit variant of Shabal hash, i.e., Shabal256. Depending on the size of the hard disk, plotting creates plot files that consist of the list of nonce values. The calculated hashes are connected into 'scoops', consisting of a group of neighboring hashes.
2. *Mining*: During the mining phase, the miner calculates a scoop number. The miner will utilize the data for scoop number 'n' from all the nonces stored in the hard drive and calculate a deadline. After calculating all the deadlines, the miner chooses the shortest deadline. The winner will be that miner who finds the block within the shortest deadline created and gets rewarded.

**Applications:** Cryptocurrencies such as Burstcoin, Btchd, Diskcoin, and LitecoinHD are based on the PoC algorithm.

#### Advantages

1. *No special hardware needed*: The hash function calculation is not compute-intensive and thus, does not require any high computation hardware. It does not require the latest hard disk as well. Any old hard drive can do the task of PoC.
2. *Less energy consumption*: Mining with hard drives saves 30 times more energy than mining with PoW.
3. *Reusability*: Once the mining is completed, the hard disk can be wiped out and reused again.

**Disadvantages** *Use of mining malware*: Mining malware [45] is used to utilize the disk space and slows down the process of mining. It makes it difficult to understand whether excess hard disk space is used for malicious purposes or not.

### 6.3.5 Proof of Activity (PoAc)

PoAc [46] is a hybrid consensus utilizing the characteristics of resource-based and economy-based mechanisms. It combines the features of PoW and PoS for generating new blocks. The mechanism of PoAc works in two phases where the first phase incorporates the mechanism of PoW and the second phase follows the PoS mechanism. Mining first begins with PoW, with miners competing to be the foremost ones to unravel mathematical problem issues on the network and claim their reward by solving them. The difference between PoW and PoAc is that the mined block does not contain transactions. They are merely templates with mining reward address and header information. As soon as mining is done for the nearly blank block, the next phase commences switching to PoS. The selection of a random group of validators for the signing block is possible due to header information. These validators are coin holders having a good amount of stake. The higher the stake a validator holds, the higher the probability of being chosen to sign the new block. The block is considered in BC once all the elected validators sign the block. The block is discarded as incomplete if it is unsigned by the elected validators. This process continues until all the validators sign a winning block. The winning miner and the validators who signed the block split the network fees.

**Applications:** Decred, Espers, and Coinbureau work on the concept of PoAc mechanism.

**Advantages** *Secure*: It is challenging to introduce a manipulated block on the network as the block needs to be signed by all the elected validators. Otherwise, it is discarded if unsigned by them.

**Disadvantages**

1. *Need of compute-intensive resources*: Since it follows the PoW mechanism, hence, PoAc also requires resources with high computational power.
2. *Need of wealth*: The nodes with high stakes always get a chance to take part in the consensus mechanism.

### 6.3.6 Proof of Weight (PoWe)

PoWe is a good alternative to PoS. It is employed in cryptocurrency Algorand. When using PoS, the stakes owned by each stakeholder determine their probability of uncovering the next block. However, the PoWe [47] assigns weight values to stakeholders as per each user's wealth in its account while the network creates a committee of random network members. It makes the consensus slightly centralized within the committee. The network is resistant to double-spending attacks until at least a two-thirds fraction of stakeholders are genuine. The system is highly scalable with PoWe. Despite the superiority of PoWe, it is challenging to reward the users as it is not developed for the generation of passive revenue streams.

## 6.3.7  Proof of Luck (PoL)

PoL built upon TEE [48], i.e., Intel SGX enabled CPUs, strives to enhance transaction throughput and decrease the computational power rendered by PoW. In PoL [49], the process of mining a new block is done by accessing a 'luck value', which is a random number between 0 and 1 within a uniform distribution. A higher value is considered lucky and a lower value is deemed to be unlucky. The block with a higher luck value is viewed in the main chain, and the chain that maintains the highest luck is favored in BC. There is lesser delay and the communication system is optimized due to this feature. It is resistant to double-spending attacks as well. However, the algorithm suffers from power deficiency, as the luck value is determined after several attempts. Synchronization between a node and the network is an essential issue in PoL, as nodes might get unlucky because of an unsynchronized timepiece.

## 6.3.8  Proof of Ownership (PoO)

PoO [50] requires a TEE for participants to participate in the network. This algorithm uses unique pseudonyms [51] generated by an Enhanced Privacy Identity (EPID) signature. The pseudonyms can track if multiple proofs are coming from a single participant to reduce Sybil attack. PoO generates unique pseudonyms so that if a faulty owner resets the owner epoch register for utilizing it numerous times, the adversary may not be able to do that. Therefore, we reach a consensus by mining a block having most proofs with unique pseudonyms. A consensus is reached when a block is generated from the user having the most proofs and unique pseudonyms. This technique can be used by artists or businesses to certify the integrity, date of publication, and ownership of their creations or contracts. Only the people with the private key associated with the signature can prove they are the owner. Applications such as Decentralized credits are based on PoO. The advantage is the use of unique pseudonyms makes it difficult to perform multiple attacks.

### 6.3.9  Proof of Existence (PoE)

Traditional models used central authorities for validating the documents causing more security breaches. A document is stored in the BC with a signature and a timestamp associated during storage. The user can validate the document anytime from any location. Privacy and security can be achieved by storing the document proof in a decentralized way where a third party cannot modify it. Cryptocurrencies such as Poex.io, Hero Node, and Dragon Chain use the PoE [52] algorithm. The advantages of PoE include maintaining integrity by document time-stamping when given by the user. Common use cases of PoE can be asset realization and proving ownership for deed transfer.

### 6.3.10  Proof of History (PoH)

PoH [33] works on one of the most challenging distributed system problems, i.e., agreement on time. The problem is identifying the sequence of transactions at particular timestamps and determining the correct sequence of events. PoH uses a high-frequency verifiable delay function to hash incoming events and transactions. Every event has a unique hash and count along with this data structure as a function of real time. This information tells us what event had to come before another event, almost like a cryptographic timestamp giving us a verifiable ordering of events as a function of time. Each node has a cryptographic clock that helps the network agree on the time and order of events without waiting to hear from other nodes. This approach executes the SHA-256 hashing algorithm consecutively to use the output of each round as the connected input to the next round. Leaders control the verification and integration of individual transactions with the prevailing hash. The confirmed transactions are the votes for the consensus algorithm. An additional layer of security is added in this network wherein an invalid hash is generated at random intervals, and those verifiers who validate it are penalized.

### 6.3.11  Proof of Believability (PoBe)

This protocol substitutes PoW and PoS as they are expensive mechanisms to be adopted. PoBe [53] is materialized by Internet of Service Token (IOST) in 2019. A node is trusted based on its previous contributions. An untradable token named SERVI is used in the IOST system to improve the fairness and decentralization of the BC. PoBe incentivizes the participants using SERVI tokens. SERVI measures the believability value associated with each user that indicates positive reviews regarding former behaviors. A higher believability value increases the chances for block creation. Tokens are awarded to the good actors to select the next validator. Each node in PoBe has a believability score based on previous node transactions, several positive reviews of a node, IOST's amount in the node, and several awarded SERVI.

## 6.4 Implementation

### 6.4.1 Proof-of-Work Implementation

We had already discussed in previous sections of this chapter the working mechanism of PoW consensus protocol used mainly in Bitcoin. This section shows the implementation of PoW consensus. For this purpose, we have implemented PoW using NODE JS in Visual Studio Code. The steps for the same are listed in Chap. 1. In this code, the hash is generated by the method *hash_calculate()*, where SHA-256 is used for the same.

```
class Data_Block {
    constructor (i, t, record, previous_hash = ''){
        this.i = i;
        this.t = t;
        this.record = record;
        this.previous_hash = previous_hash;
        this.current_hash = this.hash_calculate();
        this.nonce_value = 0;
    }

    hash_calculate(){
        return SHA256(this.i+this.t+this.previous_hash+JSON.stringify(this.record)
                +this.nonce_value).toString();
    }

    mine_block(d){
        while(this.current_hash.substring(0, d) != Array(d + 1).join("0")){
            this.nonce_value++;
            this.current_hash = this.hash_calculate();
        }
        console.log("Hash of new block " + this.current_hash);
    }
}


class Main_Blockchain{
    constructor(){
        this.current_chain = [this.compute()];
        this.d = 5;
    }

    compute(){
        return new Data_Block(0, "01/01/2022","First block i.e. genesis block","0");
    }

    getUpdatedBlock(){
        return this.current_chain[this.current_chain.length-1];
    }
```

```
newBlock(add){
    add.previous_hash = this.getUpdatedBlock().current_hash;
    add.mine_block(this.d);
    this.current_chain.push(add);
}

is_valid() {
    for(let j = 1; j < this.current_chain.length; j++) {
        const current_block = this.current_chain[j];
        const previous_block = this.current_chain[j-1];
        if(current_block.current_hash != current_block.hash_calculate()){
            return false;
        }
        if(current_block.previous_hash != previous_block.current_hash) {
            return false;
        }
    }
    return true;
}
}

let b1 = new Data_Block(1, "02/01/2022", {bal: 100});
let b2 = new Data_Block(2, "03/01/2022", {bal: 50});

let blockchain = new Main_BlockChain();

console.log("Creation of first block");
blockchain.newBlock(b1);
console.log("Creation of second block");
blockchain.addBlock(b2);
```

In order to increase the computational capability of the algorithm, our main target is to increase the difficulty for the creation of new blocks which is initially set to 3 using the statement *this.d = 3*. Let us first check the output for the above-mentioned code.

```
node proofofwork.js
Creation of first block
Hash of new block
    0002924a53652cb85d13991871563245539976d2fc6a5de94b940bea8f21ef8d
Creation of second block
Hash of new block
    000a941553c8f153c110eff06b6430039f6e566a42a3bfc4ecd455973e6afecd
{
    "current_chain": [
    {
        "i": 0,
        "t": "01/01/2022",
        "record": "First block i.e. genesis block",
        "previous_hash": "0",
        "current_hash":
            "8a71de35c0fa9f2a69b84b61ccbb51e91b2918f038ed35e602dbf0ffb899dbaa",
        "nonce": 0
    },
```

```
{
    "i": 1,
    "t": "02/01/2022",
    "record": {
        "bal": 100
    },
    "previous_hash":
        "8a71de35c0fa9f2a69b84b61ccbb51e91b2918f038ed35e602dbf0ffb899dbaa",
    "current_hash":
        "0002924a53652cb85d13991871563245539976d2fc6a5de94b940bea8f21ef8d",
    "nonce": 3052
},
    "t": "03/01/2020",
    "record": {
        "bal": 50
    },
    "previous_hash":
        "0002924a53652cb85d13991871563245539976d2fc6a5de94b940bea8f21ef8d",
    "current_hash":
        "000a941553c8f153c110eff06b6430039f6e566a42a3bfc4ecd455973e6afecd",
    "nonce": 362
}
],
    "d": 3
}
```

As per the output, three hashes are generated. Let us assume the first hash belongs to a user 'Test1'.

```
SHA256("Test1") = 8a71de35c0fa9f2a69b84b61ccbb51e91b2918f038ed35e602dbf0ffb899dbaa
    SHA256("Test2") =
0002924a53652cb85d13991871563245539976d2fc6a5de94b940bea8f21ef8d
    SHA256("Test3") =
000a941553c8f153c110eff06b6430039f6e566a42a3bfc4ecd455973e6afecd
```

If you check the hash value for 'Test2' and 'Test3', it contains three zeros in front of it. This is because we have kept the difficulty to be 3. Every time a nonce value is incremented and hash is calculated, it checks whether the hash value generated has completed the difficulty or not; if not completed, nonce is incremented and the process repeats. Let us check the output when this difficulty is set to 5.

```
Creation of first block
Hash of new block
        00000a473ded541383462c535948e7882b348f8e1e8e4ca992608c381871816b
Creation of second block
Hash of new block
        00000ca8f02f38ae2e3d1229926b3b4b155d3a7092191a99ed7bd86e11bda797
```

Now, the SHA-256 generates a hash with five zeros in front of it. The more you increase the difficulty, the more computational time will be taken to solve the problem given on the network. This is how the PoW mechanism works.

## 6.4.2  Proof-of-Stake Implementation

In this section, we will show code snippets for the implementation of PoS consensus. We have used Python to write the codes and implemented the same using Visual Studio Code. We would first create the stakes for the accounts taking part in the consensus. Then, we would be creating lots for these accounts as per the values of stakes. Later, we would be showing to select forgers, i.e., the users who validate transactions and create new blocks in the system if their stake is high [54]. Also, we would be verifying whether random and proportional stake selection works for our approach or not.

```
class ProofOfStake():

def __init__(self):
self.stakers = {}
self.setGenesisNodeStake()

def setGenesisNodeStake(self):
genesisPublicKey = open('keys/genesisPublicKey.pem', 'r').read()
self.stakers[genesisPublicKey] = 1

def update(self, publicKeyString, stake):
if publicKeyString in self.stakers.keys():
self.stakers[publicKeyString] += stake
else:
self.stakers[publicKeyString] = stake

def get(self, publicKeyString):
if publicKeyString in self.stakers.keys():
return self.stakers[publicKeyString]
else:
return None
```

Here, in *ProofOfStake* class, first a *Genesis* node is created. Later, the public key of the genesis node is used for creating the first node and is added to the list of stakers. *ProofOfStake* class keeps track of the stakes of the accounts taking part in the consensus. It keeps a mapping between the account and the corresponding stake value. Two functions are created here, i.e., *update* and *get*. *update* method takes two arguments, i.e., the account name and its corresponding stake.

```
pos.update('bob', 10)
pos.update('alice', 100)
```

We have shown a representation of *update* method, where the stake value for '*bob*' is 10 and for '*alice*', it is 100. The stake value determines how many lots can be created by the account in PoS algorithm. Hence, '*bob*' can create up to 10 lots and '*alice*' can create up to 100 lots.

The next step is to create *Lot* and corresponding *LotHashes*.

```
class Lot():
def __init__(self, publicKey, iteration, lastBlockHash):
self.publicKey = str(publicKey)
self.iteration = iteration
self.lastBlockHash = str(lastBlockHash)

def lotHash(self):
hashData = self.publicKey + self.lastBlockHash
for _ in range(self.iteration):
hashData = BlockchainUtils.hash(hashData).hexdigest()
return hashData
```

In this *Lot* class, we have initialized with public key, iteration and lastBlockHash. Iteration states how many times the account will be allowed to create a lot.

```
lot = Lot('bob', 2, 'lastHash')
```

Given the above initialization, '*bob*' will be allowed to create 2 lots.

```
hashData = self.publicKey + self.lastBlockHash
```

We are generating the *LotHashes* here, where a new hash is created with the help of the public key of an account and the *lastBlockHash* value. A stake of an account decides the amount of time hash generation is done. If the stake of an account is 2, then hash generation is done two times. If the stake is 3, hash generation is done three times and so on, and thus hash chaining is performed.

```
def validatorLots(self, seed):
lots = []
for validator in self.stakers.keys():
for stake in range(self.get(validator)):
lots.append(Lot(validator, stake+1, seed))
return lots

def winnerLot(self, lots, seed):
winnerLot = None
leastOffset = None
referenceHashIntValue = int(BlockchainUtils.hash(seed).hexdigest(), 16)
for lot in lots:
lotIntValue = int(lot.lotHash(), 16)
offset = abs(lotIntValue - referenceHashIntValue)
if leastOffset is None or offset < leastOffset:
leastOffset = offset
winnerLot = lot
return winnerLot

def forger(self, lastBlockHash):
```

```
lots = self.validatorLots(lastBlockHash)
winnerLot = self.winnerLot(lots, lastBlockHash)
return winnerLot.publicKey
```

Here, we have three methods in *ProofOfStake* class. The first method, *validatorLots()*, creates all the lots for the validator for all the stakes. You can see that it iterates the validators through all the staker public keys. The second method, *winnerLot()*, finds out which lot will be the winner lot. Initially, the *winnerLot* and *leastOffset* are set to None as there will be no winner. Now, we need to find *referenceHashIntValue* that is nearby the hash value of the account and that will be treated as forger or miner. The one with the *leastOffset* will be the winner. The third method, *forger()*, is responsible for calling the first two aforementioned methods and returns the public key of the winner.

```
pos = ProofOfStake()
pos.update('bob', 100)
pos.update('alice', 100)

bobWins = 0
aliceWins = 0

for i in range(100):
forger = pos.forger(getRandomString(i))
if forger == 'bob':
bobWins += 1
elif forger == 'alice':
aliceWins += 1

print('Bob won: ' + str(bobWins) + ' times')
print('Alice won: ' + str(aliceWins) + ' times')
```

Now, to test the implemented *ProofOfStake* class, we have used two accounts, namely 'bob' and 'alice' and both of their stakes is initialized to 100. Even though both have the same stake value, due to random selection of the miner, one should always win more times as compared to other. When we run the code, we obtain the following output:

```
> python Test.py

Bob won: 44 times
Alice won: 54 times
```

This means the random selection property of PoS is working. If we run it multiple times, we might get different outputs, but one has to win more than the other. If they both win the same amount of times, i.e., 50 times, that denotes the non-working of the random selection property. Let us check the proportional property to stake selection. If we change the stake value of 'bob' to 60, then we get the following output:

```
>python Test.py
```

```
Bob won: 34 times
Alice won: 65 times
```

This shows that stake selection is proportional to the stake value. The higher the stake value, the more times the account or node will be the winner and allow mining of the block.

### 6.4.3  Proof-of-Burn Implementation

We have not shown full implementation here. Instead, only the code snippet, which does leader election and uses the burn coins for mining the block in the network.

```
def make(mapping, block):
    dist = []
    offset_of_burn = 0
    burn_order = dict([(hash(pub_key + block.current_nonce),
                        (pub_key, burns, hash_current))
    for (pub_key, (burns, hash_current)) in mapping.items()])
        for (_, (pub_key, burns, hash_current)) in
            sorted(burn_order.items()):
                dist.append((offset_of_burn, pub_key, hash_current))
                offset_of_burn += burns
    return dist

def select(current_seed, mapping, proofs, blockHeader.current_nonce):
        if len(mapping) == 0:
            return (None, None, hash(blockHeader.current_nonce+
                current_seed))

        dist = make(mapping)
        total_burns = sum(burns for (_, (burns, _)) in mapping)
        seed = num(hash(current_seed || blockHeader.current_nonce)) /
            total_burns
        last_offset_of_burn = -1
        for (index, (offset_of_burn, pub_key, hash_current)) in
            enumerate(dist):
                if last_offset_of_burn <= seed and seed < offset_of_burn:
                    return (pub_key, hash_current,
                        hash(proofs[pub_key]))
                last_offset_of_burn = offset_of_burn
    return (dist[-1].pub_key, dist[-1].hash_current,
        hash(proofs[dist[-1].pub_key]))
```

Here, the *block* contains the PoW nonce. *mapping* include mapping of public keys to burning scores and block hashes generated during the transactions. *proofs* include the mapping of public keys with the verified proofs required for transactions of leader election. *pub_key* will be the winner public key, and *hash_current* will be the block hash of the winner.

### 6.4.4   Proof-of-Ownership Implementation

For this implementation, we have created smart contract showing PoO algorithm. We have deployed the smart contract using Truffle, Ganache, and MetaMask. The steps for using these together are shown in Chap. 1.

```
contract proof_of_ownership {
    mapping (bytes32 => add) public onwer;
    function query(bytes32 resource) view public returns (add) {
        return onwer[resource];
    }

    function register(bytes32 resource) public {
        if (add(onwer[resource]) == add(0))
        {
            onwer[resource] = message.sender;
        }
    }

    function transfer(bytes32 resource, add current_owner) public {
        if (onwer[resource] == message.sender)
        {
            onwer[resource] = current_owner;
        }
    }

    function delete(bytes32 resource) public {
        if (onwer[resource] == message.sender)
        {
            onwer[resource] = 0x0000000000000000000000000000000000000000;
        }
    }
}
```

The above-mentioned code contains four functions related to asset ownership. Firstly, *query*() function queries the asset and returns the owner details. The resource value is address passed and the address is recognized by the private key taken from Ganache. Secondly, *register*() function adds the list of addresses, i.e., owners who want to take part in the BC, in a maintained array. Thirdly, *transfer*() function transfers the ownership who got a chance to add a new block to the BC. Lastly, *delete*() function sets the owner of resource back to 0x000...000 denoting the resource does not belong to any owner.

```
register(asset_id)
{
    let user_add = this.user_add;
    this.asset.deployed().then(function(current)
    {
        current.register(assetID, {gas_value: 1400000, from:
            user_add}).then(function(i)
        {
```

```
                console.log(i.toLocaleString());
            });
        });
    }

    query(asset_id)
    {
        let user_add = this.user_add;
        this.asset.deployed().then((current)=>
        {
            current.query(asset_id, {gas_value: 1400000, from: user_add}).then((i)=>
            {
                this.set({owner: i.toLocaleString()})
            });
        });
    }

    hash_calculate(file)
    {
        this.set({value:20});
        const read = new FileReader();
        read.onload = () => {
            const binarystringfile = read.result;
            const hash = crypto.createHash('md5').update(binarystringfile).digest("hex");
            this.set({hash: hash, onwer: null, value:99});

        };
    }
```

The code shows the *register*() function, which takes the user address and allocates
some gas. *hash_calculate*() function is responsible for calculating the hash of the
ownership resource, which here is considered a file. *createHash*() function will compute the hash digest for the file ingested as an argument.

```
    Replacing 'proof_of_ownership'
    ─────────────────────────────
> transaction hash:
        0x62ef778e1b0da9c51bcad54995818ecb4979164a4d391700604d2cfa07a8b4d2
> Blocks: 0          Seconds: 0
> contract address:  0x2214dadcEfbdf15CBDD44F4CE4D10F08cC1e9efD
> block number:      3
> block timestamp:   1642877774
> account:           0x18f97d36a3d5cf30e59539349324d71FAC8931a4
> balance:           99.9883603
> gas used:          314385 (0x4cc11)
> gas price:         20 gwei
> value sent:        0 ETH
> total cost:        0.0062877 ETH
> Saving migration to chain.
> Saving artifacts
    ─────────────────────────────
> Total cost:        0.0062877 ETH
```

We get the above-mentioned output when we deploy the smart contact of PoO, which shows the account address, i.e., '0x18f97d36a3d5cf30e59539349324d71FAC8931a4', chosen for performing transactions in the network.

```
truffle(develop)> proof_of_onwership.deployed().then((current)=> current.register
     ("0x2214dadcefbdf15cbdd44f4ce4d10f08cc1e9efd", {from:
         "0x18f97d36a3d5cf30e59539349324d71FAC8931a4"}));
  {
    tx: '0x9bf9e1b2046b750abf0633afa41c771d0fed1c110d86386be8d9985a096ce1bc',
    receipt: {
        transactionHash:
            '0x9bf9e1b2046b750abf0633afa41c771d0fed1c110d86386be8d9985a096ce1bc',
        transactionIndex: 0,
        blockHash:
            '0x5060908f59c534f643dfc5231b55c4e818a6d3e42d9818beec8c3291a0c74cc3',
        blockNumber: 5,
        from: '0x18f97d36a3d5cf30e59539349324d71fac8931a4',
        to: '0x2214dadcefbdf15cbdd44f4ce4d10f08cc1e9efd',
        gasUsed: 43328,
        cumulativeGasUsed: 43328,
        contractAddress: null,
        logs: [],
        status: true,
        rawLogs: []
    },
    logs: []
}
```

The function listed will transfer the asset to account '0x2214dadcefbdf15cbdd44f4ce 4d10f08cc1e9efd'. The keys are private keys taken from the Ganache. Without the association of private keys, ownership cannot be claimed. We can open MetaMask and check the ETH transfer from the owner.

## 6.5  Comparison

We have done a comparison analysis of consensus algorithms in a permissionless environment on the basis of some chosen parameters as shown in Table 6.1.

## 6.6  Summary

BC is a trending technology for security and immutable databases. A private BC is secured as it allows only authenticated users to join the BC, but a third party maintains trust. While working with a public BC, there is no third-party involvement, and it is public, anyone can join the BC. So, malicious activities can be involved. Consensus mechanisms are there to protect the system from malicious users. The consensus mechanism can be proof-based or voting-based. PoW is one of the proof-based

**Table 6.1** Comparative analysis of consensus algorithms in permissionless environment

| Sr. no. | Algorithm | Scalability | Cost | Energy efficiency | Transaction rate | Throughput | Decentralization | Computing overhead | Finality |
|---------|-----------|-------------|------|-------------------|------------------|------------|------------------|--------------------|----------|
| 1. | PoW | H | H | No | L | L | H | H | Probabilistic |
| 2. | PoS | H | M | Yes | H | L | H | M | Probabilistic |
| 3. | DPoS | H | L | Yes | L | H | M | M | Probabilistic |
| 4. | LPoS | H | M | Yes | M | H | M | M | Probabilistic |
| 5. | PoI | H | H | Yes | H | L | H | M | Probabilistic |
| 6. | PoSV | H | H | Yes | H | L | H | M | Probabilistic |
| 7. | PoB | H | L | Yes | H | L | H | M | Deterministic |
| 8. | PoC | H | L | Yes | M | L | H | L | Probabilistic |
| 9. | PoAc | H | L | No | L | L | H | M | Probabilistic |
| 10. | PoWe | H | L | M | M | L | H | M | Instant |
| 11. | PoL | L | M | M | M | L | H | M | Deterministic |
| 12. | PoO | H | L | H | L | M | M | L | Deterministic |
| 13. | PoE | M | L | H | L | M | M | L | Probabilistic |
| 14. | PoH | H | M | M | H | L | M | M | Deterministic |
| 15. | PoBe | L | M | M | M | L | H | M | Probabilistic |

H: High, M: Medium, L: Low

consensus mechanisms where miners need to perform complex tasks formulated using the SHA-256 hashing algorithm. These consensus mechanisms solve security attacks such as double spending, 51% attack, and Sybil attack. Each consensus algorithm has its own advantages and disadvantages. They can be compared based on consensus parameters such as power consumption scalability, security, and throughput. Many cryptocurrencies are created by various companies based on these consensus algorithms and perform specific use cases in real estate, supply chain, banking, health care, etc.

## 6.7   Practice Questions

### 6.7.1   MCQ Questions

1. Which of the following is used to ensure consensus in only a Permissionless Blockchain Environment?

(a)  Proof of Work.
(b)  Paxos consensus.
(c)  Proof of Stake.
(d)  Byzantine fault tolerance.

2. Which of the following consensus algorithms consider virtual resources or digital coins for participating in the mining activity?

(a)  Proof of Stake.
(b)  Raft consensus.
(c)  Proof of Burn.
(d)  Proof of Work.

3. What is the default setup time for the miners to collect all the transactions that occurred in the 'Bitcoin' network?

(a)  6 min.
(b)  15 min.
(c)  10 min.
(d)  16 min.

4. Considering the cryptographic sortition function for the selection procedure, which of the following arguments of the function ensures that the subset of the random committee change in every iteration?

(a)  *seed*: A publicly known random value.
(b)  *role*: User for proposing a block/committee member.
(c)  $w_i$: The weight of the user i.
(d)  $W$: Weight of all users.

5. Which of the following challenges make the application of traditional distributed consensus algorithms to blockchain-based systems difficult over the Internet?

(a) Global accessibility of the Internet.
(b) Large number of users.
(c) Asynchronous nature of the Internet.
(d) All of the above.

6. Which of the following strategy is used by the Bitcoin network to control the inflow of bitcoins in the network?

(a) Lower the mining reward.
(b) Increasing the nonce.
(c) Increasing the block size.
(d) All of the above.

7. With mining pools in effect, which of the following problem or problems may appear in the Bitcoin network?

(a) Participation of smaller miners with lesser resources.
(b) Discouragement among miners to fully complete the task posed by the network.
(c) There are disadvantages to this scheme.
(d) Monopoly in mining.

8. Which consensus algorithm requires the need of unique pseudonyms?

(a) Proof of Work.
(b) Proof of Ownership.
(c) Proof of Existence.
(d) Proof of Luck.

9. Proof of Stake is

(a) How private keys are made.
(b) A password needed to access an exchange.
(c) A transaction and block verification protocol.
(d) A certificate needed for the blockchain.

## 6.7.2  Fill in the Blanks

1. _____ and _____ is ensured in a traditional distributed environment by applying consensus.
2. In _____ attack, all the chains are followed so that participant gets rewards either way, and one chain out of them will be picked as a winner.
3. The PoWe assigns _____ values to stakeholders as per _____ in user's account.
4. In PoE, a document is stored in the BC with a signature and a _____ associated during storage.

5. _____ mechanism demands nodes to compete in a lottery and the final value is proposed by the node that wins.
6. PoAc combines the features of _____ and _____ for generating new blocks.
7. The _____ property states that the correct individual must not agree on an incorrect value.
8. _____ overcomes the high energy consumption drawn by PoW and favors the rich issue drawn by PoS and PoB.
9. In _____, if forgers decide to leave the network, they can take the coins back and sell them to the market. This will not create market scarcity.
10. PoH uses a _____ function to hash incoming events and transactions.

### 6.7.3   Short Questions

1. What is distributed consensus blockchain? Why could consensus be crucial to your business?
2. Discuss how issues with interoperability and scaling affect the wider use of blockchain.
3. Examine and justify whether a public blockchain needs to issue its native cryptocurrency to incentivize its validator network.
4. Distinguish between a permissionless and a permissioned blockchain.
5. Discuss whether the SHA-256 hash is appropriate for most blockchains.

### 6.7.4   Long Questions

1. List some of the popular consensus algorithms in a permissionless environment. Describe their properties. Why do we need different consensus mechanisms?
2. PoW has been criticized for its high and continually rising mining cost. Discuss how mining cost impacts the tamper resistance attribute of public blockchains.
3. Perform a comparative analysis of consensus algorithms with respect to security and its corresponding attacks.
4. Discuss how high power consumption and high stake issue are solved by consensus algorithms.
5. Elaborate on the factors you will consider when you want to develop your own consensus algorithm and justify the same.

# References

1. Bodkhe U, Mehta D, Tanwar S, Bhattacharya P, Singh PK, Hong W-C (2020) A survey on decentralized consensus mechanisms for cyber physical systems. IEEE Access 8:54371–54401. https://doi.org/10.1109/ACCESS.2020.2981415
2. Zheng Z, Xie S, Dai H, Chen X, Wang H (2017) An overview of blockchain technology: architecture, consensus, and future trends. IEEE Int Congr Big Data (BigData Congr) 2017:557–564. https://doi.org/10.1109/BigDataCongress.2017.85
3. Bach LM, Mihaljevic B, Zagar M (2018) Comparative analysis of blockchain consensus algorithms. In: 41st international convention on information and communication technology, electronics and microelectronics (MIPRO), pp 1545–1550. https://doi.org/10.23919/MIPRO.2018. 8400278
4. Gupta R, Nair A, Tanwar S, Kumar N (2020) Blockchain-assisted secure UAV communication in 6G environment: architecture, opportunities, and challenges. IET Commun. https://doi.org/10.1049/cmu2.12113
5. Zou W, Lo D, Kochhar PS, Le XBD, Xia X, Feng Y, Chen Z, Xu B (2019) Smart contract development: challenges and opportunities. IEEE Trans Softw Eng
6. Cachin C, Vukolić M (2017) Blockchain consensus protocols in the wild. arXiv:1707.01873
7. Shaik VA, Malik P, Singh R, Gehlot A, Tanwar S (2020) Adoption of blockchain technology in various realms: opportunities and challenges. Secur Priv 3:e109. https://doi.org/10.1002/spy2. 109
8. Berman P, Garay JA, Perry KJ (1989) Towards optimal distributed consensus. In: FOCS, vol 89, pp 410–415
9. Amelchenko M, Dolev S (2017) Blockchain abbreviation: implemented by message passing and shared memory. In: 2017 IEEE 16th international symposium on network computing and applications (NCA). IEEE, pp 1–7
10. Wright A, De Filippi P (2015) Decentralized blockchain technology and the rise of lex cryptographia. Available at SSRN 2580664
11. Baudet M, Ching A, Chursin A, Danezis G, Garillot F, Li Z, Malkhi D, Naor O, Perelman D, Sonnino A (2019) State machine replication in the libra blockchain. The Libra Assn, Technical report
12. Fan K, Sun S, Yan Z, Pan Q, Li H, Yang Y (2019) A blockchain-based clock synchronization scheme in IoT. Futur Gener Comput Syst 101:524–533
13. Fischer MJ (1983) The consensus problem in unreliable distributed systems (a brief survey). In: International conference on fundamentals of computation theory. Springer, Berlin, pp 127–140
14. Barborak M, Dahbura A, Malek M (1993) The consensus problem in fault-tolerant computing. ACM Comput Surv (CSur) 25(2):171–220
15. Shah MA, Hellerstein JM, Brewer E (2004) Highly available, fault-tolerant, parallel dataflows. In: Proceedings of the 2004 ACM SIGMOD international conference on management of data, pp 827–838
16. Yanovich Y, Ivashchenko I, Ostrovsky A, Shevchenko A, Sidorov A (2018) Exonum: byzantine fault tolerant protocol for blockchains. bitfury. com, pp 1–36
17. Ferdous MS, Chowdhury MJM, Hoque MA, Colman A (2020) Blockchain consensus algorithms: a survey. arXiv:2001.07091
18. Hoffman RS, Hoffman R (2000) Does consensus equal correctness? J Toxicol: Clin Toxicol 38(7):689–690
19. Mostefaoui A, Raynal M (2001) Leader-based consensus. Parallel Process Lett 11(01):95–107
20. Zhang L, Li Q (2018) Research on consensus efficiency based on practical byzantine fault tolerance. In: 2018 10th international conference on modelling, identification and control (ICMIC). IEEE, pp 1–6
21. Mingxiao D, Xiaofeng M, Zhe Z, Xiangwei W, Qijun C (2017) A review on consensus algorithm of blockchain. In: IEEE international conference on systems, man, and cybernetics (SMC), pp 2567–2572

22. Helliar CV, Crawford L, Rocca L, Teodori C, Veneziani M (2020) Permissionless and permissioned blockchain diffusion. Int J Inf Manag 54:102136
23. Gupta R, Kumari A, Tanwar S (2020) A taxonomy of blockchain envisioned edge-as-a-connected autonomous vehicles. Trans Emerg Telecommun Technol. https://doi.org/10.1002/ett.4009
24. Rizal BF, Ubacht J, Janssen M (2019) Unraveling transparency and accountability in blockchain. In: Proceedings of the 20th annual international conference on digital government research, pp 204–213
25. Mitani T, Otsuka A (2020) Traceability in permissioned blockchain. IEEE Access 8:21573–21588
26. Gupta R, Aparna K, Sudeep T, Neeraj K (2020) Blockchain-envisioned softwarized multi-swarming UAVs to tackle COVID-19 situations. IEEE Netw. https://doi.org/10.1109/MNET.011.2000439
27. Nakamoto S (2008) Bitcoin: a peer-to-peer electronic cash system. Decentralized Bus Rev 21260
28. Stinson DR (2006) Some observations on the theory of cryptographic hash functions. Des, Codes Cryptogr 38(2):259–277
29. Natoli C, Gramoli V (2016) The blockchain anomaly. In: 2016 IEEE 15th international symposium on network computing and applications (NCA). IEEE, pp 310–317
30. Karame GO, Androulaki E, Capkun S (2012) Double-spending fast payments in bitcoin. In: Proceedings of the 2012 ACM conference on computer and communications security, pp 906–917
31. Douceur JR (2002) The sybil attack. In: International workshop on peer-to-peer systems. Springer, Berlin, pp 251–260
32. Jamal T, Haider Z, Butt SA, Chohan A (2018) Denial of service attack in cooperative networks. arXiv:1810.11070
33. Zhou X, Dong J, Zhang X, Zhang P (2018) Application of blockchain technology in the financial industry and its legal norms. In: 2018 2nd international conference on man, education and social science. Atlantis Press
34. Nguyen CT, Hoang DT, Nguyen DN, Niyato D, Nguyen HT, Dutkiewicz E (2019) Proof-of-stake consensus mechanisms for future blockchain networks: fundamentals, applications and opportunities. IEEE Access 7:85727–85745
35. King S, Nadal S (2012) PPCoin: Peer-to-peer crypto-currency with proof-of-stake. self-published paper, 19 Aug, no 1
36. Ye C, Li G, Cai H, Gu Y, Fukuda A (2018) Analysis of security in blockchain: case study in 51%-attack detecting. In: 2018 5th international conference on dependable systems and their applications (DSA). IEEE, pp 15–24
37. Li W, Andreina S, Bohli J-M, Karame G (2017) Securing proof-of-stake blockchain protocols. In: Data privacy management, cryptocurrencies and blockchain technology. Springer, Cham, pp 297–315
38. Azouvi S, McCorry P, Meiklejohn S (2018) Betting on blockchain consensus with fantomette. arXiv:1805.06786
39. Larimer D (2014) Delegated proof-of-stake (DPoS). Bitshare whitepaper 81:85
40. Salimitari M, Chatterjee M (2018) An overview of blockchain and consensus protocols for IoT networks, pp 1–12. arXiv:1809.05613
41. Bamakan SMH, Motavali A, Bondarti AB (2020) A survey of blockchain consensus algorithms performance evaluation criteria. Expert Syst Appl 154:113385
42. Ren L (2014) Proof of stake velocity: building the social currency of the digital age. Self-published white paper
43. Karantias K, Kiayias A, Zindros D (2020) Proof-of-burn. In: International conference on financial cryptography and data security. Springer, Cham, pp 523–540
44. Bach LM, Mihaljevic B, Zagar M (2018) Comparative analysis of blockchain consensus algorithms. In: 2018 41st international convention on information and communication technology, electronics and microelectronics (MIPRO). IEEE, pp 1545–1550

45. Azab A, Layton R, Alazab M, Oliver J (2014) Mining malware to detect variants. In: 2014 fifth cybercrime and trustworthy computing conference. IEEE, pp 44–53

46. Bentov I, Lee C, Mizrahi A, Rosenfeld M (2014) Proof of activity: extending bitcoin's proof of work via proof of stake [extended abstract] y. ACM SIGMETRICS Perform Eval Rev 42(3):34–37

47. Goldin D, Raisch J (2013) On the weight controllability of consensus algorithms. In: 2013 European control conference (ECC). IEEE, pp 233–238

48. Sabt M, Achemlal M, Bouabdallah A (2015) Trusted execution environment: what it is, and what it is not. In: 2015 IEEE Trustcom/BigDataSE/ISPA, vol 1. IEEE, pp 57–64

49. Milutinovic M, He W, Wu H, Kanwal M (2016) Proof of luck: an efficient blockchain consensus protocol. In: Proceedings of the 1st workshop on system software for trusted execution, pp 1–6

50. Abreu PW, Aparicio M, Costa CJ (2018) Blockchain technology in the auditing environment. In: 2018 13th Iberian conference on information systems and technologies (CISTI). IEEE, pp 1–6

51. Bao S, Cao Y, Lei A, Asuquo P, Cruickshank H, Sun Z, Huth M (2019) Pseudonym management through blockchain: cost-efficient privacy preservation on intelligent transportation systems. IEEE Access 7:80390–80403

52. Chopra K, Gupta K, Lambora A (2019) Proof of existence using blockchain. In: 2019 international conference on machine learning, big data, cloud and parallel computing (COMITCon). IEEE, pp 429–431

53. Bada AO, Damianou A, Angelopoulos CM, Katos V (2021) Towards a green blockchain: a review of consensus mechanisms and their energy consumption. In: 2021 17th international conference on distributed computing in sensor systems (DCOSS). IEEE, pp 503–511

54. Yakovenko A (2018) Solana: a new architecture for a high performance blockchain v0. 8.13. Whitepaper

# Chapter 7
# Mining Procedure in Distributed Consensus

**Abstract** The rapid growth of blockchain technology over the preceding decade has attracted the interest of both scholars and businesses. Bitcoin and cryptocurrency are the foundations of blockchain technology, widely recognized as a decentralized, immutable ledger system for transaction data ordering and an open network structure. Anyone may join the network because it is an accessible network framework. It might also be a legal or illegal entity. This involves the implementation of a blockchain that is both secure and tamper-proof. Mining is a core of a distributed consensus mechanism that maintains transaction consistency, making the network secure and efficient. Every miner must establish their identity in an access network to join the blockchain network. The miner must complete numerous tasks and allocate their time and resources to become a member of the network and earn a reward. This process has been designed in such a manner that it requires a large number of processing resources and power. With the help of an example, this chapter explains the entire process of mining a block. This mining process creates an incentive for miners to contribute more toward the network. Difficulty requirements are an important factor to consider during the mining procedure of a block. As a result, the connection between mining and hash cash is thoroughly examined. Moreover, the growing demand for blockchain technology highlights the need for mining pools and their many sorts, as well as comparative research. In addition, this chapter explores the permissionless and permissioned consensus algorithm and its working process. This chapter explores the types of sources, cryptocurrency, and required energy to execute any specific consensus algorithm.

**Keywords** Mining · Distributed consensus · Mining pools

## 7.1 Introduction

In Chap. 2, we have already discussed how blockchain technology has evolved since the invention of bitcoin by a group of individuals named Satoshi Nakamoto in 2008. We have mentioned various generations of blockchain and its development, starting from bitcoin utilizing Blockchain 1.0 to Blockchain 5.0 with the property of cross-chain function, enhanced security, scalability, and cost-effectiveness. Blockchain 5.0 is designed so that institutions in varied fields can deploy it in their applications integrated with advanced technologies. The innovation of blockchain technology as a decentralized system is mainly based on two aspects, i.e., immutability and tamper resistance. Chapter 1 discusses the basic working of blockchain and how data transactions are added to the network to make it immutable, secure, and tamper-resistant using applied key technologies such as cryptographic hash functions, smart contract, and consensus mechanism. It provides a sustainable and highly reliable environment to the nodes participating in the networks. For example, two users want to transfer some money that involves a transaction that needs to be secured against any malicious attack. For that, blockchain uses various cryptographic hash functions and message authentication protocols to preserve the network's immutability, further achieving data integrity.

Blockchain consists of a chain of blocks to store the data transactions securely, which is connected to the Merkle tree [1]. It is used to keep track of all the transactions in the network by associating them with their hash. So, if data has been tampered with in the network, it automatically changes the hash of that corresponding block; that change will be reflected in the Merkle root at the top of the Merkle tree. Thus, blockchain can easily detect the tampered data from the Merkle root without accessing all the transactions in the Merkle tree. The basic concepts and technologies related to blockchain technology have been discussed in detail in Chap. 1. One of the key technology of blockchain, i.e., consensus mechanisms, which include Proof of Work (PoW), Proof of Stake (PoS), and Proof of Burn (PoB), play an essential role in validating the transactions between nodes, i.e., users in the blockchain network [2]. Every node plays a different role in the network, but if we want to get insights into the consensus mechanism, then we need to consider different consensus nodes such as lightweight nodes and full nodes that have already been discussed in Chap. 1.

Consensus nodes are responsible for declaring the transactions as valid, i.e., distributed consensus mechanism can be used to authorize the transactions if all the consensus nodes in the network agree to it. For example, Satoshi Nakamoto had used PoW to transfer bitcoins between users in the blockchain network. The PoW involves a mining mechanism in which miners, i.e., network consensus nodes, play an important role in validating the transactions by computing a cryptographic puzzle. However, the mining mechanism can differ depending on the consensus mechanism used in the network. But, before understanding the mining procedure in various consensus mechanisms, we need to understand the type of environment in which blockchain technology works, i.e., permissionless and permissioned environments. A permissionless environment can be defined as an open environment that provides transparency and anonymity to the nodes or users interacting with the other nodes in the network [3]. However, various consensus mechanisms are used to authenticate the arriving nodes as any node can deviate from the decision of adding transactions or can alter the transactions. It is applied to ensure that all the nodes should agree on the same decision about the addition of transactions to the blockchain network. On the other hand, permissioned blockchain is managed through the known group of users in a closed environment. It allows only a limited number of users or nodes to access information about the transactions in the network. It does not provide transparency and anonymity to the users as an authority or a team of users handling the transactions in the network. Therefore, any adversary can easily access the data and exploit the confidential information [4]. In this chapter, we will discuss the mining procedure in a blockchain-based permissionless environment on two types of consensus mechanisms, such as proof-based consensus and voting-based consensus.

## 7.2 Proof-Based Consensus Mechanism

Consensus mechanisms are considered as the backbone of the blockchain network. As discussed earlier, different consensus protocols are adapted to verify and add the data transactions to the blockchain network. For that, all the participating nodes in the network need to be in consensus on a particular decision about the authenticity of new transactions that need to be appended to the network. Proof-based consensus mechanisms provide security to the network by adapting the mining mechanism, which includes a number of participants, i.e., miners in the network who compete to verify the authenticity of the transactions. Once the transactions are valid, a smart contract can be implemented to append them to the blockchain. But miners work differently in different proof-based consensus mechanisms, which will be explained in the next section initiating with PoW.

### 7.2.1 Proof of Work

As mentioned above, Satoshi Nakamoto has first released a cryptocurrency bitcoin, that can be transferred between users participating in the blockchain network in which they have applied the PoW consensus mechanism, which mainly computes a complex cryptographic puzzle so that transactions can further be authenticated in the Peer-to-Peer (P2P) network [5]. The detailed procedure of mining in the PoW consensus mechanism can be explained in the following steps [6, 7]:

- Figure 7.1 shows how transactions can be verified by the miners using the PoW consensus mechanism. As we are talking about a permissionless environment of blockchain, anyone can become a miner to verify the transactions to check their authenticity for the blockchain network.
- First, the bitcoin transaction gets validated using the sender's private key to further send it to the group of participating nodes in the network.
- Then, miners get involved in the network to group transactions into the block so that the SHA-256 hash algorithm can be applied to the block to generate the desired hash while providing security to the block. As discussed in Chap. 1, it considers string as an input to generate a hash value of a 256-bit fixed length.

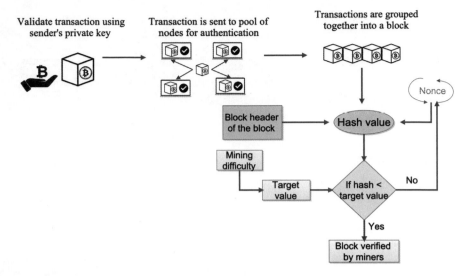

**Fig. 7.1** Verification of block using PoW with mining

- After that, miners have to guess the hash value by adding the different values of the nonce to the block header of the blockchain, which contains a hash of the previous block, nonce, version, and timestamp.
- Nonce is a 32-bit random string that can be tried different times to add it to the block header of the data block to generate the desired hash value with the help of the SHA-256 algorithm.
- If the generated hash value tends to be less than the target value, then it reflects that miners have solved the complex mathematical puzzle so that the validated block can be passed to the consensus nodes in the network.
- The target value is calculated with the mining difficulty which depends upon the number of miners competing to solve the cryptographic hash puzzle as whoever solves the puzzle first will get the reward.
- As a huge number of miners will be competing to solve the puzzle, it will lead to high energy consumption in the PoW mechanism.
- After verification of the block by miners, the block will be passed to the consensus nodes in the network to further add the validated block to the blockchain network if all the nodes are in consensus about the addition of the block.

### 7.2.2   Proof of Stake

Blockchain-based platforms bitcoin and Ethereum utilize the PoW consensus mechanism to solve a complex mathematical puzzle, which incurs high network energy consumption. But, due to the wastage of increased computing power in PoW, Vitalik

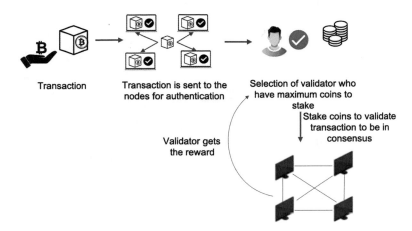

**Fig. 7.2** Verification of block using PoS with validator

Buterin in 2018 suggested a gradual move toward the PoS consensus mechanism from PoW. The aim of the PoS is similar to the PoW, i.e., to achieve the consensus between participating nodes in the network, but there are no miners involved in the PoS consensus mechanism. PoS consensus works on the principle that any user can verify the transactions based on the coins they have in their wallet. If a user owns the maximum number of coins in their wallet, that user is eligible to validate the transaction and get the reward in the form of transaction fees as shown in Fig. 7.2, although the reward obtained in the case of PoS will be less than the reward of mining in PoW [8]. PoS consensus incurs less energy consumption in the network as it works based on the staking of coins instead of mining. But there can be an issue associated with this consensus mechanism as it may favor the prosperous person more, leading to partiality to the other users. Other users may not even get a chance to win the reward for validating the transaction [9].

### 7.2.3 Proof of Burn

Another consensus mechanism, i.e., PoB, mainly works to reduce the energy consumption incurred in the PoW consensus. PoB works on the principle that participating nodes who own coins in their wallet have to send them to an address where that particular coin cannot be used again. Iain Stewart first proposed this consensus mechanism to provide users with a creative incentive scheme to encourage them to invest their coins for validating the transactions. But what are the chances of getting the incentive to validate the transaction by using their coins? It depends on how many coins the users can send to that particular address, but it can also cause loss for them as their coins get wasted if they do not get a chance to validate the transaction.

**Fig. 7.3**  Verification of
block using PoB with miner

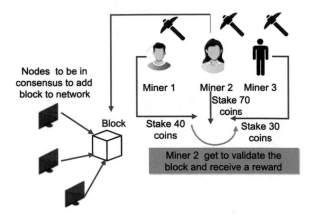

Figure 7.3 shows how miners can validate the transaction using the PoB consensus
mechanism, which can be explained in the following steps [10]:

- In PoW, we have already explained how a transaction is passed to the nodes for
  authentication using a digital signature with the sender's private key.
- In PoB, after the transaction gets authenticated in the permissionless environment,
  miners get involved in the network for verifying the transaction using a consensus
  mechanism.
- Figure 7.3 depicts three miners, which are Miner 1, Miner 2, and Miner 3, and
  they are willing to stake their coins so that miner with the highest staking coin can
  get the reward.
- As shown in Fig. 7.3, Miner 2 with the 70 stake coins will get to validate the data
  block of transactions and will receive the reward in the form of coins only.
- Then, nodes in the network need to be in consensus to append the verified block
  to the blockchain network.
- But, staking coins of Miner 1 and Miner 1 will not be of any use after staking them
  leading to the loss of other miners.

So, we have discussed mining procedure in three types of proof-based consensus
mechanisms, which include PoW, PoS, and PoB consensus mechanisms. But there
is one more category, i.e., voting-based consensus mechanisms such as Byzantine
Fault Tolerance (BFT) and Crash Fault Tolerance (CFT) consensus. These consensus
mechanisms will be explained in the next section. But before that, we need to get a
basic understanding of the mining pools and their benefits.

## 7.3  Mining Pools

We have seen different types of proof-based consensus mechanisms that work based
on different mining mechanisms. But there is a question that needs to be answered by
miners that they mine individually or join the mining pool. With the increase in the
difficulty of solving the cryptography puzzle, it may be a challenging and costly task

for miners to mine the block individually. Therefore, the idea behind the mining pool is to combine multiple miners together and then generate the hash in a distributed way. Every miner starts generating the hash for a block and shares their pricing over the network to mine a new block. If miners work in a cooperative way to mine the block, power consumption can be reduced, and rewards get distributed among the miners.

### 7.3.1 Benefit of Mining Pools

There can be some benefits associated with the mining pools, which encourage miners to work cooperatively instead of mining the block individually for appending the block to the blockchain network. Also, miners can validate the block with reduced power consumption, making mining profitable for them. Furthermore, due to the low energy consumption, scalability of the network can be maintained, leading to more blocks mined in the network. But a third-party authority involved in the mining pool manages the reward distribution among the deserving miners. But it can cause security issues as the third-party system may become faulty and can get the rewards for themselves only, leading to the loss for miners. Therefore, it is not always appropriate for miners to join the mining pool for validating the transactions [11, 12].

### 7.3.2 Mining Pools Methods

In a mining pool, hundreds or thousands of miners give rise to a very genuine problem, such as how the pool manager evaluates the amount of actual work done by the pool members against the claimed work. There can be a scenario where the miners in the pool are sending a lot of validated blocks to the nodes in the network to check for consensus; then, this can give the pool manager a statistical idea of the work done by miners. Also, they cannot fake the mining procedure as there can be cryptographic hash functions, smart contracts, and consensus mechanisms associated with ensuring confidentiality and security in the network. Moreover, there's no way one can impersonate it because of the robust properties of the hash function.

Pool managers collect the transactions needed to be validated and group them into a block to send them to the miners further. Merkle tree should also be included with the transaction to maintain the security of the network. Then, one of the miners can show the validated block sent to the network to append it to the blockchain. According to the mining work done by each miner, a reward gets distributed among them, which the pool manager obtained from the validated block. It is important for miners to get the reward equally according to their contribution to mining the block. There can be several mining pool payment methods that the pool manager should adapt to distribute the reward equally among the users, which is explained in the next section [13].

- **Pay per Share (PPS)** The PPS model can be defined as a payment method where the pool manager announces that he pays a flat fee for every share above a certain difficulty that miners are willing to send based on a particular block. In some ways, it is the best for miners due to the timely and quick payout to the miner. Moreover, it is favorable for miners to exit anytime from the mining pool by withdrawing their reward. Nonetheless, it can be a tedious task for the pool manager to manage the mining pool if many transactions appear for mining, but miners are unwilling to stay in the pool for a long time.
- **PROP** Another mining pool payment method can be a proportional model. Rather than paying a level expense for each share, the amount per share depends on whether the pool found a valid block. Therefore, at each mining round, the pool finds out a block that calculates the total share of the individual miner. The distribution of rewards to the members is in proportion to their performed work. Formally, the risk is proportional to the risk of the pool.
- **Pay Per Last N Shares (PPLN)** The PPLN payment model is similar to the proportional share model. There is only a slight difference that pool managers in PPLN consider, that is, the last N shares of the miner for the work done while validating the transactions. On the other, in PROP, they used to consider all the shares of work done by miners. This method is beneficial for miners when transactions appear quickly as it leads to more blocks mined for the blockchain network.

## 7.4   Voting-Based Consensus Mechanism

Any node can join the blockchain network in proof-based consensus mechanisms to verify the transactions. However, in a voting-based consensus mechanism, participating nodes have to communicate with each other to access the particular block to add to the blockchain network or not. Therefore, they need at least a threshold (Th), i.e., several nodes that agree to append a block to the network. Furthermore, some of the nodes in the network may be defined as faulty or crashed; for that, we need to categorize the voting-based consensus mechanism into two more mechanisms, i.e., BFT and CFT. There is an assumption in this type of consensus mechanism, i.e., N number of nodes out of Th number of nodes is working correctly. In the case of BFT, N is approximately two-thirds of the Th number of nodes, i.e., 2Th/3+1, and for CFT, N is approximately one-half the Th number of nodes, i.e., Th/2+1. These two consensuses are explained in the next section, which deals with the failure of nodes in the system.

### 7.4.1   Byzantine Fault Tolerance Consensus

Practical Byzantine Fault Tolerance (PBFT) has been used by many industries, which is a type of BFT. PBFT consists of two types of nodes, i.e., leader node and validators

**Fig. 7.4** Verification of block using PBFT with validator and leader

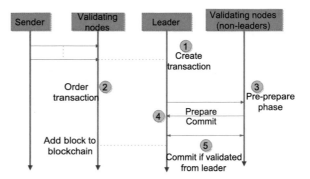

nodes. In PBFT, the procedure to validate the transactions can be explained in the following steps [14]:

- As shown in Fig. 7.4, the sender sends a request to validating nodes about the transactions which they want to add to the network.
- After that, the requests are sent to other nodes in the network, including the leader node.
- As we have already mentioned about the Th number of nodes, the leader considers the transactions till Th number of nodes, then after that leader arranges them according to their timestamp to group them into a block.
- Then, it initiates the three phases of PBFT, i.e., pre-prepare, prepare, and commit phases.
- In the pre-prepare phase, the validator nodes send the arranged block to other nodes in the network to store the block in their local storage,
- After that, they need to ensure the authentication of the received block. The block is sent to the prepare phase and commit phase.
- Then, in prepare and commit phases, other nodes in the network receive the block of local storage and satisfy the condition of PBFT, i.e., two-thirds of the nodes in the network, then that block will be valid to add it to the blockchain network.

## 7.4.2 Crash Fault Consensus Tolerance

Raft quorum is an Ethereum-based blockchain platform mainly designed for industrial purposes. Raft is a CFT-based consensus mechanism responsible for working properly in the network if it satisfies the criteria of Th number of nodes, i.e., one-half of the Th number of nodes. We get the understanding of the CFT consensus mechanism in the following steps as shown in Fig. 7.5 [15]:

- Each validator node is assigned three roles comprising of follower, leader, and candidate.

**Fig. 7.5** Selection of leader
for verifying the transactions

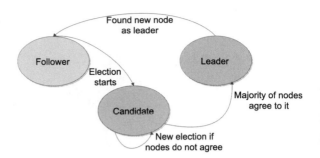

- The leader's main role is to interact with the senders, and the follower node has to follow the leader node.
- Candidate can become the leader node depending on that particular time at which leader nodes want to interact and make changes to the transactions in the network.
- All the candidate nodes hold an election to choose the leader node with the help of the RequestVote message. If the majority of the nodes agree to it, a node can be the next leader; else, that node is considered a follower node.
- To validate the transactions, the leader sends AppendEntries to all the followers with information about the transactions, such as the hash of the previous transaction.
- Followers make sure that received transactions are only recent and received according to their timestamp in sequence order.
- Further, the leader considers the verified transactions and groups them into a block so that they can be added to the blockchain network.

We have discussed how miners can validate the transactions based on the two types of consensus mechanisms, i.e., proof-based consensus and voting-based consensus mechanisms. Miners work differently based on the consensus mechanism, i.e., some may consider the validator nodes or miners for verifying the transactions as mentioned above. Now, we have to understand how to implement the consensus mechanism to show the mining procedure in the network. For that, we consider a scenario in which PoW consensus is implemented to understand the mining procedure in the permissionless environment of blockchain.

## 7.5  Mining Procedure Implemented with PoW

We understood mining procedures in different proof-based consensus mechanisms and how miners validate the transaction using these consensus mechanisms. This section explains the practical implementation of mining using the PoW consensus mechanism, which is categorized into three parts, i.e., creation of blockchain, mining using PoW, and how miners get the reward for validating the transaction. Thus, we show the creation of the blockchain. We use the JavaScript programming language that can be deployed in Visual Studio (VS) Code for the implementation purpose.

### 7.5.1 Create Blockchain

Firstly, a chain of blocks has to be created so that users can request to add their data transactions to the network. For that, miners will get involved in the network to validate the transactions using PoW. Some steps need to be followed to create the chain of blocks to form a blockchain network.

```javascript
const SHA256 = require ('crypto-js/sha256');
class Block{
constructor(index, timestamp, data, previousHash = ''){
this.index = index;
this.timestamp = timestamp;
this.data = data;
this.previousHash = previousHash;
this.hash = this.computehash();
}
computehash(){
return SHA256(this.index + this.previousHash +
this.timestamp + JSON.stringify(this.data)).toString();
}
}
class Blockchain{
constructor(){
this.chain = [this.creategenesisblock()];
}
creategenesisblock(){
return new Block(0, "01/01/2022", "Genesis block", "0");
}
getrecentBlock(){
return this.chain[this.chain.length - 1];
}
  addBlock(newBlock){

newBlock.previousHash = this.getrecentBlock().hash;
newBlock.hash = newBlock.computehash();
this.chain.push(newBlock);
}
}
let Coin = new Blockchain();
Coin.addBlock(new Block(1, "10/01/2022", {amount : 4}));
Coin.addBlock(new Block(2, "10/01/2022", {amount : 8}));
console.log(JSON.stringify(Coin, null, 4));
```

- Firstly, create a file 'Blockchain.js' in VS code. Further, the chain of blocks can be built using these information indexes, timestamps, data, and previous hash. We have created a class 'Block' that contains all the necessary information needed to create the block.

- As we already discussed, a hash can be generated using the SHA-256 cryptographic algorithm to implement the bitcoin-based PoW consensus.
- So, we have imported the SHA-256 to compute the transaction's hash using the function 'computehash'.
- Then, we have created a genesis blockchain block using the function 'creategenesisblock' in class 'Blockchain' according to its timestamp.
- After that, the next block has been created in function 'addBlock' using the information about the previous hash of the block and the amount is considered as 4 and 8.
- Finally, the output of a blockchain with a chain of blocks connected with each other is shown below with the information such as timestamp, the hash generated, and a previous hash of the block with the help of the command 'node Blockchain.js' in the terminal of VS Code.

```
OUTPUT:
PS H:\Mine> node Blockchain.js
{
"chain": [
{
"index": 0,
"timestamp": "01/01/2022",
"data": "Genesis block",
"previousHash": "0",
"hash": "e43ca773cc4de3e7cc04271c088648dfb5a8ad668080ed
8f11fcf16ebec6fdb1"
},
{
"index": 1,
"timestamp": "10/01/2022",
"data": {
"amount": 4
},
"previousHash":
"e43ca773cc4de3e7cc04271c088648dfb5a8ad6680
80ed8f11fcf16ebec6fdb1",
"hash":
"44401c87323d8be67e28d46729b865059ede6d96a3f11754ef
c823b4be09365c"
},
```

## 7.5.2 Block Mining Using PoW

After creating the blockchain, now we have to understand how PoW can be implemented to mine the transactions of block in which miners solve the cryptographic
puzzle, which is explained in the following steps:

```
const SHA256 = require ('crypto-js/sha256');
class Block{
constructor(index, timestamp, data, previousHash = ''){
this.index = index;
this.timestamp = timestamp;
this.data = data;
this.previousHash = previousHash;
this.hash = this.computehash();
this.nonce = 0;
}
computehash(){
return SHA256(this.index +
this.previousHash + this.timestamp
+ JSON.stringify(this.data)+ this.nonce).toString();
}
Blockmining(difficulty){
while(this.hash.substring
(0, difficulty) !== Array(difficulty + 1).
join("0")){
this.nonce++;
this.hash = this.computehash();
}
console.log("Block mined: " + this.hash);
}
}
```

- As explained in the creation of the block, class 'Block' is created, which contains
  an index, timestamp, data, previous hash, and nonce that is initialized as 0. A
  nonce is a random number that miners will guess to get the hash value through the
  SHA-256 algorithm.
- Then, we also have to define the difficulty based on the number of miners competing
  to try the different nonce values using the function 'Blockmining'.
- Then, in function 'addBlock', depending on the difficulty, i.e., miners can verify
  the transactions to check if they can be added to the network or not.
- Then, nodes in the network with their previous hash and hash should be in consen-
  sus to further add the block of transactions to the network defined in the function
  'ValidChain'.
- Finally, the mined blocks are shown below as output with the help of the command
  'node first.js'.

```
class Blockchain{
constructor(){
```

```
this.chain = [this.creategenesisblock()];
this.difficulty = 2;
}
creategenesisblock(){
return new Block(0, "01/01/2022", "Genesis block", "0");
}
getrecentBlock(){
return this.chain[this.chain.length - 1];
}
addBlock(newBlock){
newBlock.previousHash = this.getrecentBlock().hash;
newBlock.Blockmining(this.difficulty);
this.chain.push(newBlock);
}
ValidChain(){
for(let i=1; i< this.chain.length; i++){
const recentBlock = this.chain[i];
const previousBlock = this.chain[i-1];
if(recentBlock.hash != recentBlock.computehash()){
return false;
}
if(recentBlock.previousHash != previousBlock.hash){
return false;
}
}
return true;
}
}
let Coin = new Blockchain();
console.log('Mining block 1');
Coin.addBlock(new Block(1, "10/01/2022", {amount : 4}));
console.log('Mining block 2');
Coin.addBlock(new Block(2, "10/01/2022", {amount : 8}));
```

```
OUTPUT:
 PS H:\Mine> node first.js
Mining block 1
Block mined: 007eda6d5f6318803fc5e8ae9f8ac16039f751ad18
95e0eaf6f7c21e4a938825
Mining block 2
Block mined: 00970b563d65b7dcecbd45d3c109905f4248412574
01f9f30fa6fb2db5c388e7
```

## 7.5.3 Rewards for Miners

- In the previous section, miners verify the block of transactions to add them to the blockchain using PoW. Then, a miner guesses the correct nonce value to get the hash less than the target determined by the difficulty, as explained in the previous section.

- Miner who validates the transaction gets the reward. For that, we have created another class, 'class Transaction', which considers the 'fAdd', 'tAdd', and amt to transfer that particular reward from one address to another, i.e., will be transferred to the miner's address.
- Now, the amount transferred to the miner can be determined with function 'get-BalanceOfAddress' and also a balance of the miner can be displayed, but with pending transactions in function 'minePendingTransactions', it will not show the balance of miner.
- Therefore, we have to start the miner again to display the desired balance associated with the mining, as shown below.

```
class Transaction{
constructor(fAdd, tAdd, amt){
this.fAdd =fAdd;
this.tAdd = tAdd;
this.amt= amt;
}
}
class Blockchain{
constructor(){
this.chain = [this.creategenesisblock()];
this.difficulty = 2;
this.PendingTransactions = [];
this.miningReward = 110;
}
creategenesisblock(){
return new Block("01/01/2022", "Genesis block", "0");
}
getrecentBlock(){
return this.chain[this.chain.length - 1];
}
minePendingTransactions(miningRewardAddress)
{
let block = new Block(Date.now(),
this.PendingTransactions);
block.Blockmining(this.difficulty);
console.log('Block successfully mined:');
this.chain.push(block);
this.PendingTransactions = [
new Transaction(null, miningRewardAddress,
this.miningReward)
];
}
createTransaction(transaction){
this.PendingTransactions.push(transaction);
}
getBalanceOfAddress(address){
let balance = 0;
for(const block of this.chain){
```

```
for(const tran of block.transactions){
if(tran.fAdd == address){
balance -= tran.amt;
}
if(tran.tAdd == address){
balance += tran.amt;
}
}
}
return balance;
}
```

```
isChainValid(){
for(let i=1; i< this.chain.length; i++){
const currentBlock = this.chain[i];
const previousBlock = this.chain[i-1];
if(currentBlock.hash != currentBlock.computeHash()){
return false;
}
if(currentBlock.previousHash != previousBlock.hash){
return false;
}
}
return true;
}
}
let Coin = new Blockchain();
Coin.createTransaction(new Transaction
('address1', 'address2', 80));
Coin.createTransaction(new Transaction
('address2', 'address1', 120));
console.log('\n Starting the miner...');
Coin.minePendingTransactions('Riya-address');
console.log('\n Starting the miner once again...');
Coin.minePendingTransactions('Riya-address');
console.log('\nBalance of riya is',
Coin.getBalanceOfAddress('Riya-address'));
```

```
OUTPUT:
 Starting the miner once again...
Block mined: 009465f74c6696ff444ae30711a64b87221fd678243f3
dcc670093ec6850b172
Block successfully mined:

Balance of riya is 110
```

Finally, we have got an understanding of the mining procedure using PoW so that miners receive the reward for validating the transaction. Similarly, we can also implement mining procedures using different consensus mechanisms.

## 7.6 Summary

Blockchain is a developing technology that is gaining attraction in every business. The mining process is at the core of the blockchain. This chapter goes through the entire mining process in depth. This chapter also describe the process of becoming a miner. This is an important step since miners must complete a difficult challenge in order to find the nonce. They are encouraged to engage in the blockchain because of the reward system. The miners can add the transaction to the public ledger. This chapter also explored the permissionless consensus algorithms and their process to verify the new nodes of the network. The information on resources, power, energy, and cryptocurrency provides to execute various consensus algorithms. The requirement of mining pools and their sorts are also explored. Finally, statistics are used to explain some of the difficulties and challenges of mining.

## 7.7 Practice Questions

### 7.7.1 Multiple Choice Questions

- What is the abbreviation of PoB?

  - Proof of Burn.
  - Proof of Blockchain.
  - Proof of Blocks.
  - None of the above.

- What is Merkle tree is known for?

  - Dictionary tree.
  - Hash tree.
  - Leaf node.
  - None of the above.

- The Merkel tree is constructed in an _____

  - top-down approach.
  - bottom-up approach.
  - both a and b.
  - None of the above.

- What is the size of the 'Nonce' in the PoW with mining?

  - 64 bit.
  - 32 bit.
  - 128 bit.
  - 1024 bit.

- The name of the inventor _____ who first proposed the PoB consensus mechanism is

  – Wences Casares.
  – Gavin Andresen.
  – Iain Stewart.
  – Charles Hoskinson.

- The practical Byzantine fault tolerance needs two types of nodes, mark all the suitable options.

  – Leader node.
  – Byzantine node.
  – validator node.
  – follower node.

- What is the 'genesis block' in the blockchain?

  – Fixed hash inserted into the blocks.
  – First block of the blockchain.
  – Second block of the blockchain.
  – None of the above.

### 7.7.2  Fill in the Blanks

1. The innovation of blockchain technology as a decentralized system is mainly based on two aspects, that is, _____ and _____.
2. Blockchain consists of a chain of blocks to store the data transactions securely, which is connected to the _____.
3. _____ had used PoW to transfer the bitcoins between users in the blockchain network.
4. _____ model can be defined as a payment method where the pool manager announces that he pays a flat fee for every share above a certain difficulty that miners are willing to send based on a particular block.
5. Name the payment method _____ that instead paying a level expense for each share, it depends on whether the pool found a valid block or not.

### 7.7.3  Short Questions

1. Explain Peer-to-Peer network in the viewpoint of Blockchain.
2. Define Nonce in the cryptography.
3. Compare Proof of Burn and Proof of Work.
4. Explain the benefits of mining pools.
5. What is Byzantine fault tolerance?

## 7.7.4 Long Questions

1. Enlist and explain the mining pool method.
2. Explain in great detail the working of the voting-based consensus mechanism.
3. Write a source code that shows a mining procedure using PoW with SHA-512.
4. How can rewards be calculated using miners?

## References

1. Kakkar R, Gupta R, Tanwar S, Rodrigues JJPC (2021) Coalition game and blockchain based optimal data pricing scheme for ride sharing beyond 5g. IEEE Syst J 1–10
2. Bodkhe U, Mehta D, Tanwar S, Bhattacharya P, Singh PK, Hong W-C (2020) A survey on decentralized consensus mechanisms for cyber physical systems. IEEE Access 8:54371–54401
3. Gupta R, Tanwar S, Tyagi S, Kumar N, Obaidat MS, Sadoun B (2019) Habits: blockchain-based telesurgery framework for healthcare 4.0. In: 2019 international conference on computer, information and telecommunication systems (CITS), pp 1–5
4. Permissioned blockchain vs. permissionless blockchain: key differences (2022). https://cointelegraph.com/blockchain-for-beginners/permissioned-blockchain-vs-permissionless-blockchain-key-differences. Accessed 27 Jan 2022
5. Shaik VA, Malik P, Singh R, Gehlot A, Tanwar S (2020) Adoption of blockchain technology in various realms: opportunities and challenges. Secur Priv 3:e109
6. What is bitcoin mining: how does it work, proof of work, mining hardware and more (2022). https://www.simplilearn.com/bitcoin-mining-explained-article. Accessed 28 Jan 2022
7. Level up coding- bitcoin proof of work (2022). https://levelup.gitconnected.com/bitcoin-proof-of-work-the-only-article-you-will-ever-have-to-read-4a1fcd76a294. Accessed 28 Jan 2022
8. Forkast: what is proof of stake? (2022). https://forkast.news/what-is-proof-of-stake/. Accessed 28 Jan 2022
9. Ledger academy: what is proof-of-stake? (2022). https://www.ledger.com/academy/blockchain/what-is-proof-of-stake. Accessed 28 Jan 2022
10. Proof of burn- does it work? (2022). https://medium.datadriveninvestor.com/proof-of-burn-9e348725953c. Accessed 28 Jan 2022
11. What is a cryptocurrency mining pool? (2022). https://academy.bit2me.com/en/what-is-cryptocurrency-mining-pool/. Accessed 29 Jan 2022
12. How bitcoin mining pools work (2022). https://river.com/learn/how-bitcoin-mining-pools-work/. Accessed 29 Jan 2022
13. Different bitcoin mining pool payment methods (pps vs fpps vs pplns vs pps+) (2022). https://medium.com/luxor/mining-pool-payment-methods-pps-vs-pplns-ac699f44149f. Accessed 29 Jan 2022
14. Nguyen G-T, Kim K (2018) A survey about consensus algorithms used in blockchain. J Inf Process Syst 14(1):101–128
15. Podgorelec B, Keršič V, Turkanović M (2019) Analysis of fault tolerance in permissioned blockchain networks. In: 2019 XXVII international conference on information, communication and automation technologies (ICAT). IEEE, pp 1–6

# Chapter 8
# Distributed Consensus for Permissioned Blockchain

**Abstract** Consensus protocols are used to manage large databases in a distributed environment without trust among participating nodes in the network. The choice of a specific protocol relies on the characteristics and purpose of the system itself. This chapter aims to identify the practical advantages and disadvantages of consensus protocols. Seven consensus protocols for permissioned environment are covered in this chapter. All the consensus algorithms solve either crash faults or byzantine faults. PAXOS and RAFT, falling under crash fault tolerance, are intended for systems with no apprehension of unreliable users. The notion of byzantine fault was developed from the Three General Byzantine Problems and Lamport Shostak Pease Algorithm. Practical Byzantine Fault Tolerance (DBFT) is intended for systems with fewer nodes. Federated Byzantine Fault Tolerance (FBFT) shows better scalability and is more appropriate for large-scale systems but can resist fewer malicious nodes. Proof of Authority (PoA) and Proof of Elapsed Time (PoET), even though used in both permissionless and permissioned environments, its characteristics make it more suitable for the permissioned environment. PoA can resist the most significant number of malicious nodes without hindering the system's functioning. Various companies have developed applications that are based on consensus algorithms. While working with extensive transactions and a large chain of blocks, scalability, efficiency, and malicious attacks are significant issues. This chapter also present a comparative analysis of all the consensus algorithms based on such issues.

**Keywords** Blockchain · Consensus · Permissioned · Voting-based · Crash Fault · Byzantine Fault

## 8.1 Introduction

The introduction chapter discussed that Blockchain (BC) is of two main types—private and public. The first operates on a closed network and the latter is open to let anyone join with an internet connection. Consensus algorithms [1] are used in both private and public BC to validate transactions. They both store the transactions on a distributed ledger with a synchronized copy available with every participant

© The Author(s), under exclusive license to Springer Nature Singapore Pte Ltd. 2022       211
S. Tanwar, *Blockchain Technology*, Studies in Autonomic, Data-driven and Industrial Computing, https://doi.org/10.1007/978-981-19-1488-1_8

in BC. The only difference between both is private BC, which requires special permission to interact, whereas anyone can freely enter a public BC and interact with it. Thus, public BC is also called '*permissionless*' [2] BC. Chapter 6 already discussed permissionless BC and the consensus algorithms under permissionless BC. There is a third type of BC known as '*permissioned*' [3] or '*consortium*' BC. Many different permutations are possible for permissioned BC as it combines public and private BC. This chapter will discuss permissioned BC, its use cases, its design limitations, its advantages, various consensus algorithms falling under this category, implementation of a few of the algorithms, and at the end, a comparison between permissioned consensus algorithms. In permissioned BC, as the name suggests, participants require permission to participate in the network or the consensus process. This does not mean the private and permissioned BC are the same. Where private BC requires an isolated network to work with, permissioned BC, along with being private, can also be a public network allowing participation based on diverse access levels. For instance, a bank can run a private BC that operates via a defined number of internal bank nodes. In comparison, permissioned BCs can allow anyone to enter a network once they have established their identity and function [4]. Permissioned BC has an additional layer of security, i.e., an access control layer that permits only identifiable participants to conduct transactions. The participants are known to a network operator who allows them to enter a permissioned network [5]. The network operator limits participants' access and assigns different roles to them. For example, the network operator can enable a node or miner participation without providing them access to the whole transaction record or providing any additional functions. As far as technical characteristics are concerned, permissioned BC has the same technology as the underlying BC protocol, excluding the access control layer. They maintain all the positive characteristics of the original BC but differ significantly among each network. For example, suppose the network is employing a Bitcoin [6] implementation, in that case, the new network will be an open-source, PoW-based BC just like Bitcoin, with an access control layer. Based on the needs of the network operators, the level of decentralization, consensus architecture, and governance can be diverse for each implementation.

Some examples of permissioned networks include The Energy Web Chain, Ripple, IBM Food Trust, and Nokia Data Marketplace. All these network belong to different application domains where permissioned BC can be employed. There are many popular permissioned BC frameworks out there. It includes Hyperledger, Quorum, Corda, and others.

## 8.1.1   Provenance Tracking Use Case

Let us look into certain use cases of this permissioned environment. It is helpful for business applications where you want to execute specific contracts among a closed set of participants. One such interesting use case of this can be like this 'provenance tracking of an asset' [7]. When an asset moves from one particular supplier to the distributor to a vendor and goes to the market, certain locks are maintained at every stage. Inside that lock, you make an entry that this particular asset has now moved from location A to location B. The interesting fact is that why do you not want to use a centralized server to have this entire data lock, and then anyone can look into that central lock and verify. Nowadays, whenever you send a specific courier service via, say, blue dart or any courier platform, we usually track when the courier is moving from one particular location to another specific location. The courier agency makes an entry in their central database when the courier reaches location A, verifies it, and sends it to location B. Whenever it moves from one location to another, the entire pass from source to destination through several agents, an entry is made to the central database. Everyone working in blue dart or any courier service platform can look into the lock and find out where the particular asset is moving or how it is moving between source and destination. In this centralized architecture, any user with authenticated login and password can assess the data. This centralized database is only under the control of that specific courier company.

Let us have a typical use case where we send postal mail from India to the USA. Multiple authorities will be involved between the source and destination, with numerous policies to be adapted. Due to numerous authorities trying to access the central database, there will be trust issues. This is a typical use case of a permissioned BC environment. Hence, you have a closed setup of participants participating in the entire BC environment, but the trust relationship is still not there. So, you have to maintain a certain kind of security, or you have to ensure that the data is not getting tampered with while transferring from one authoritative domain to another authoritative domain.

A similar case happens when you transfer certain goods between the suppliers and distributors. The supply chain includes multiple suppliers, distributors, and vendors. Every individual supplier, distributor, or vendor will have the authoritative domain and have their policy of data entering. Regardless, the access to the entire data is with a third-party auditor. This should reliably verify what data is passing through a supplier, distributor, or vendor and the final market, to identify the correctness of data. To ensure this kind of environment in a distributed setting, we want a permissioned model.

## 8.1.2 Design Limitations of Permissioned Environment

Following are the design limitations of a permissioned environment:

1. *Sequential execution*: The transactions verified in a specific order will be executed first, and then later transactions are performed in a sequence. This gives an effective throughput in the system due to certain ordering of transactions.
2. *Non-deterministic execution*: The execution of smart contracts requires a deterministic approach. But generally, smart contracts are written in general-purpose programming language wherein the approach becomes non-deterministic.

3. *Execution on all nodes*: Generally, smart contracts are executed at all nodes and the state is propagated to others to reach consensus. But to do that, there should be a sufficient number of trusted nodes in the system to validate the execution of smart contracts.

### 8.1.3 Challenges Faced by Permissioned Environment

1. *Inconsistent security*: The security in a permissioned environment depends entirely on members' integrity. A malicious member can spoil the integrity of the network, making it compromised. The system should have proper permissions to prevent fraudulent effects by bad actors.
2. *External data storage*: A permissioned BC requires external data storage space depending upon the degree of decentralization used. This puts on-chain data storage at risk.
3. *Control and regulation*: A permissioned BC works like a public BC but with regulations. There is authority control that can restrict or control transactions from taking place. This control, regulation, and censorship threaten using permissioned BC.

### 8.1.4 Advantages of Permissioned Environment

1. *Performance*: The performance of permissioned BC is much better compared to permissionless BC due to pre-determined nodes and thus performing only necessary computations to reach consensus.
2. *Governance*: Permissioned BC are more organized compared to permissionless ones due to governance. Any rule update takes time to get updated on the network by administrators, whereas in the case of permissionless, all nodes had to work together to update the rules on the network.
3. *Decentralization*: Permissioned BC can choose their level of decentralization [8] wherein they can be fully centralized or partly decentralized.

## 8.2 Taxonomy of Consensus Algorithms for Permissioned Environment

Figure 8.1 shows the taxonomy of consensus algorithms for permissioned environment. The taxonomy is divided into proof-based and voting-based algorithms. They are explained as follows:

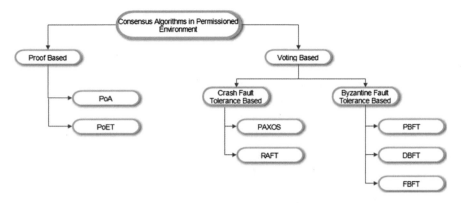

**Fig. 8.1**  Taxonomy of consensus algorithms for permissioned environment

1. *Proof-based consensus algorithms:* In these algorithms [9], nodes can freely join and leave the network. They are more inclined toward permissionless environment. But, Proof of Authority and Proof of Elapsed Time, due to its properties and working are more useful in permissioned environment.
2. *Voting-based consensus algorithms:* In these algorithms [10], nodes should be known in the network and are adjustable. All the participating nodes in the network will have to verify the transactions or the blocks together. These algorithms suffer from crash faults and byzantine faults.

## 8.3  State Machine Replication

In the context of permissioned BC, the concept of state machine replication helps you to achieve a consensus in a permissioned model. This idea is to execute a smart contract (SC) on a subset of nodes instead of every node in the network. The state of the contract is propagated to the neighboring nodes, and the states get further spread in the network. In that way, the same states of the contract are available at every node in the system and can be on the same page as your smart contract. What if you execute on all the nodes in the network and some of the nodes are faulty? In that case, the propagation of the states in the network will stop. Hence, in state machine replication, the contract is executed on a subset of nodes and is ensured that the state of the contract is getting propagated to all the nodes in the network. There are specific consensus mechanisms that will ensure the states in which multiple state machines have propagated or by the contract executor are on the same page and are consistent and correct. Hence, by applying this kind of distributed state machine replication [11] technology, you can ensure consensus in a permission BC environment.

A state machine is defined as a function of six parameters which are as follows:

- A set of states ($S$) based on system design.
- A start state.
- A set of inputs ($I$).
- A set of outputs ($O$).
- A transition function $S \times I \rightarrow S$.
- A output function $S \times I \rightarrow O$.

In the following context, it is assumed that the state machines are deterministic [12], i.e., for a given state, when the same input is given for the same declare, it always generates the same output values in the same declare. At any point, the state machine stores the state of the system. When it receives a set of *inputs* or commands associated with that state, the sequential input generates an declared *output* using a *transition function*. It goes to the *next state* while maintaining the synchronization. An example of a state machine operation is shown below in an algorithm.

```
s = first
log = []
while true:
    on receiving command from a client:
        log.append(command)
        s, o = perform(command, s)
        send o to the client
```

The state machine at an initial state *first*. While receiving a *command* from the client, input is added to the log. By applying the *perform* state transition function, it executes the command. The state transition function identifies whether a transaction is valid or not. If yes, then a next state is obtained and the result $o$ is sent to the client, otherwise not.

All the server replicas at the beginning start from the same state, but when it receives multiple requests from the client simultaneously, all the replicas must collectively decide on the sequence of client commands they are receiving. This problem is called log replication. After the replicas have agreed upon a single sequence, it is then used to transit to the next state. Specifically, to make a fault-tolerant system using state machine replication, it needs to guarantee the following:

1. *Safety*: Two non-faulty replicas should have the same log entry.
2. *Liveness*: Non-faulty replicas will certainly execute a command at some point.

### 8.3.1  Working of Distributed State Machine Replication

In a typical distributed architecture, the distributed state machine replication [13] mechanism works in this way. You have multiple servers, which work in a distributed fashion. The advantage of using distributed servers in place of a centralized server was discussed earlier. The idea is that first, you have multiple servers here. You place the copies of the state machine on each of these servers. Each of these servers has a copy of this entire state machine, then severs get the client request. Figure 8.2

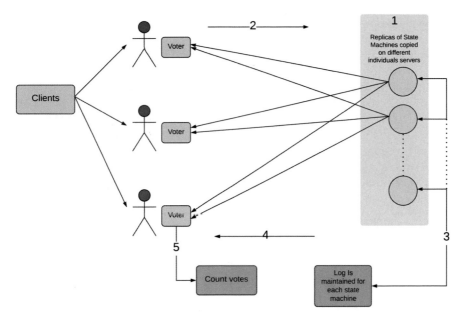

**Fig. 8.2**   Distributed state machine replication

the working of a typical sate machine replication in a distributed network with the following steps:

1. Place copies of the state machine on multiple independent servers.
2. Receive client requests as an input to the state machine.
3. Propagate the input to all the servers.
4. declare the inputs based on some declaring algorithm.
5. Sync all the state machines across the servers to avoid failure.
6. If the output state is produced, inform the clients about the output.

If Bob and Alice submit their request, the request goes to different servers. If you execute the state independently at its server, one server's state will be different from the state in another server. On one side, you will reach a state that Bob has transferred and on another side, you reach the state that Alice has transferred the money. But, collectively, you have to ensure that all the three servers are in a state after a specific time that Bob and Alice have transferred their share of the money. To achieve this, the state machine replication mechanism works in the principle that you propagate these inputs to all the servers. All the servers have the input of Bob and Alice transferring money. Later, you declare the inputs based on a certain declaring algorithm. You can have an associated timestamp with every individual transaction. Based on the timestamp, you can have declaration of the transactions. Once this declaring is done, every individual system executes the inputs based on the declares that have been decided. Once this transaction is done, check the state machines across all the servers to avoid failure. There is a chance of failure of servers during the execution, but they may recover after some time. So, once it recovers, it should get this updated copy, that both Bob and Alice have transferred their share of the money. That update is done through the synchronization of the state machine.

In this entire procedure, there are two glitches; the first one is that you need to maintain declare in service. The second is that you need to ensure that all the individual servers are on the same page in the presence of a failure. To ensure the transactions are successful, you need to have a consensus algorithm in the system. Why do we apply this kind of state machine replication-based consensus algorithm in a permissioned model, in contrast to the challenge response-based consensus model, which we apply for the permissionless settings? So, there are specific natural reasons to use state machine-based consensus algorithms over permissioned BC, which are as follows:

- Firstly, the network is closed. In this case, every node knows each other. Hence, state replication is possible among the nodes. If peers are known, one can constantly replicate the state machine with the current state to one's peers.
- Secondly, the state machine replication-based scenario avoids mining overhead. So, mining has considerable overhead regarding the system power used and the time provided.

But as mentioned, a consensus is still required on top of this state machine replication because the machines can be faulty or behave maliciously. Hence, the later section will discuss the consensus algorithms falling under permissioned BC.

## 8.4    Voting-Based Consensus

There are two broad categories of consensus algorithms: proof-based consensus and voting-based consensus. Proof-based consensus where nodes can freely join and leave the network was under a permissionless environment. In contrast, in voting-based consensus, nodes should be known to the network and are adjustable under a permissioned environment. In voting-based consensus, all the participating nodes in the network will have to verify the transactions or the blocks together. It means that before resolving to append their proposed block to the BC, they communicate about their decision. In this mechanism, at least $T$ nodes should have the same proposed block as you to append the block where $T$ is a given threshold value. Not all the nodes in a consensus will act in a non-fraudulent way. Some nodes will act as malicious nodes and create negative behavior in the network. Thus, a voting-based consensus is divided into two types described as follows:

1. *Crash fault tolerance-based consensus*: A crash fault-tolerant consensus [14] prevents crash faults that occur when a node unexpectedly crashes or the nodes become unreachable in between an ongoing communication. Anyone will not receive any message from the crashed node. This is one type of typical fault that can be a kind of hardware fault or a software fault due to which the node or the process, which is communicating with another one that particular process, fails.

2. *Byzantine fault tolerance-based consensus*: A Byzantine fault-tolerant consensus [15] prevents byzantine faults [16], which is more challenging to handle in a distributed environment where a node starts acting maliciously. Whenever a node starts acting maliciously, you do not know what the node's action would be. The node can send a positive vote or sometimes send a negative vote. In case of the crash fault or a partition fault, you can find out what will be the fault's effect, but the impact of a byzantine fault is challenging to guess because it entirely depends on how maliciously the node is acting and what the node is doing. Sometimes the node can give a vote against a consensus, or sometimes, the node can vote for the consensus. Therefore, handling byzantine nodes becomes difficult in a typical distributed system.

During a crash fault, the crashed nodes are unreachable. Hence, nodes wait for messages from other reachable nodes. To decide to execute a transaction in the network, to prevent a crash with $N$ nodes, at least $2N + 1$ nodes should be operating normally. In the case of Byzantine fault tolerant-based consensus, where nodes start acting maliciously, with $N$ malicious nodes, at least $2N + 1$ nodes should be acting normally.

## 8.5  Crash Fault Tolerance-Based Consensus

A distributed system encounters several threats. Operations may crash, machines at some locations may fail, or a network connection may cease functioning. In any case, the consensus algorithm must have resilience against various threats. Crash fault tolerance constructs an extent of resiliency in the protocol so that the algorithm can precisely take the process ahead and reach a consensus, even if specific components fail. Two consensus algorithms were developed to solve crash faults: PAXOS and RAFT. The following section will describe these algorithms in detail.

### 8.5.1  PAXOS

PAXOS [17] was the first consensus algorithm that Leslie Lamport proposed in 1989, and the objective of PAXOS is to come to a consensus under crash or network fault.

Let us look into the PAXOS in a simplified view and we will discuss how PAXOS works in a real system to ensure consensus. Let us take an example of a group of friends choosing between going for a movie or dinner at a restaurant. The people in the group can collectively decide that all of them want to go to either movie or a restaurant. How will they select if they wish to go to the movie and the restaurant? The interesting part is that they all want to go together; otherwise, the party would not be fun. That is why the friends want to make some decisions collectively. There is no central leader here. The only way to communicate is that everyone will propose

**Fig. 8.3** PAXOS: type of nodes

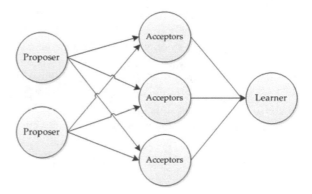

specific values and from that value proposal, they will try to reach a consensus. Every friend will either propose a value or wait for another person to propose a value. If someone else proposes a value, you may accept it or reject it. If you propose a value, you will see how many friends are accepting the value or not. That way, the information get is propagated in the entire network and, they try to have some certain consensus based on a majority decision that well. If the majority of the nodes proposes that or agrees that we want to go to a movie, then all of them will go for a movie, or else if the majority of the friends propose to go for dinner, then all of them go to the restaurant. So, that is the broad idea behind PAXOS.

Now, let us look at the working of the PAXOS algorithm. PAXOS consists of three types of nodes as shown in Fig. 8.3 and described as follows:

1. *Proposer Node*: The proposer node proposes the values that the consensus algorithm should choose.
2. *Acceptor Node*: Acceptor nodes form the consensus and accept the values. If the acceptor nodes are reviewing proposal number 1 and another proposal number 2 comes, reject proposal number 1.
3. *Learner Node*: Learner nodes learn which value was chosen by each acceptor and collectively accept that particular value. Every node is a learner in the network.

**Working of PAXOS algorithm** PAXOS algorithm as shown in Fig. 8.4 is elaborated in detail in following steps:

1. *Making a Proposal—Proposer Process*: A proposal number forms a timeline, the biggest number considered up-to-date. For example, if two proposals are coming from; P1—100 and P4—102, we will accept the proposal coming from P4, as it has the biggest proposal number.
2. *Making a Proposal—Acceptor's Decision Making*: The role of each acceptor is to compare the received proposal number with the present known values for all proposers' prepare message. If the proposal number is less than the $acceptor_i$ proposal number, you accept it; otherwise, reject it.
3. *Making a Proposal—Acceptor's Response Message*: The acceptor's response will be based on the following points:

- *Accept/decline*: Whether prepare accepted or not.
- *Proposal number*: The biggest number the acceptor has seen.
- *Accepted values*: Already accepted values from other proposer and this accepted value is informed to the proposer.

4. *Accepting a Value—Proposer's Decision Making*: The proposer node receives a response from a majority of the acceptor nodes before proceeding. Whenever the majority of the acceptors send some accepted values, if they have accepted your value, the value you have shared is coming to be the consensus.
5. *Accepting a Value—Accept Message*: Proposer sends the accept message to all acceptors, which includes the following:

- *Proposal number*: Same as prepare phase value.
- *Value*: A single value proposed by the proposer.

6. *Accepting a Value—Notifying Learner*: Now, whenever the acceptor accepts values from the proposer, it informs the learner about this majority voted value. So, everyone in the network learns what the majority voting in the environment is.

**Handling Failure: Acceptor Failure** If you have a single proposal in the system, then the system is straightforward. With a single proposal, every acceptor will accept the proposal because that will be the biggest, so the proposal does not get rejected if the acceptors are correct. But what if the acceptors fail or crash? Following cases are possible when acceptors fail:

- *Acceptor failing during prepare phase*: If the acceptor fails during the prepare phase, there is no issue as there are other acceptors who can hear the proposal and vote either for the proposal or against the proposal.
- *Acceptor failing during accept phase*: If the acceptor fails during the accept stage, there is no issue because other acceptors have already voted for the proposal. The only thing that one has to ensure is that up to $N/2$ number of acceptors can fail simultaneously because you are going for the majority voting principle. Whenever you are going for the majority voting principle in a synchronous environment, you need to ensure that will majority of the nodes are correct. If more than $N/2$ number of acceptors fail, then no proposer gets a reply and you cannot come to a consensus algorithm.

**Handling Failure: Proposer Failure** Let us look into the scenario, what happens when a proposer fails:

- *Proposer failing during prepare phase*: The proposer can fail during the prepare phase or during the say accept phase. So, if the proposer fails during the prepare phase, it is like no one is proposing any value. The acceptor will wait for some specific time and, if they do not hear any proposal, one of the acceptors becomes the proposer and propose a new value.
- *Proposer failing during accept phase*: If the proposer fails during the accept phase, they have already agreed upon whether to choose or not to choose the proposal

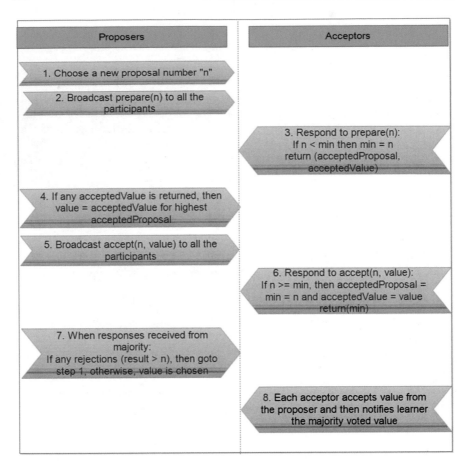

**Fig. 8.4** PAXOS consensus algorithm

based on the majority port. So, they have shared the majority port among themselves and they can find out whether the proposal has been accepted or not.

**Handling Failure: Dueling Proposers Attack** There is an interesting attack called a dueling proposers [18] attack as shown in Fig. 8.5 . Suppose proposer received confirmations to prepare message from a majority, but yet to send accept messages. Some other proposer sends a prepare message with a higher proposal number. To break the tie between the two proposers, block the first proposer's proposal from being accepted and use the 'leader election algorithm' to handle this attack to select one of the proposers as a leader. PAXOS can be used for leader election.

Now, this simple view of the PAXOS is easier to understand. It is just like making one selection, but if you look into the real system, you may go for multiple sequences of choices rather than having a single choice. That kind of system is called a multi PAXOS [19] system. Multi PAXOS system makes a sequence of choices by

**Fig. 8.5** Two nodes sending prepare message simultaneously

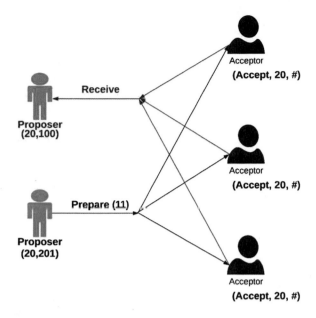

applying repeated PAXOS protocol. Using repeated PAXOS is complicated because you have all these messages that need to be exchanged repeatedly. This increases the complexity of the PAXOS protocol. That is why all the PAXOS gives an excellent theoretical idea about why the system can reach a consensus or how are distributed system can lead to a consensus. Hence, in the next section, we will discuss a more simplified consensus algorithm, which is widely adopted as an alternative to PAXOS called a RAFT consensus mechanism.

## 8.5.2   RAFT

RAFT [20] was coined by Ongaro & Ousterhout in 2014. RAFT, as shown in Fig. 8.6, is an algorithm of consensus designed to be easy to understand and is resilient to crash faults. In terms of fault tolerance and efficiency, it is similar to PAXOS. Whenever the system starts up in the RAFT algorithm, it has a set of follower nodes. These follower nodes check whether there is already a leader or not. If this validation times out, it implies there is no leader in the system you start the election. In the election, you choose some of the candidates. The participant who wants to become a leader will send a message on the network. The sender participant will receive votes from the majority of the network if agreed on as leader. Among the candidates who win the majority of the votes, that person becomes the leader, and thus the leader election is done. So, the proposed value is figured by this particular leader. Later, the followers either vote for the proposal arriving from the leader or

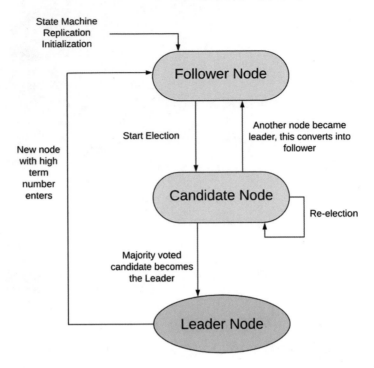

**Fig. 8.6** RAFT consensus algorithm

deviate against the leader. If no consensus is reached, the condition is called a split vote, and the term ends without a representative. RAFT consensus algorithm uses the concept of the replicated database. There are multiple replicated servers and you want to build up a consensus among these multiple replicated servers. So, whenever some transactions are coming up from the clients, then replicated servers are needed. They collectively take some decisions about this consensus and based on that, they decide whether to commit those transactions or not. Let us move to how the RAFT algorithm works.

**1. Electing the Leader—Voting Request:** The initial part of RAFT is to elect a leader. The question comes that how you will elect the leader. To elect a leader, you need to be certain leader candidates. Among the candidates, someone can wait for a certain amount of time to see whether some leader is already there. If this wait times out, it implies there is no leader yet. Now, one can decide to be a leader or not. If one chooses to be a leader, one can announce their candidacy. Once these leader candidates are there, these leader candidates request the votes. The votes contain two parameters: one is called the *term*, and the second one is called the *index*. RAFT algorithm runs in multiple rounds like in PAXOS. In every round, you need to take one decision and the term parameter denotes which particular round you are in. The term is calculated as the last calculated number known to the candidate. In the current round, say round 20, the term value will equal 20. You want to have a new vote to elect a new leader. Hence, the term value is now 21. The second parameter is the index parameter. The index parameters represent the committed transaction available to the candidate. The index parameters validate up to $x$ transactions committed until now. Now, this request vote message is passed to all the nodes in the network.

**2. Electing the Leader—Follower Node's Decision-Making:** Now, the nodes in the network receive the request vote message. Once the nodes receive a message, their task is to elect a leader. It is done by comparing the term and index value. When node $j$ received the term value from one of the leader candidates, it looks into its term value. If the received term is more than the node $j$ term value, then the corresponding index value is checked. We vote for the candidate if the index value is more than the node $j$ index. If the received term value or index value is less than node $j$ values, then the vote is declined. In that case, because node $j$ has the highest term value or index value, it will send the request vote message declaring it is a leader candidate.

**3. Electing the Leader—Majority Voting:** Now, the leader is getting a vote. Every node sends its vote and the concept of majority voting is used for leader election. Later, commit the corresponding log entry. It is certain that if a certain leader candidate receives a majority of the vote from the nodes, then that particular candidate becomes the leader and the other becomes the follower of that node.

**4. Multiple leader candidates: Current leader failure** Let us look into an example of the presence of multiple leader candidates and the current leader fails. There is a leader with three followers, the current term is 10, and a committed index is 100. Now, say the leader node has failed. Once the leader node has failed, a new leader with term value 11 will be elected. That means you need to move from round 10 to round 11 because your leader is going to change. At this particular time, if the old leader recovers, the old leader will receive a heartbeat message from the new leader with greater term value. Thus, the old leader drops to the follower state.

**5. Multiple leader candidates: Simultaneous request vote** If two nodes send request vote messages with the same term simultaneously, it implies there are two leader candidates. In that case, majority voting is the only option. One of them will get majority voting and will be elected as a leader. The elected leader sends a heartbeat signal with an updated term value. Another leader candidate automatically drops it to the follower state.

**6. Log replication** For now, it is presumed that the client requests are written-only. Each request consists of a command preferably executable by all the servers 'replicated state machines. When a leader receives a request from a client, it adds it as a new entry to its own log. Each log entry contains the following:

- Contains commands defined by the client
- An index to identify the log entry location.
- A term number to logically define when the entry was written.

To keep logs consistent, the entry must be repeated to all follower nodes. The append entry is parallelly issued to all other servers by the leader. The leader repeats this until all followers reproduce the new entry securely. When the leader replicates the entry to a majority of servers, it is committed, including all the previous entries.

**7. Handling failures—Follower failed** It may happen that a follower has crashed. So, the system can tolerate up to the failures of $N/2 - 1$ number of nodes. It does not affect the system because we rely on majority voting. In that case, the majority of the followers are non-faulty. They can send a vote and the leader can take the majority decision whether to accept or reject a particular transaction or not now.

The key difference between PAXOS and RAFT algorithms is that RAFT allows leader selection which is best-updated nodes, while any node can be a leader with the PAXOS algorithm. The well-known BC platforms using the RAFT consensus algorithm are R3 Corda [21] and Quorum [22]. PAXOS and RAFT both can tolerate up to a number of crash faults. But what if the nodes start behaving maliciously? How is the system going to reach a consensus? This scenario will be explained in the further section, where we will discuss Byzantine Generals Problem.

## 8.6  Notion of Byzantine Fault

We have already discussed PAXOS and RAFT consensus protocols. Both protocols efficiently handle crash faults. But there might be a case where a node or a group of nodes behave maliciously and this kind of behavior is called byzantine behavior. The PAXOS and RAFT consensus algorithms can't handle this kind of scenario. For the distributed system, we need different types of fault tolerance protocols known as byzantine fault tolerance protocols. We will first discuss two problems from where the byzantine fault concept which are Three Byzantine Generals Problem [23] and Lamport Shostak Pease Algorithm [24].

### 8.6.1  Three Byzantine Generals Problem

If there are two lieutenants and one commander, the system becomes faulty if the commander sends out two different messages, namely 'attack' to one soldier and 'retreat' to another soldier. It is difficult for the system to find out what to do. We will discuss two separate scenarios where the lieutenant or commander is faulty.

**Lieutenant Faulty:** As shown in Fig. 8.7, in round 1, the commander correctly sends the same message to lieutenants. In round 2, lieutenant 1 (L1) correctly echoes to lieutenant two (L2), but L2 incorrectly resonates to L1. We assume L2 is faulty. Here, the commander is correct, but L2 is faulty. We will now frame out the consensus. L1 received a differing message. In a typical military scenario, the lieutenant must obey the commander's declare. If the commander is not faulty, then the entire system works perfectly, and by integrity condition, L1 is bound to decide on the commander's message.

**Fig. 8.7**  Message passing
when Lieutenant is faulty

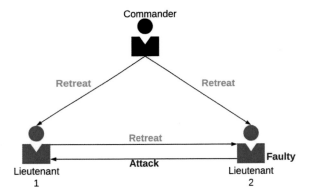

**Fig. 8.8**  Message passing
when commander is faulty

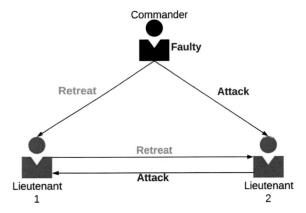

**Commander Faulty:** As shown in Fig. 8.8, in round 1, the commander sends differing messages to lieutenants, and in round 2, both the lieutenants correctly echo the
message to each other. We assume that both the lieutenants are correct and the commander is faulty. By message passing, the entire system won't work. L1 received
the differing message because the commander was faulty. By the integrity condition, both lieutenants conclude with the commander's message. This contradicts the
agreement condition. No solution is possible for three generals, including one faulty
general. Suppose we don't have any way to finalize whether to go for majority voting or follow the commander instructions. Then, the entire system is in a dilemma
that which particular instructions need to be followed. We can solve this byzantine
problem with the principle of majority voting. In a byzantine system, if the leader
(commander) sends different messages to different peers (lieutenants), coming to a
consensus on this principle is very difficult. But the Three Generals problem with
one commander and two lieutenants is still unsolvable through the voting principle.

In a general principle with the normal byzantine general problem with f number
of faulty nodes. If a lieutenant is faulty, we need to ensure that $2f + 1$ number of
correct lieutenants should be present in the system. If e have $2f + 1$ lieutenants and

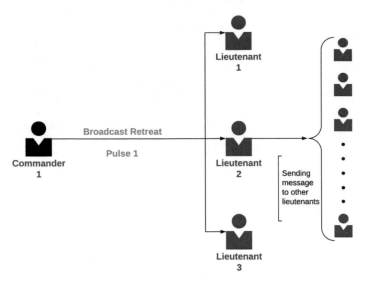

**Fig. 8.9** Commander sends a message to all lieutenants and all lieutenants send message to other lieutenants

1 commander, then we will be again able to correctly apply this majority voting principle to find out the byzantine nodes in the system.

### 8.6.2   Lamport Shostak Pease Algorithm

As shown in Fig. 8.9, the commander broadcasts a message to all the lieutenants in the network. The base condition for the commander includes two parameters, i.e., Broadcast (N, t = 0), where N is the number of processes and t is the algorithm parameter which denotes the individual rounds. If t = 0 means you are in pulse 0 when the commanders send the message to all lieutenants. Lieutenants receive the message from the commander, so we have to check whether it is coming from the commander (pulse-1(first message in the system)). If not, then don't take any decision because the source is not genuine. If the source is genuine, then accept the commander's message. The commander sends the message to all lieutenants. Now broadcast the received message to all other processes in the system. As the system progresses in rounds, at every individual round, say commander T sends the message to all in the $T_{th}$ round. Every lieutenant broadcasts its value to the other lieutenants except the sender. Similarly, every lieutenant broadcasts whatever message they have received from the commander to all other lieutenants except the sender of that particular message. If you have N numbers of lieutenants in the system, then after $N_{th}$ round, you will get messages from all the individual lieutenants. We can apply the majority principle to take a decision (whether to follow the commander or the majority of lieutenants).

In this case, if we have $f$ lieutenants and one commander as faulty, we can achieve consensus in a synchronous system. If $f$ number of lieutenants is faulty, you will reach a consensus with $2f + 1$ number of lieutenants. Otherwise, if the commander is faulty, but all the lieutenants are correct, then also you will be able to reach a consensus under Lamport's algorithm.

This is a consensus algorithm for a synchronous environment [25], but as we understand, our practical networking systems environments are asynchronous [26], which does not behave like a synchronous environment. And in that asynchronous network, it may happen that you will not be able to receive a message within a predefined timeout. Hence, in the next section, we will look into another class of consensus problems for an asynchronous network. We call this algorithm the practical byzantine fault-tolerant algorithm widely used in a BC environment. We will look at the variants of a practical byzantine fault-tolerant algorithm.

## 8.7    Byzantine Fault Tolerance-Based Consensus

A Byzantine fault-tolerant consensus prevents byzantine faults, which are more challenging to handle in a distributed environment where a node starts acting maliciously. Whenever a node starts acting maliciously, you do not know what that node's action would be. The node can send a positive vote or sometimes send a negative vote. In case of the crash fault or a partition fault, you can find out what will be the fault's effect, but the impact of a byzantine fault is challenging to guess because it entirely depends on how maliciously the node is acting and what the node is doing. Sometimes the node can give a vote against a consensus, or sometimes, the node can vote for the consensus. Therefore, handling byzantine nodes becomes difficult in a typical distributed system. Subsequent sections will describe consensus algorithms for handling byzantine fault in a distributed environment.

### 8.7.1    Practical Byzantine Fault Tolerance (PBFT)

Since Lamport raised Byzantine General's Problem in 1982, many discussions on Byzantine Fault Tolerance (BFT) [27] solutions were accomplished, but the solutions were inefficient, complex, and slow. In 1999, Castro and Liskov presented Practical Byzantine Fault Tolerance (PBFT) [28] which improved the situation. This algorithm is termed 'practical' because it ensures safety over an asynchronous network but not liveness on a pure asynchronous network; otherwise, it will inviolate the impossibility theorem or the impossibility principle. To ensure liveness [29], we assume a weak asynchronous where we deviate from a pure asynchronous system. The PBFT algorithm is applied for many real applications. It is widely used for the permissioned BC model like the standard meant IBM's OpenChain, Hyperledger Fabric [30], Tendermint [31], Diem, etc. To ensure consensus among the participants in a closed environment, we apply the PBFT algorithm. We will go through the details of this PBFT algorithm.

We consider a state machine replication concept in the system model that we discussed earlier in Section 2 of this chapter. We have a state machine that is replicated across different nodes, and we have $3f + 1$ replica, where $f$ is the number of faulty nodes. In a synchronous environment, we discussed that if you have a $f$ number of faulty nodes, you require $2f + 1$ number of nodes apart from the commander to ensure consensus. But in the case of an asynchronous network, we need $3f + 1$ replicas. These replicas move through a successive configuration known as 'views'. Like in Byzantine General Model, we had a few lieutenants and a commander. Similarly, in these replicas, we have one primary and multiple other backups. That way, we consider this setup of primary and a backup as a single view. These views are changed when a primary is detected as faulty. Whenever the backups collectively detect that a primary is faulty, they will look into this view change mechanism. These views are changed and every view is identified by a unique integer number $v$, using which only the messages from the current views are accepted. If there is a view change, then the messages which have been broadcasted in the previous view are discarded. Because of the asynchronous nature of the system, it may always happen that you are receiving a message which is delayed or out of declare; such messages are also discarded. Only the messages from the current views are accepted.

In BC, a transaction is valid if all the backups or replicas decide that the current transaction is correct. It sends a success or a commit message to the client; otherwise, it will send a failure message. The consensus process adopts the three-phase protocol [29]. Figure 8.10 shows that five phases are experienced, from the client launching requests to receiving responses. The following content briefly describes the five phases:

1. *Initiate*: The client $c$ sends a request to invoke a service operation $rec$ to the primary.
2. *Pre-prepare*: After verifying the validity of the request message $rec$, the leader node, i.e., replica 0, assigns a sequence number $n$ to the request $rec$ in the view and broadcasts the message $rec$ to all the backup nodes. The message looks like $<< PRE - PREPARE, v, n, d >_{\sigma_p}, rec >$ where $v$ is the current view number, $n$ is the message sequence number, $d$ is the message digest, $\sigma_p$ is the private key of primary, and $rec$ is the message to transmit. $\sigma_p$ works like a digital signature [32]. Let us assume that one node is faulty. We have a $3f + 1$ number of different replicas. We are considering here $f$ equal to 1, so we have 1 faulty replica. We have a total of 4 different replicas, 1 primary and 3 backups. So, in the pre-prepare phase, the client assigns a unique sequence number to the request message and broadcasts that request message to all the replicas.
3. *Prepare*: In this phase, the secondary, i.e., replicas 1, 2, and 3, sends a prepare message to all the replicas. If the backup accepts the PRE-PREPARE message, it enters prepare phase by multicasting [33] a message. A replica accepts the

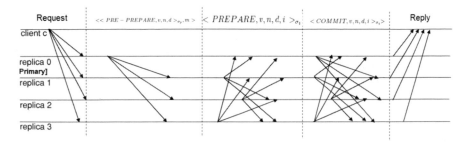

**Fig. 8.10**  PBFT mechanism

PREPARE message if signatures are correct, view number is equal to the current view, and the sequence number is within a threshold. The message looks like $< PREPARE, v, n, d, i >_{\sigma_i}$ to all other replicas where $i$ is the backup number from where multicast is originating and $\sigma_i$ encrypts with the private key of backup $i$ acting like a digital signature. A node commits a message transferred in the pre-prepare message if you have received a prepare message with the same message and the same message digest and comes from at least $2f$ number of prepare messages. If a total of $2f + 1$ votes are received where one vote is from the primary and the rest $2f$ number of votes from other replicas, commit takes place.

4. *Commit*: All the nodes, i.e., primary and backup (secondary), multicast a commit message to all other nodes once the assignment agreement message is received from the cluster. All the replicas have agreed on the declare and confirmed the received request in this phase. The commit message looks like $< COMMIT, v, n, d, i >_{\sigma_i}$. Commit a message when a replica has sent a commit message its and has received $2f + 1$ number of such commit messages, including its own.

5. *Reply*: After receiving all the commits from the network, all the nodes directly return messages to the client.

A question that arises here is why do you require $3f + 1$ replicas to ensure safety in an asynchronous system where there are $f$ faulty nodes? The answer is if you have $2f + 1$ replicas, you need all the votes to decide the majority boils down to a synchronous system. You may not receive votes from certain replicas due to delay in the case of an asynchronous system. $f + 1$ votes do not ensure a majority. Maybe you have received $f$ votes from Byzantine nodes and just one vote from a non-faulty node. Byzantine nodes can vote for an against, which is not known as apriori. If you did not receive a vote, either the node is faulty and not forwarded the vote at all, or the node is non-faulty, forwarded a vote, but the vote got delayed. A majority can be decided once $2f + 1$ votes have arrived, even if $f$ are faulty, and you know $f + 1$ are correct nodes, so we do not care about the remaining $f$ nodes.

**Handling failures in PBFT with View Change:** We can tolerate up to $f$ number of such failures. Within $f$ number of failures, if $f$ number of secondary's or the backups are faulty nodes, the system can tolerate that, but what if the primary is faulty? The non-faulty nodes detect the fault. The backups detect the fault collectively and together start a view change [34] operation. They remove the primary from the system or consider that the replica designated as the primary in view $v$ will get changed. Another replica from the backup will be designated as the primary. The view change protocol ensures 'liveness', but we must receive the message from all the nodes within a timeout duration to confirm this view change. To ensure that the view change is done at a proper time interval, you have to ensure that all the backups are able to detect that the primary is not sending any message within some timeout duration. Here we put a timeout in the system. When you set the timeout in a system, you assume that if you are not receiving the message within this duration, you consider that message to be lost and the primary to be faulty. The backups start a timer when it receives a request. The timer stops if a request is executed. Everything is complete within that timeout duration, and the timer gets restarted when some new request comes. If the timer expires at view $v$, then the backup starts a view change to move the system to view $v + 1$. On timer expiry, a backup stops accepting any new messages; except the standard checkpointing message, the view change message, and a new view message.

When a primary has a timeout, it multicasts a view change message to all the replicas. It initiates a view change for a new view $v + 1$. The view change message looks like $< VIEW\_CHANGE, v + 1, n, C, P, i >_{\sigma_i}$ where $n$ is the sequence number of the last stable checkpoint $s$ know to $i$. $C$ is a set of $2f + 1$ valid checkpoint messages proving the correctness of $s$. $P$ is a set containing a set $P_{rec}$ for each request $rec$ that prepared at $i$ with a sequence number higher than $n$. Each set $P_{rec}$ contains a valid pre-prepare message and $2f$ matching. The new view is initiated after receiving $2f$ view change messages. Hence, the view change operation takes care of the synchronization of checkpoints across the replicas and all the replicas are ready to start at the new view $v + 1$.

**Correctness of PBFT Algorithm** Correctness [35] in PBFT algorithm is observed by the following parameters:

- *Safety*: If all non-faulty backups decide on the sequence numbers of requests that commit locally, then the algorithm provides safety.
- *Liveness*: Replicas must shift to a new view if they cannot perform a request. A backup waits for $2f + 1$ view change messages and then begins a timer to start a new view. If a replica acquires a set of $f + 1$ valid view change messages for views greater than its current view, it sends a view change message. Faulty replicas are unable to hinder progress by compelling regular view change.

## 8.7.2   Delegated Byzantine Fault Tolerance (DBFT)

Delegated Byzantine Fault Tolerance (DBFT) [36] is a modification of the PBFT consensus algorithm. It was coined during the design of the NEO blockchain [37], which is the first public BC project in China. NEO is also known as the 'Ethereum

of China'. Along with being a cryptocurrency, it also delivers a code for constructing smart contracts. DBFT facilitates large-scale participation by performing proxy
voting in the consensus. By voting, the NEO token holder can select a bookkeeper
that it supports. Employing the BFT algorithm, the chosen bookkeepers reach a consensus and yield new blocks in the BC. DBFT works similarly to Delegated Proof of
Stake (DPoS) [38]. First of all, NEO token holders need to vote for finding delegates
who are their proxies. For becoming a delegate, the node needs to fulfill specific
requirements such as a reliable and stable internet connection, a validated identity,
the proper equipment, and 1000 GAS [39]. A delegate gets rewarded in terms of
GAS. A new block is created by a random speaker selected from the chosen delegates. A consensus can be reached only if two-thirds of the chosen delegates validate
the new block for adding the validated block to the existing BC. Otherwise, it leads
to rejection of the proposal, the election of a new random speaker, and the process
continues. The chosen delegates will validate the new block. The chosen delegates
can track and record all the transactions on the network and share and compare the
proposals. DBFT promises total finality in the BC.

The question is, how will you find if the random speaker or any of the delegates
is a corrupted one? The answer lies in comparing blocks. When a block is received
from a random speaker for validation, at least 2/3rd of the delegates should approve
it. If the delegates compare their own versions of the block proposal, they can decide
its validity. If the block is corrupted, 2/3rd of the delegates will agree to the block's
invalidation and the replacement of the random speaker. If any delegate is corrupted,
then 2/3rd of delegates validating a block will not be possible. Hence, in any case,
corrupted nodes will be handled. The advantage of DBFT is the fast creation of a
new block and high throughput. It can support commercial applications due to the
presence of large-scale participation. DBFT is criticized as nodes have to reveal their
true identity for the voting process of delegates. Also, DBFT supports increased
centralization. BC cannot ensure privacy due to lack of anonymity and the demand
for centralization.

### 8.7.3  Federated Byzantine Fault Tolerance (FBFT)

A Federated Byzantine Fault Tolerance (FBFT) or Federated Byzantine Agreement
(FBA) [40] is also a modification of the PBFT consensus algorithm. FBFT is known
for improved throughput, high network scalability, and lower transaction costs. To
overcome the issue of extensive communication, the FBFT indicates that nodes
should exchange messages and communicate with only the nodes they trust. We
will discuss two variants of FBFT, i.e., Stellar [41] and Ripple [42].

Stellar uses a group of nodes called 'quorum slice' [43] for intercommunicating
between verifying nodes. This variant of FBFT requires each node to decide which
other nodes they trust and the group of trusted nodes create a quorum slice. The
quorum slices of each verifier node connect the whole network to create a quorum
for the entire network to reach a consensus on a transaction. Each quorum slice is a

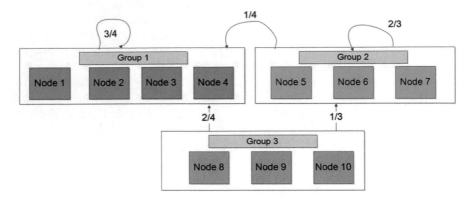

**Fig. 8.11** Quorum slices in Stellar

subset of a quorum helping a node reach consensus. Each verifier node can belong to one or many quorum slices. A node belonging to multiple quorum slices represents quorum intersection [44]. When a node wants to validate a transaction, the other nodes in the quorum slice need to agree on the validating transaction. The transaction is valid if all the nodes in the quorum slice successfully verify the transaction. Quorum slices are organized in a hierarchical structure. The top tier in this hierarchy will get the first chance to verify the transactions.

Let us take an example and understand the working of quorum slices and validation of transactions in BC. Consider the example shown in Fig. 8.11. Consider 'Node 1' creates quorum slices with some nodes of 'Group 1' and some nodes of 'Group 2'. The two quorum slices created are nodes 1, 2, 3, and 5 and nodes 1, 2, 4, and 6. Group 1 and 2 belong to the top, and Group 3 belongs to the lower levels in this hierarchical structure. Group 1 conducts authentication services, Group 2 conducts banking services, and Group 3 conducts payment services. Suppose Node 10 from Group 3 wants to verify a transaction and requires services from the other two groups, then he has to convince nodes such as node 1, node 3, and node 6 to agree to the transaction, and once agreed, the transaction is considered valid. Without needing a central authority to decide the list of nodes in a quorum slice, the nodes can spin the slices and participate in the consensus by adding new nodes they trust and require services from.

As a quorum in Stellar, Ripple uses Unique Node List (UNL) [45] for transaction validation. Every node has its own list called UNL, which contains nodes that it trusts. To verify a transaction, at least 80% nodes should agree to the same. That means at least $1/5$ part of all the nodes in a UNL should be non-malicious for a transaction to be validated. To ensure correctness in the system, only $(n-1)/5$ nodes can be malicious, where $n$ is the total nodes within a UNL. Even if the malicious nodes increase, the system ensures that if malicious nodes are less than or equal to $(4n+1)/5$, the system shows weak correctness, which means it will not be able to validate the transactions. Still, at least it will node allow introducing malicious transactions. The advantage of FBFT is that it will enable forking, unlike the previous algorithms of FBFT, which would allow finality in the system. Depending on the quorum size, forking is allowed using FBFT if quorums partially overlap and there is no lower limit on quorum size.

## 8.8 Proof-Based Consensus Algorithm

We discussed permissionless BC and the consensus algorithms used in permissionless BC in Chapter 6. Although proof-based consensus algorithms are part of permissionless BC, three consensus algorithms, i.e., Proof of Authority (PoA) and Proof of Elapsed Time (PoET), are discussed under the category of permissioned BC. These algorithms can be used under both permissionless and permissioned environments but are more applicable to permissioned BC. Hence, we will be discussing the same in the next section under the permissioned category.

### 8.8.1 Proof of Authority (PoA)

Proof of authority (PoA) [46] is a consensus protocol based on identifying nodes and their reputation. It was created in 2017 based on the Proof of Stake (PoS) [47], with the requirement that instead of stake with monetary value, i.e., coins, a node's identity is at stake. Identity means the resemblance between a node's personal identification on the network with officially issued documentation for the same person, i.e., assurance that a node is precisely who that individual depicts to be. Nodes that append blocks in the BC are called validators. To become a validator, the node's identity must not only be publicly known. In PoA, in case of disfavored behavior, the validator earns a negative reputation and loses the prospect of participation in the future block creation process. Nodes with a good reputation can take up the role of validating the block on a round-robin basis.

PoA provides high performance, high transaction rate, and fault tolerance. The transaction rate is high as blocks are generated in sequence at regular time intervals by only authorized nodes and the speed at which transactions are validated also increases. PoA does not require any high computing hardware like in Proof of Work (PoW). PoA tolerates malicious nodes as long as 51% nodes are not compromised. PoA can also withstand denial of service (DoS) attacks since the nodes are verified and block creation occurs at regular intervals. Since it does not confirm maintaining the user's anonymity, it is not helpful for public BC and is useful for private BC only. Also, PoA is more inclined toward centralization rather than decentralization as it promises high throughput and scalability. PoA is used in Aura, Clique, PoA Network, and Vechain platforms.

## 8.8.2   Proof of Elapsed Time (PoET)

Proof of Elapsed Time (PoET) [48] was invented in 2016 by Intel. It is used to solve the 'Random Leader Election' problem and is more helpful in permissioned BC. PoET reaches consensus by implying a random timer at every node in the network. The random timer decides which node gets to validate the block on BC and gets rewarded. PoET focuses on efficiency and ensures there is an equal probability of every node to become the miner and validate the block. The working of PoET is divided into the following two stages:

1. *Verification and Joining Phase*: PoET uses a sophisticated technology called Software Guard Extension (SGX) [49] which is a security-related instruction code to be executed on a central processing unit that isolates certain trusted regions of data and code. SGX enables a trusted execution environment (TEE) [50] that allows only trusted code and data to be executed on the system applications. A node wanting to join a permissioned network needs to be trusted. The node needs first to download the PoET code and run on SGX. SGX renders an attestation and a new private/public key pair. The node signs the attestation and forwards this signed attestation (with the public key generated) with a request message to join the network. The already verified nodes in the network verify the attestation. The network can either accept or reject the attestation based on the verification. If accepted, the node is allowed to be part of the BC network.
2. *Mining lottery based on random timer*: For the mining process to start, a node needs to be elected for validating a block. Each trusted node receives a timer object from the trusted code using SGX. The timer is entirely randomized and prevents malicious users from fooling the network by continuously acquiring the shortest timer. All the nodes go into a waiting state when their timer starts and the node whose timer expires first gets a chance to validate the block and get the reward. This leader election is broadcasted to the entire network to announce the winner node waited for a specific amount of time before the mining process started. Once the transactions are validated and the block is mined, the random timer starts again to choose the next node to participate in the mining process.

Applications like Hyperledger Sawtooth [51] developed by Intel Corporation work on the concept of PoET. The randomized timer used in PoET makes it more efficient and eliminates the need for any compute-intensive mining process. The randomized timer enables an equal probability of all the trusted nodes to get a chance to mine the block and claim the reward. Also, PoET does not suffer from any scalability issues. PoET has a dependency on TEE for establishing the trust to take part in the network. Also, it is susceptible to a Sybil attack where the attacker creates a lot of fake identities and tries to influence the network.

## 8.9 Implementation

**Three Byzantine Generals Problem** We have shown the implementation of Leslie Lamport's algorithm for The Byzantine Generals Problem. The algorithm works in multiple recursions that you can specify when running it. When the recursion is 0, in Lamport's algorithm, the commander sends his proposal to every lieutenant. After receiving the proposal from the commander of either 'ATTACK' or 'RETREAT', each lieutenant employs the proposal or uses the proposal 'RETREAT' by default if they did not receive any proposal from the commander.

```
class Commander:
    def __init__(s, id, is_faulty=False):
        s.id = id
        s.other_commanders = []
        s.declares = []
        s.is_faulty = is_faulty

def __call__(s, rec, declare):
    s.byz_algo(leader=s, rec=rec, declare=declare,)

def _next_declare(s, is_faulty, declare, i):
    if is_faulty:
        if i % 2 == 0:
            return "ATTACK" if declare == "RETREAT" else "RETREAT"
    return declare

def byz_algo(s, leader, rec, declare):
    if rec < 0:
        s.declares.append(declare)
    elif rec == 0:
        for i, l in enumerate(s.other_commanders):
            l.byz_algo(leader=s, rec=(rec - 1),
                declare=s._next_declare(s.is_faulty, declare, i))
    else:
        for i, l in enumerate(s.other_commanders):
            if l is not s and l is not leader:
                l.byz_algo(leader=s, rec=(rec - 1),
                    declare=s._next_declare(s.is_faulty, declare, i))

def decision(s):
    c = Counter(s.declares)
    return c.most_common()

def init_commanders(commanders_spec):
commanders = []
    for i, spec in enumerate(commanders_spec):
        commander = Commander(i)
        if spec == "l":
            pass
        elif spec == "t":
            commander.is_faulty = True
```

```
        else:
            print("Error, bad input in commanders list:
                {}".format(commanders_spec))
            exit(1)
        commanders.append(commander)
    for commander in commanders:
        commander.other_commanders = commanders
    return commanders

def print_decision(commanders):
    for i, l in enumerate(commanders):
        print("Commander {}: {}".format(i, l.decision))

def main():
        print_decision(commanders)

if __name__ == "__main__":
main()
```

When the recursion value is more than 0, the commander sends his proposal to every lieutenant. For each $i$, let $p[i]$ be the proposal Node $i$ acquires from the commander, or else considers 'RETREAT' by default if no proposal is obtained. Node $i$ acts as the commander in the algorithm and sends the proposal $p[i]$ to $n - 2$ other lieutenants. For each $i$, and $j! = i$, let p[j] be the value Node $i$ acquired from Node $j$ or considers 'RETREAT' by default if no proposal is obtained. Node $i$ uses the proposal majority $(p[0].....p[n])$.

```
python byz_algo.py -m 3 -N L,L,L,M,M,L,L,M,L,L -D ATTACK
Node 0: [('ATTACK', 303), ('RETREAT', 273)]
Node 1: [('ATTACK', 403), ('RETREAT', 173)]
Node 2: [('ATTACK', 303), ('RETREAT', 273)]
Node 3: [('ATTACK', 403), ('RETREAT', 173)]
Node 4: [('ATTACK', 303), ('RETREAT', 273)]
Node 5: [('ATTACK', 403), ('RETREAT', 173)]
Node 6: [('ATTACK', 303), ('RETREAT', 273)]
Node 7: [('ATTACK', 403), ('RETREAT', 173)]
Node 8: [('ATTACK', 303), ('RETREAT', 273)]
Node 9: [('ATTACK', 403), ('RETREAT', 173)]
```

Here, -rec stands for the number of recursions, -N stands for node values where L is loyal and M is malicious. The first node is the commander. -D stands for the declaration from the commander, i.e., to ATTACK or RETREAT. In the above mentioned output, there are 10 nodes, out of which 3 are malicious and 7 are loyal. All the nodes issue the proposal as 'ATTACK', as it is the commander's proposal and other lieutenants have to obey the commander. When the commander changes his decision to 'RETREAT', the output changes as follows:

```
python byz_algo.py -m 3 -N L,L,L,M,M,L,L,M,L,L -D RETREAT
Node 0: [('RETREAT', 303), ('ATTACK', 273)]
Node 1: [('RETREAT', 403), ('ATTACK', 173)]
```

```
Node 2: [('RETREAT', 303), ('ATTACK', 273)]
Node 3: [('RETREAT', 403), ('ATTACK', 173)]
Node 4: [('RETREAT', 303), ('ATTACK', 273)]
Node 5: [('RETREAT', 403), ('ATTACK', 173)]
Node 6: [('RETREAT', 303), ('ATTACK', 273)]
Node 7: [('RETREAT', 403), ('ATTACK', 173)]
Node 8: [('RETREAT', 303), ('ATTACK', 273)]
Node 9: [('RETREAT', 403), ('ATTACK', 173)]
```

When we reduce the number of nodes to 9 with 3 faulty nodes and 6 loyal nodes, since at least $2f + 1$ non-malicious nodes condition is not satisfied, hence the output is as follows wherein there is no guarantee of the final decision:

```
python byz_algo.py —m 3 —N L,L,L,M,M,L,L,M,L,L —D ATTACK
Node 0: [('RETREAT', 210), ('ATTACK', 182)]
Node 1: [('ATTACK', 244), ('RETREAT', 148)]
Node 2: [('RETREAT', 210), ('ATTACK', 182)]
Node 3: [('ATTACK', 244), ('RETREAT', 148)]
Node 4: [('RETREAT', 210), ('ATTACK', 182)]
Node 5: [('ATTACK', 244), ('RETREAT', 148)]
Node 6: [('RETREAT', 210), ('ATTACK', 182)]
Node 7: [('ATTACK', 244), ('RETREAT', 148)]
Node 8: [('RETREAT', 210), ('ATTACK', 182)]
```

### 8.9.1  Proof of Authority (PoA) Implementation

The following code shows three nodes that can be chosen as a validator. We have chosen a validator randomly in this code for showing the working of PoA. Also, the addresses mentioned will be used to transfer the amount when a block of transactions is validated.

```
class Candidates {
    static nodes() {
        return [
        ["node—address—1", 0],
        ["node—address—2", 0],
        ["node—address—3", 0]
        ];
    };

    static accounts() {
        return [
        ["account—address—1", 0],
        ["account—address—2", 0]
        ];
    };
}
module.exports = Candidates;
```

The following code shows the creation of three blocks where each block chooses a candidate as a validator randomly. *createTran()* takes three arguments, i.e., from Address, to Address, and amount. When a validator is chosen randomly and successfully validates a transaction, the amount is transferred to his account.

```
console.log('New Blockchain started with PoA')

console.log('\Authorities selected...');
let authorities = Candidates.nodes();

let bc = new BC('poa');
console.log('Genesis Block 1 created')

console.log('\nFirst Transactions created...');
bc.createTran(new
    Tran(Candidates.accounts()[0][0],Candidates.accounts()[1][0],100));
bc.createTran(new
    Tran(Candidates.accounts()[1][0],Candidates.accounts()[0][0],50));

console.log('Creating Block 2...');
let authority = authorities[Math.floor(Math.random() *
    Candidates.nodes().length)][0];
console.log('\nAuthority randomly chosen: ' + authority);
bc.generateBlock(authority);

console.log('\nNew Transactions created...');
bc.createTran(new
    Tran(Candidates.accounts()[1][0],Candidates.accounts()[0][0],130));

console.log('Creating Block 3...');
authority = authorities[Math.floor(Math.random() *
    Candidates.nodes().length)][0];
console.log('\nAuthority randomly chosen: ' + authority);
bc.generateBlock(authority);

console.log('Validation check...')
bc.validationCheck();

Candidates.accounts().forEach(function(account){
    console.log('Balance of '+account[0]+':\t'+
        bc.getBalanceOfAddress(account[0]))
});

console.log('Blockchain')
console.log(JSON.stringify(bc.chain,'', 4));
```

As you can see in the output, at first, *node-address-1* was chosen as the validator, and later, *node-address-3* was selected as the validator. After successfully validating transactions, the mentioned amount was transferred from corresponding accounts.

```
New Blockchain started with PoA
Authorities selected...
Genesis Block 1 created

First Transactions created...
Creating Block 2...

Authority randomly chosen: node-address-1
```

```
Block generated:6cafad98737d252fcf36e63dbc137cb5be1972e3a2053b55ffc1d187f47640ee

New Transactions created...
Creating Block 3...

Authority randomly chosen: node-address-3
Block generated:6dbea6abcd0ee59ecd80c95a6d05fbb3f023e2d9bfaae7ae87e676bed166609b
Validation check...
Blockchain is valid.
Balance of account-address-1: 80
Balance of account-address-2: -80
Blockchain
[
{
    "block": 0,
    "previousHash": "0",
    "timestamp": "01/01/2022",
    "transactions": "GenesisBlock",
    "hash": "c089b45f1bd2411f615e2d75eb6578488385199477c18c4022f8273ca557abfe",
    "current_nonce": 0
},
{
    "block": 1,
    "previousHash":
        "c089b45f1bd2411f615e2d75eb6578488385199477c18c4022f8273ca557abfe",
    "timestamp": 1643560539650,
    "transactions": [
    {
        "fromAddress": "account-address-1",
        "toAddress": "account-address-2",
        "amount": 100
    },
    {
        "fromAddress": "account-address-2",
        "toAddress": "account-address-1",
        "amount": 50
    }
    ],
    "hash": "6cafad98737d252fcf36e63dbc137cb5be1972e3a2053b55ffc1d187f47640ee",
    "current_nonce": 0,
    "validator": "node-address-1"
},
{
    "block": 2,
    "previousHash":
        "6cafad98737d252fcf36e63dbc137cb5be1972e3a2053b55ffc1d187f47640ee",
    "timestamp": 1643560539651,
    "transactions": [
    {
        "fromAddress": "account-address-2",
        "toAddress": "account-address-1",
        "amount": 130
    }
    ],
    "hash": "6dbea6abcd0ee59ecd80c95a6d05fbb3f023e2d9bfaae7ae87e676bed166609b",
    "current_nonce": 0,
```

```
      "validator": "node—address—3"
  }
  ]
```

## 8.10   Comparison

A comparative analysis of consensus algorithms in permissioned environment on the basis of some chosen parameters as shown in Table 8.1.

## 8.11   Summary

This chapter covered the concept of the permissioned model, which employs a control layer that executes on blockchain and verifies the tasks carried out by the nodes enabled. Permissioned blockchain requires a trust-building mechanism when nodes want to participate in the blockchain network. A permissioned environment is susceptible to Crash and Byzantine faults, where nodes become unavailable suddenly or start acting maliciously. All the consensus mechanisms covered in this chapter solve the mentioned faults. The consensus dealing with Byzantine fault requires some number of non-malicious nodes to be present in the system to reach consensus. The ones dealing with crash fault require a certain number of nodes to be available to reach consensus. Each consensus algorithm has its advantages and disadvantages. They can be compared based on consensus parameters such as power consumption scalability, security, throughput, etc. Many cryptocurrencies are created by various companies based on these consensus algorithms and perform specific use cases in the field of real estate, supply chain, banking, healthcare, etc.

## 8.12   Practice Questions

### 8.12.1   Muliple Choice Questions

1. Suppose, 15 trustworthy nodes are performing some task distributedly. As per the process, at a certain interval, every node of the team shares the results for making the consensus. After starting the task, 7 trustworthy nodes drop the plan and they are replaced by 7 other nodes whose trustworthy information is unknown. After joining the new nodes, some discrepancy occurs in the system, although all the nodes are running correctly without any software or hardware error. What is the type of fault it is in the context of distributed consensus?

**Table 8.1** Comparative analysis of consensus algorithms in permissioned environment [52]

| Sr. No | Algorithm | Scalability | Cost | Energy efficiency | Transaction rate | Throughput | Decentralization | Computing overhead | Finality |
|--------|-----------|-------------|------|-------------------|------------------|------------|------------------|--------------------|----------|
| 1. | PoA | H | N/A | Yes | H | L | M | N/A | Deterministic |
| 2. | PoET | H | M | No | M | H | M | L | Probabilistic |
| 3. | PAXOS | H | M | Yes | H | M | H | N/A | Deterministic |
| 4. | RAFT | H | N/A | Yes | L | H | M | L | Probabilistic |
| 5. | PBFT | L | L | Yes | H | H | M | L | Immediate |
| 6. | DBFT | H | L | Yes | H | H | M | L | Absolute |
| 7. | FBFT | H | L | Yes | H | H | H | L | Immediate |

H: High, M: Medium, L: Low

(a) Crash Fault
(b) Network Fault
(c) Byzantine Fault
(d) None of the above

2. In distributed consensus, all the correct individuals either reach a value or null. What is the name of the property?

(a) Termination
(b) Validity
(c) Integrity
(d) Agreement

3. Which of the following is considered as the primary assumption for a permissioned blockchain?

(a) Closed network
(b) Chosen miners
(c) No malicious miners
(d) All of the above

4. Considering the Proof of Elapsed Time (PoET) adapted in Hyperledger Sawtooth framework, which of the following mechanisms is used to ensure that the miner (or block leader) is a legitimate participant and not an attacker and has waited for the random amount of time assigned by the network?

(a) By verifying the acquired stake that the user has obtained by consuming the given random amount of time
(b) By verifying the amount of bitcoins sent to a verifiable un-spendable address
(c) By ensuring that the trusted regions of the code are run in Trusted Execution Environment (TEE) and the user cannot tamper it
(d) By ensuring that the nonce is very difficult to obtain and the user wastes enough time for it

5. Suppose in a distributed network, running PAXOS as the underlying consensus algorithm, has 3 proposers and 5 acceptors and 1 learner. Say, 3 of the acceptors have failed, which of the following is true about the network?

(a) A value gets accepted by default
(b) Someone else becomes proposer
(c) None of the proposers get a reply
(d) A new value gets proposed

6. Suppose, your system identifies that the leader is Byzantine using RAFT consensus mechanism. What will the system do now?

(a) New Leader Selection will be initiated
(b) This scenario will not happen
(c) The other nodes in the system will reject the leader based on majority voting
(d) Continue with the old leader and leader with change in the next term

7. Which are the examples of synchronous consensus techniques?

(a) RAFT
(b) PAXOS
(c) Byzantine General Model
(d) Practical Byzantine General Model

8. Suppose, you have developed an asynchronous system which can tolerate at most f faulty nodes. What is the minimum number of the similar response needed for concluding the decision?

(a) $f$
(b) $f + 1$
(c) $2f + 1$
(d) $3f + 1$

9. Suppose you execute your tasks distributedly from four different systems at four different locations. For maintaining the consensus among the systems, you are using the BFT model. You found that one system is permanently failed due to a hardware fault and another system is compromised by an attacker. Does your system correctly work at all?

(a) No
(b) Yes with the remaining nodes
(c) Yes with all the nodes
(d) This situation cannot arise

10. Alice wants to develop a secure distributed system where she wants to keep track of the node identity. Additionally, she wants fixed message content representation although any node in the system can transfer the message of any size. You as a system consultant, suggest a consensus protocol to Alice which is extremely suitable for her system.

(a) RAFT
(b) PAXOS
(c) BFT
(d) PBFT

## 8.12.2 Fill in the Blanks

1. If the timer expires at view $v$, then the backup starts a view change to move the system to view _____.
2. In PoA, instead of coins at stake, a node's _____ is at stake.
3. In three general byzantine problem, if a lieutenant is faulty, we need to ensure that _____ number of correct lieutenants should be present in the system.
4. In the pre-prepare phase of PBFT, the client assigns a _____ to the request message and _____ that request message to all the replicas.

5. PoET uses a sophisticated technology called _____ that enables _____ for allowing only trusted code and data in the system.
6. A state machine is defined as a function of parameters which include a set of _____, set of _____, _____ state, _____ function, and _____ function.
7. In DBFT, a delegate gets rewarded in terms of _____.
8. Stellar uses a group of nodes called _____.
9. In PAXOS, whenever the acceptor accepts values from the proposer, it informs the _____ about this majority voted value.
10. In Ripple, to verify a transaction, at least _____ nodes should agree to the same.

### 8.12.3  Short Questions

1. In PBFT, why do you require $3f + 1$ replicas to ensure safety in an asynchronous system where there are $f$ faulty nodes?
2. Distinguish between PAXOS and RAFT consensus mechanism.
3. Why PoA consensus algorithm is more suitable for permissioned environment?
4. Why log replication is required in RAFT consensus mechanism?
5. How does PoET justify the equal probability of every node in BC to get a chance to mine the block?

### 8.12.4  Long Questions

1. How liveness property is guaranteed by a asynchronous environment?
2. Discuss in detail how state machine replication is helpful in distributed consensus?
3. Discuss the failure of leader and follower in RAFT consensus and how consensus is achieved in such a scenario.
4. Explain three general byzantine problems and discuss any consensus mechanism that overcomes the problem.
5. Discuss in detail how PAXOS handles the failure of acceptor and proposer node?

### References

1. Azbeg K, Ouchetto O, Andaloussi SJ, Fetjah L (2021) An overview of blockchain consensus algorithms: comparison, challenges and future directions. Adv Smart Soft Comput 357–369
2. Wang H, Ge C, Liu Z (2021) On the security of permission less blockchain systems: challenges and research perspective. In: 2021 IEEE conference on dependable and secure computing (DSC). IEEE, pp 1–8

3. Tomić NZ (2021) A review of consensus protocols in permissioned blockchains. J Comput Sci Res 3(2)
4. Gupta R, Tanwar S, Kumar N (2021) B-iomv: blockchain-based onion routing protocol for d2d communication in an iomv environment beyond 5g. Vehicul Commun 100401. https://doi.org/10.1016/j.vehcom.2021.100401, https://www.sciencedirect.com/science/article/pii/S221420962100070X
5. Gupta R, Tanwar S, Kumar N (2021) Blockchain and 5g integrated softwarized uav network management: architecture, solutions, and challenges. Phys Commun 101355. https://doi.org/10.1016/j.phycom.2021.101355, https://www.sciencedirect.com/science/article/pii/S1874490721000926
6. John K, O'Hara M, Saleh F (2021) Bitcoin and beyond. Ann Rev Financ Econ 14
7. Yiu NC (2021) Toward blockchain-enabled supply chain anti-counterfeiting and traceability. Future Internet 13(4):86
8. Maier M, Ebrahimzadeh A (2021) Decentralization via blockchain
9. Rebello GAF, Camilo GF, Guimarães LC, de Souza LAC, Thomaz GA, Duarte OC (2021) A security and performance analysis of proof-based consensus protocols. Ann Telecommun 1–21
10. Popov S, Müller S (2021) Voting-based probabilistic consensuses and their applications in distributed ledgers. arXiv preprint arXiv:2104.05313
11. Alqahtani S, Demirbas M (2021) Bottlenecks in blockchain consensus protocols. arXiv preprint arXiv:2103.04234
12. Distler T (2021) Byzantine fault-tolerant state-machine replication from a systems perspective. ACM Comput Surv (CSUR) 54(1):1–38
13. Gupta S, Hellings J, Sadoghi M (2021) Fault-tolerant distributed transactions on blockchain. Synthesis Lect Data Manage 16(1):1–268
14. Ranchal-Pedrosa A, Gramoli V (2021) Rational agreement in the presence of crash faults. In: 2021 IEEE international conference on blockchain (Blockchain). IEEE, pp 470–475
15. Barger A, Manevich Y, Meir H, Tock Y (2021) A byzantine fault-tolerant consensus library for hyperledger fabric. In: 2021 IEEE international conference on blockchain and cryptocurrency (ICBC). IEEE, pp 1–9
16. Chondros N, Kokordelis K, Roussopoulos M (2012) On the practicality of practical byzantine fault tolerance. In: ACM/IFIP/USENIX international conference on distributed systems platforms and open distributed processing. Springer, pp 436–455
17. Lamport L et al (2001) Paxos made simple. ACM Sigact News 32(4):18–25
18. Mocanu, A, Bădică C (2016) Paxos-based weighted argumentation framework approach to distributed consensus. In: 2016 international symposium on innovations in intelligent systems and applications (INISTA). IEEE, pp 1–6
19. Du H, Hilaire DJS (2009) Multi-paxos: an implementation and evaluation. Department of Computer Science and Engineering, University of Washington, Technical report, UW-CSE-09-09-02
20. Hu J, Liu K (2020) Raft consensus mechanism and the applications. In: Journal of physics: conference series, vol 1544. IOP Publishing, p 012079
21. Mohanty D (2019) Corda architecture. In: R3 corda for architects and developers. Springer, pp 49–60
22. Baliga A, Subhod I, Kamat P, Chatterjee S (2018) Performance evaluation of the quorum blockchain platform. arXiv preprint arXiv:1809.03421
23. Lamport L, Shostak R, Pease M (2019) The byzantine generals problem. In: Concurrency: the works of leslie lamport, pp 203–226
24. Pease M, Shostak R, Lamport L (1980) Reaching agreement in the presence of faults. J ACM (JACM) 27(2):228–234
25. Gupta S, Hellings J, Rahnama S, Sadoghi M (2019) An in-depth look of bft consensus in blockchain: challenges and opportunities. In: Proceedings of the 20th international middleware conference tutorials, pp 6–10
26. Pass R, Seeman L, Shelat A (2017) Analysis of the blockchain protocol in asynchronous networks. In: Annual international conference on the theory and applications of cryptographic techniques. Springer, pp 643–673

27. Driscoll K, Hall B, Sivencrona H, Zumsteg P (2003) Byzantine fault tolerance, from theory to reality. In: International conference on computer safety, reliability, and security. Springer, pp 235–248
28. Castro M, Liskov B et al (1999) Practical byzantine fault tolerance. OSDI 99:173–186
29. Kindler E (1994) Safety and liveness properties: a survey. Bull Europ Assoc Theor Comput Sci 53(268–272):30
30. Androulaki E, Barger A, Bortnikov V, Cachin C, Christidis K, De Caro A, Enyeart D, Ferris C, Laventman G, Manevich Y et al (2018) Hyperledger fabric: a distributed operating system for permissioned blockchains. In: Proceedings of the thirteenth EuroSys conference, pp 1–15
31. Buchman E (2016) Tendermint: Byzantine fault tolerance in the age of blockchains. PhD thesis
32. Kaur R, Kaur A (2012) Digital signature. In: 2012 international conference on computing sciences. IEEE, pp 295–301
33. Amiri MJ, Agrawal D, El Abbadi A (2019) Parblockchain: leveraging transaction parallelism in permissioned blockchain systems. In: 2019 IEEE 39th international conference on distributed computing systems (ICDCS). IEEE, pp 1337–1347
34. Li Y, Wang Z, Fan J, Zheng Y, Luo Y, Deng C, Ding J (2019) An extensible consensus algorithm based on pbft. In: 2019 international conference on cyber-enabled distributed computing and knowledge discovery (CyberC). IEEE, pp 17–23
35. Onireti O, Zhang L, Imran MA (2019) On the viable area of wireless practical byzantine fault tolerance (pbft) blockchain networks. In: 2019 IEEE global communications conference (GLOBECOM). IEEE, pp 1–6
36. Bach LM, Mihaljevic B, Zagar M (2018) Comparative analysis of blockchain consensus algorithms. In: 2018 41st international convention on information and communication technology, electronics and microelectronics (MIPRO). IEEE, pp 1545–1550
37. Elrom E (2019) Neo blockchain and smart contracts. In: The blockchain developer. Springer, pp 257–298
38. Yang F, Zhou W, Wu Q, Long R, Xiong NN, Zhou M (2019) Delegated proof of stake with downgrade: a secure and efficient blockchain consensus algorithm with downgrade mechanism. IEEE Access 7:118541–118555
39. Zarir AA, Oliva GA, Jiang ZM, Hassan AE (2021) Developing cost-effective blockchain-powered applications: a case study of the gas usage of smart contract transactions in the ethereum blockchain platform. ACM Trans Softw Eng Methodol (TOSEM) 30(3):1–38
40. Yoo J, Jung Y, Shin D, Bae M, Jee E (2019) Formal modeling and verification of a federated byzantine agreement algorithm for blockchain platforms. In: 2019 IEEE international workshop on blockchain oriented software engineering (IWBOSE). IEEE (2019), pp 11–21
41. Mazieres D (2015) The stellar consensus protocol. A federated model for internet-level consensus, Version, p 14
42. Schwartz D, Youngs N, Britto A et al (2014) The ripple protocol consensus algorithm. Ripple Labs Inc White Paper 5(8):151
43. García-Pérez Á, Gotsman A (2018) Federated byzantine quorum systems. In: 22nd international conference on principles of distributed systems (OPODIS 2018). Schloss Dagstuhl-Leibniz-Zentrum fuer Informatik
44. Lachowski Ł (2019) Complexity of the quorum intersection property of the federated byzantine agreement system. arXiv preprint arXiv:1902.06493
45. Cachin C, Vukolić M (2017) Blockchain consensus protocols in the wild. arXiv preprint arXiv:1707.01873
46. De Angelis S, Aniello L, Baldoni R, Lombardi F, Margheri A, Sassone V (2018) Pbft vs proof-of-authority: applying the cap theorem to permissioned blockchain
47. Deirmentzoglou E, Papakyriakopoulos G, Patsakis C (2019) A survey on long-range attacks for proof of stake protocols. IEEE Access 7:28712–28725
48. Dhillon V, Metcalf D, Hooper M (2017) The hyperledger project. In: Blockchain enabled applications. Springer, pp 139–149
49. Chakrabarti S, Hoekstra M, Kuvaiskii D, Vij M (2019) Scaling intel® software guard extensions applications with intel® sgx card. In: Proceedings of the 8th international workshop on hardware and architectural support for security and privacy, pp 1–9

50. Sabt M, Achemlal M, Bouabdallah A (2015) Trusted execution environment: what it is, and what it is not. In: 2015 IEEE Trustcom/BigDataSE/ISPA, vol 1. IEEE, pp 57–64
51. Aggarwal S, Kumar N (2021) Hyperledger. In: Advances in computers, vol. 121. Elsevier, pp 323–343
52. Bodkhe U, Mehta D, Tanwar S, Bhattacharya P, Singh PK, Hong WC (2020) A survey on decentralized consensus mechanisms for cyber physical systems. IEEE Access 8:54371–54401. https://doi.org/10.1109/ACCESS.2020.2981415

# Chapter 9
# Consensus Scalability in Blockchain Network

**Abstract** The consensus mechanisms are used in a blockchain to achieve the essential agreement on a single value among various multiagent and distributed processes. The consensus algorithm is categorized into two networks, i.e., permissionless and permissioned networks of blockchain. The permissionless network consists of Proof of Work (PoW), Proof of Stake (PoS), Bitcoin-NG, and others. The permissioned network contains PBFT-based consensus mechanisms. This mechanism performs the roles in a leader selection and in a mining procedure to enhance the system's transaction throughput. First, we identify the issues of PoW consensus and find out the problems such as scalability in terms of transaction throughput and consensus finality. These two parameters have been required to enhance the consensus mechanism so that it can improve the system's performance. Then, this chapter described the bitcoin-NG, which improves the scalability of the PoW consensus mechanism. We also highlight steps of scaling the byzantine to improve the transaction throughput and performance of the system. Finally, we consider various scenarios of leader and process to solve the issue of consensus scalability with different PBFT-based consensus mechanisms. All mechanism runs various transactions within a specific time and avoids the forks that occurred during the insertion process of the new block. Thus, the chapter presents multiple research directions to improve the scalability of the consensus by identifying various issues in the current consensus mechanism.

**Keywords** Consensus · Blockchain · Smart contract · Byzantine fault tolerance

## 9.1 Introduction

Blockchain is a collaborative network where nodes operate autonomously and require collaboration and coordination to establish an agreement on shared data to communicate information. The consensus issue [5] states that the coordination should exponentially settle to a shared value. The consensus problem specifies characteristics such as validation, agreement, and termination. Validity ensures that every valid process can converge to the same result if all valid processes offer the same value. The agreement principle says that no two valid procedures may make different decisions [8]. The termination attribute assures that each right procedure decides at some

© The Author(s), under exclusive license to Springer Nature Singapore Pte Ltd. 2022     251
S. Tanwar, *Blockchain Technology*, Studies in Autonomic, Data-driven and Industrial
Computing, https://doi.org/10.1007/978-981-19-1488-1_9

point. The first two are also known as security qualities, and the third is known as liveness. According to consensus security, a specific instance is picked from a list of possible values, and an appropriate node knows that value. The liveness attribute indicates that a value is finally picked among the offered values, and the nodes can acquire it. However, the incidence of byzantine failures makes it difficult to develop consensus methods since nodes depart from anticipated behavior in various ways [1]. Even in an uncertain environment, it is difficult to analyze the behavior of a node, and a blockchain network must establish distributed consensus.

Consensus algorithms are the basis of a blockchain network. It has a massive impact on the network's entire throughput, latency, and fault tolerance. Proof-based and voting-based consensus are two standard consensus algorithms. PoW, PoS, and delegated proof of stake (DPoS) are examples of proof-based consensus mechanisms. To add a new block to the chain, nodes entering the network should demonstrate that they are more qualified than other nodes. The two voting-based consensus mechanisms are practical byzantine fault tolerance (PBFT) and tendermint. This algorithm requires sharing the recent new block or transaction confirmation findings across network nodes. Aside from that, there are a lot of diverse opinions. Their basic concept is to build on each other's strengths in order to increase the robustness and efficiency of the actual consensus.

The bitcoin blockchain and its proof-of-work (PoW) method were founded to establish a framework for a secure and decentralized financial accounting system based on a peer-to-peer transaction, and further use cases have subsequently emerged [7]. They use a technique in which principle should operate in a huge, globally distributed system like the Internet. Because it combines participation with security, PoW is incredibly successful in enabling an open membership by securing the network against Sybil assaults [2]. In a consensus, find the probability of nodes selecting the next block with their resource consumption. For an attacker, it is expensive to vandalize the structure of PoW. It also grows effectively across many nodes, but it fails in terms of scalability and throughput. Moreover, the PoW system has faults such as (i) PoW wastes a large number of resources and energy, (ii) It performs poorly in terms of latency and throughput, and (iii) It does not ensure consensus finality [4] which means a determined block might be changed in the future.

To address the issue of the PoW consensus mechanism, byzantine fault-tolerant (BFT) consensus is used in well-known mechanisms like PBFT, which might be utilized as a substitute solution for ordering transactions in DLTs [9]. BFT techniques may be used as a proof-of-stake variation, Nodes are randomly allocated to the authority to apply blocks, and all nodes agree on the specific block that is being added into the blockchain to ensure consensus finality. It has certain characteristics such as (i) BFT contains efficient energy for work, (ii) It can handle ten thousand transactions per second within a few seconds, and (iii) it has verified liveness and security features. On the other hand, traditional BFT mechanisms restrict PoW's capability for an open membership and scalability for high numbers of nodes up to thousands [11]. We may dispute that permissioned blockchains are justified or that DLTs should look at different techniques to assure membership, but scalability remains a significant issue for massive blockchain-based infrastructures.

Existing PoW and BFT-based blockchain mechanisms are being explored to overcome scalability limitations and a method that evaluates terms of reference for improving PoW based blockchain's performance. In addition, comprehensive and systematic research of consensus in blockchain mechanisms analyzes security, performance, design aspects, and scalability. We give a complete overview in this study that focuses on techniques used in recent schemes to increase scalability [10].

## 9.1.1  Bitcoin-NG

The performance and scalability of the consensus mechanism are crucial aspects. The PoW based consensus mechanism gets a maximum of around 7 to 8 transactions per second, which is limited transaction throughput. Another issue of PoW is a consensus finality, in which various miners mine the block subsequently at the same time, so there is a possibility of a fork. If two nodes try to mine a block simultaneously, the other node must wait for the following blocks to find the longest chain between all forks. Due to the problem of consensus finality, miners cannot know the information about the last mined block. So the two problems of PoW are consensus finality and transaction throughput or scalability. To improve the scalability and performance of the PoW system, design the bitcoin-NG consensus mechanism. It aims to provide better transaction throughput and reduce the forks in the PoW system. This process creates a block after every 10 min and estimates the subsequent creation of the block based on mining difficulty, controlling the average block creation frequency. The second parameter is a block size, which is limited upto 1 MB in PoW. If the size of the block is increased, it results in a fork creation. So, this is not a promising solution to increase the block size. To improve the scalability of consensus, initially, bitcoin-NG decouples the characteristics of bitcoin PoW for the miners. The first consideration in this mechanism is a leader election, where it is necessary to find the leader in a specific round. In a bitcoin-NG, at every 10 min, the new leader was elected and applied to serialize the transactions. The frequent selection of leaders in this mechanism improves the performance of the consensus process. The leader serializes the transaction till the next leader is elected. When a leader gets enough transactions, it can serialize the transaction and insert the transaction details into the current blockchain. Figure 9.1 shows two sorts of blocks, the first is a key block, and the other blocks are called microblocks. These blocks generate using the PoW consensus mechanism. After every PoW round, create one key block containing a miner's public key, which can solve the puzzle challenge for some time. Now the specific leader creates various microblocks containing the transactions. The leader can create various microblocks which can be encrypted using the private key of the leader. So in this, a leader can encrypt all microblocks using their private key, and others can use the public key to verify the microblocks. The key block contains a reference of the last block, current Unix time, coinbase transaction for paying a reward, targeted hash value, and a nonce field. The coinbase transaction provides information on how much reward the specific miners give. Thus, a coinbase transaction can pay the reward to

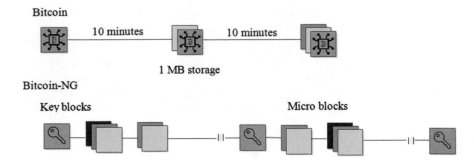

**Fig. 9.1**  Bitcoin-NG based consensus mechanism

the miner. The targeted hash value is either equal or less than as compared to standard bitcoin. Each miner subsequently tries to find the nonce, which helps get an equal hash value as a targeted hash value. If any miner successfully finds out a nonce, they are elected as a leader, creating a related key block and various microblocks. This contains the serializability of the transactions. The key block must be required to be valid in the bitcoin-NG mechanisms, and the cryptographic hash of its header is smaller than a targeted value. The key block contains a public key of a miner. At the end of a specific key block round, other rounds of key block creation, miners can try to solve the challenge and create a new key block. Now, this new block can use the following microblocks to the process of transaction serializability. So in this Bitcoin-NG, after every 10 min, a miner can create a new key block. These key blocks are infrequently created, so the interval time between the two key blocks is distributed exponentially.

When the node creates a new key block, it becomes the next leader. This leader permits creating microblocks to set a rate that is smaller than the maximum predefined. The generated rate is higher compared to the key block creation rate. If a key block is created every 10 min, then microblocks are created every 1 min. It means, within a time of 1 min, the transaction has arrived at that specific miner working as a leader of this specific round. The miner can create a new microblock with the serialization process of the transactions append into the existing blockchain. So, in this case, a leader can generate various microblocks frequently and add several transactions at a time to the system, which improves the system's scalability. A microblock contains ledger entries, and the header contains a reference of the last block, current Unix time, a cryptographic hash of ledger entries, and the signature of the header. This signature is based on a private key related to the key block and its public key. This public key and all microblocks are called a header of microblock. This header contains a cryptographic signature, which generates a standard digital signature using the private key related to the public key. As a verifier, a node can validate this signature. This signature comes from a miner who creates a specific key block. The confirmation time of this mechanism is to check the possibility of a fork. When a miner creates a new key block, they do not know about the microblocks created previously by the last leader. The issue is solved by assuring the reception of the key block, and when a node can see a microblock, it can wait for the propagation time of the system to make sure that a new key block does not clip it. Thus, Bitcoin-NG improves the transaction throughput and scalability, and solves the issue of consensus finality of PoW-based mechanism [4].

## 9.1.2 Scaling Byzantine

Scaling Byzantine consensus is a never-ending process that needs the research, development, and integration of various approaches and procedures, and approaches. The following questions were recognized as being important to improve the scalability and efficiency of byzantine consensus:

- Who has a reason to communicate? Is the whole consortium actively involved in reaching an agreement, or is it decided by one or more individually selected representation committee?
- What is the flow of communication? To improve communication efficiency, an appropriate communication topology, such as gossip-based/hierarchical or overlay, may be required.
- How may parallel transactions be organized and committed?
- Which new paths may be created by the use of appropriate cryptographic primitives?
- How can we transfer mission-critical consensus stages to hardware made up of reliable hardware devices?

### 9.1.2.1 Communication Topology

To enhance the efficiency of the communication flow and eliminate the exposure of bottlenecks, it distributes a load of communication as widely as feasible. For example, the well-known PBFT protocol cannot grow successfully due to its all-to-all broadcast stages and the enormous burden on the leader, who should suggest large-size messages to all nodes.

In flat communication, the first step is to lower the communication complexity. Then, PBFT can use signatures for the distribution of messages. These messages are lower from $O(n^2)$ to $O(n)$ in the network. The next step is to aggregate various signatures in one. In tree communication, arrange the flow of communication in a tree-based structure where the leader is at the root position, and other nodes can forward the messages to their child nodes. In a communication tree, scalability emerges when the node can receive integrated $O(1)$ size instead of $O(n)$ size message and requires only $O(1)$ instead of $O(n)$ computation with verification of collective signatures rather than n personal signatures in commitment stage [2].

In an entire network, the leader starts to send a message m to r and other arbitrarily selected nodes. At each hop, the node receives m and forwards it to k of its connected neighbors. The speed relies on the parameter k, and the total communication load is moved from the leader to the other $O(1)$ nodes, which improves scalability. The gossip method is used to select randomness and the probabilities. Communication without a leader uses a gossip technique; each node requests k to other nodes in a loop. It accepts the value that is being returned by an updated majority of nodes, such that accurate nodes are finally directed to the same consensus value. Even in a wide network, nodes in the probabilistic technique may quickly converge to an irreversible state, but there is no assurance.

The precise architecture of the trust and dependency connections established by the consensus slice and periodic intersections determines scalability in federated byzantine agreement. Creating a layered, horizontal consensus structure is being recommended, in which nodes build a consensus slice including itself and a fixed number of random nodes from the higher adjacent levels. While this approach saves time and effort when compared to traditional BFT methods. Fault tolerance no longer linearly scales the overall number of control nodes, at least from the view of the individual node. This results in an open and scalable system, in which a few terrible trust decisions can prevent a single valid node from achieving a consensus with the federation.

### 9.1.2.2  Cryptographic Primitives

Collective signatures (CoSi) is a cosigning mechanism for large groups of attendees. An authoritative statement (i.e., a leadership proposal message) is verified and publicly documented by the observers and signed by the authority and all or most of the observers. The CoSi mechanism scales by building a schnorr multisignature across a spanning tree in 4 communication stages (basically two top-down- and then-bottom-up passes over the tree) because it can easily reduce hundreds of distinct signatures into one unique signature of the same size using blockchain technology. In the byzantine scenario, the adoption of this mechanism is tentative and can only enhance optimist performance because defective nodes may not deliver information to their siblings. As demonstrated by ByzCoin, the leader only gathers an $O(1)$-size message instead of an $O(n)$-size message, it has less verification difficulty since it only wants to verify a single collective signature rather than n individual signatures, and saving both compute and network resources.

It has been shown that threshold signatures can decrease communication costs of various BFT mechanisms by a factor of $O(n)$ when the adopted approach is used correctly. Threshold signatures, like multisignatures, reduce the size of numerous signatures to a unique one, with the condition that at most t members out of n is required to establish a valid signature for a (t,n)-threshold signature. For example, PBFT employs $O(n)$-sized commit certificates to ensure that at most $2f + 1$ nodes agree on a commitment decision. These certificates may be incorporated into an individual commit certificate of size $O(1)$ employing $t = 2f + 1$ threshold signatures, making the system more scalable.

In order to decipher an encrypted communication in a (t,n) threshold cryptography method, participants must collaborate at most t out of n overall. Setting t = f + 1 in the BFT adversarial model means that overall, as no valid node releases its decryption part, an attacker never read encrypted messages. The scalability emerges from the reality that (i) committee selected by verifiable random functions in a pre manner without impact directly, regardless of the overall the network size, and (ii) only members of the committee engage actively in the consensus mechanism (which needs sharing messages), while the rest only gain knowledge of the agreed value, results in scalability. Moreover, the selection process' anonymity, randomization, and lack of interaction protect committee members against cyberattacks.

### 9.1.2.3 Representative Committees

The announcement of a committee representative with active responsibilities (e.g., acceptors and proposers), while many nodes keep silent (e.g., they just know about the agreed value), is crucial for reaching scalable consensus in more extensive networks. This committee is best described as a random and representative group of actors that interact with one another to achieve an agreement. Because the delegate's choice is binding on everybody, the whole system comes to an agreement as to the decision and spreads over the network. Moreover, because the committee has a fixed volume, this approach scales effectively; in fact, it is practically independent of such a total number of nodes, and it can operate securely as far as a few key requirements are met, (i) The selection process should not necessarily require coordination between nodes; (ii) delegates are selected at random, but in a usual manner, such as by using stake-based graded probabilities, so that an intruder cannot release a Sybil attack; and (iii) delegates are being selected secretively to ignore being the attacker's target.

### 9.1.2.4 Parallelizing Transactions

The concept of committees may be used to parallelize the transaction validation and ordering method. Sharding is a strategy in which the entire system is divided into smaller, (nearly) equal-sized committees. An attacker should not be able to modify the distribution; for example, defective nodes should distribute uniformly over all committees with a high likelihood. Every committee then runs byzantine consensus on its own and processes separate sets of transactions simultaneously. With an increasing total number of nodes, the general approach should allow for nearly constant scaling of throughput. Equally sized committee can refer to have the same hash power, stake, or the same number of nodes with verified identification.

The significance relies on how identities are handled and voting power is shared through the mechanism. To provide security, the equally scattered nodes between committees must be random, and the necessary seed can be produced, for example, by utilizing proof of work to generate epoch randomness or by using a distributed power generation mechanism. Moreover, rather than relying on all-to-all broadcasts, the discovery for participants of the similar committee, and hence the formation of an efficient network must be handled effectively. Two transactions can be saved in separate blocks by various committees, and thus parallelized if people don't contradict or rely on one another, for example, if people don't try to double-spend (pay with the similar unspent transaction output UTXO two times) or if one transaction generates a UTXO as an output that is input to the other. A block-based directed acyclic graph can be used to efficiently analyze conflict transactions and connections between transactions (like A must execute before B).

### 9.1.2.5   Trusted Hardware Components

The removal of the expenses of reaching consensus, in particular the explosive broadcast primitive, of the protocol's main chain and moving it to hardware is an exciting approach for increasing efficiency. Field programmable gate arrays were employed in their crash fault-tolerant approach to deliver consensus as a solution to applications, such as an instance of Zookeeper explosive broadcast with a relevant key. Although the need for specific hardware may limit the transparency and decentralization of DLTs, we believe the basic concept of hardware consensus merits, such as for BFT blockchain protocol versions. We can see a promising approach in using trusted execution environments (TEEs), i.e., built on Intel SGX. By showing immutable originality inside a network, hardware components might effectively prevent Sybil assaults [12].

Recent research suggests a proof-of-luck (PoL) consensus that uses Intel SGX (but may be used for other TEEs). PoL integrates the concepts of TEE-based proof of ownership and proof of time to enable better mining with sufficient time and higher energy; therefore, the massive issues of PoW are being addressed. TEEs are also employed in FastBFT, an equipment secret sharing strategy for reaching scalable byzantine consensus. FastBFT offers an effective message aggregation method by incorporating TEEs with lightweight key exchange. It also includes other optimizations such as tree topology and hopeful execution to improve scalability and performance, making it an attractive choice for combination in DLTs.

## 9.1.3   Helix Byzantine Fault Tolerance Scalable Algorithm

In helix [3], the transactions are encrypted via a threshold encryption scheme to hide information from the ordering nodes, limiting censorship and front-running. The encryption is further used to realize a verifiable source of randomness, which is used to elect the committees unpredictably and introduce a correlated sampling scheme of transactions included in a proposed block. The correlated sampling scheme restricts nodes from promoting their transactions over others. Nodes are elected to participate in committees in proportion to their relative reputation. Reputation, attributed to each

**Fig. 9.2** Helix-based
consensus mechanism

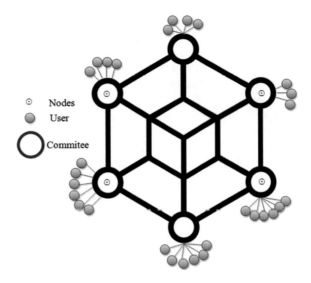

node, is a measure of obedience to the protocol's instructions. Committees are thus chosen in a way that is beneficial to the protocol. Figure 9.2 shows the consensus process among various nodes, users, and committee. For byzantine fault tolerance, helix depends on PBFT. In reality, the helix may be regarded as a blockchain implemented using PBFT, with the arbitrary committee that allows it to operate several PBFT instances. Each being used for an individual block value. It maintains PBFT's finality and eliminates the chances of common forks in the blockchain. To reduce PBFT's inherent issue to scale big networks, it limits the agreement procedure to a confined committee chosen from a broader collection of all consensus nodes. The committee's size is based on a high estimation of the number of defective nodes f. When the set of network participants $n > 3f + 1$, performance is compromised in service of redundant security in PBFT. Even when n is big, helix's performance improves by making the elected committee size $m = 3f + 1$, which may fulfill m $< n$, the smallest committee size that keeps PBFT's features. The re-election of the committee is frequent and unexpected to offer resistance from DoS assaults.

Helix specifies the interests that drive its consensus nodes explicitly and promotes fairness with them in their application of these principles. It focuses on maintaining equal power distribution between the nodes and guaranteeing that an individual node cannot change the sequence of the blockchain network. Helix also considers the protocol's end-users to prevent them from becoming controlled. The ordering of transactions is the focus of helix. It organizes entirely indifferent to any state changes that these transactions may cause. As a result, threshold encryption may be used to disguise transactions between nodes. The substance of a transaction is being exposed once the transaction's position is finalized. Inside this protocol, the threshold encrypting approach is applied to generate verified randomization. Helix creates a randomization flag without the need for extra communication rounds. Afterward, it uses the variations in two ways: first, to elect a leader (and committee, which are used to discuss scalability concerns, as PBFT sometimes does not scale well to large-scale networks); and second, to implement correlated selecting method, which enables nodes to extract their group of recent transactions when constructing blocks arbitrarily. The correlated selection technique allows for novel charge structures (for example, continuous charges) that are different from existing techniques and protect the establishment of undesirable charge markets.

Helix works with a known set of consensus nodes in charge of creating and verifying blocks. In the permissioned environment of blockchain [21], the content of the list is openly determined by an authority; and in a permissionless environment, the list's structure is being determined by a stake in the system, PoW, or other factors. For convenience, we will assume that a set of consensus nodes is provided throughout this research. Since consensus nodes are identifiable, helix can add a reputation metric, which evaluates a node's compatibility with the protocol's instructions. The reputation measurement is being used to reward or penalize a node for appropriate conduct. Helix comes to a consensus on a balanced transaction ordering, in which each block contains an impartial collection of transactions. To look at it the other way, a node cannot give priority to transactions and get to choose which transactions are being required to insert in the new block. Such a constraint allows for circumstances in which transaction charges only cover processing expenses and generate zero revenues. Nodes prioritize transactions based on other standards in such cases, and this form of equality becomes significant. Second, with helix, end-users who do not participate actively in the consensus process are protected from being discriminated against with the network nodes. It is accomplished by requiring users to encrypt individual transactions before sending them to nodes. Finally, although various consensus mechanisms depend on a steady leader to move quickly and incur significant delays from leader replacement, helix efficiently changes leaders under normal patterns (and committee members). The leader cycle has a small communication cost and, unlike other round-robin protocols, is unexpected and can be graded non-equally among the nodes.

## 9.1.4   Various Byzantine-Based Scalable Consensus Algorithms

Table 9.1 shows the comprehensive survey of various scalable consensus mechanisms.

### 9.1.4.1   Proof of Work and Proof of Stake

**ByzCoin** [13]- this approach integrates various consensus methods, for example, use proof-of-work method to solve the issue of defining membership in the absence of the method being vulnerable to Sybil attacks and while using byzantine con-

**Table 9.1** A comprehensive survey of existing scalable consensus mechanisms designed for blockchain structures [2]

| | Fast BFT | Stellar | ByzCoin | Algorand | HoneyBadger BFT | Omni Ledger | Gosig |
|---|---|---|---|---|---|---|---|
| Evaluation Scalability | 199 | currently running 100 | 1004 | Up to 500k | 104 | 1800 | Up to 10k |
| Transactions throughput | 370 (n+199) | 1000 (n = 100) | 700 (n+1004) | 1000 | 1200 (n = 140) | ≥4000 (n = 1800) | 4000 (n = 140) |
| Latency (in time) | <1s (1 Gbps LAN) | Few seconds | 30s | 1 min | 100s | <2s | <1 min |
| Synchrony | Weak synchronous | Asynchronous, but progress rely on the process of synchrony | Weak synchronous | Weak synchronous | Asynchronous | Asynchronous | Asynchronous, but achievable liveness only under weak synchrony |
| Determination of consensus | Deterministic | Deterministic | Deterministic | Probabilistic | Probabilistic | Probabilistic | Probabilistic |
| Scaling consensus | Hardware-based TEE + tree topology, secret exchange | Federated byzantine consensus with hierarchical infrastructure | Collective signatures + communication tree | Gossip + committee (cyptographic sortition) | Novel ACS reduction with thresold encryption, and efficient RCB with erasure codes | Parallelizing transactions, collective signature, and communication tree | Gossip + multi-signatures |

sensus for the transaction process to improve its performance. ByzCoin provides a proof-of-work algorithm on a different identification chain to implement this approach. Miners gain hash proportional consensus rights, which affect the likelihood of being the leader of either an epoch on the keyblock chain and, therefore, the ability to organize transactions employing BFT consensus. Traditional scalability advancements over conventional BFT protocols such as PBFT, interaction, and collaborative signatures lower the interaction difficulty from quadratic to logarithm (linear in adverse circumstance) and signature authentication charges from linear to continuous time

**Tendermint**- Tendermint [15] is a BFT replication technique that employs proof of stake (PoS), a voting power distribution method in which members (nodes that allow executing the consensus) place risk that they potentially lose if they are not trustworthy. In this method, more than 2/3 of active, risk-weighted members perform effectively; the protocol can securely get the experience of open membership. Tendermint's primary consensus mechanism is a different round of voting, similar to PBFT, but with a few key modifications for scalability and decentralization, (i) Round-robin based leadership rotations system, and (ii) a unique termination technique that effectively exploits media communication. On the other hand, the leader rotation works very sincerely; therefore, is expected for an assault.

### 9.1.4.2 Sharding BFT

**ELASTICO** [16]- Sharding protocols can improve a system's performance by splitting transaction processing across shards and processing those in parallel by discontinuous sets of verifiers committee. ELASTICO is a permissionless sharding mechanism, (i) establishes identities using proof-of-work, (ii) securely splits the task into little committee, correlated processing of transactions, and (iii) grows linearly with existing computation capacity (that consistent with existing verified identities). In certain ways, sharding is an orthogonal solution that may be built using any byzantine consensus process. It uses a PBFT mechanism to execute the intra-committee consensus mechanism. In tests, ELASTICO raised the number of blocks per epoch from 1 to 16 as the total number of nodes starts from 100 and goes upto 1600). However, the time it takes to locate evidence of work grows slightly from 600s to 711s, while the time it takes to gain consensus within the committee remains at 100s.

**OmniLedger** [17]- OmniLedger can use a bias-resistant mechanism to generate a random distribution system rather than PoW. It also includes Atomix, a fast cross-shard commit mechanism that atomically manages transactions to affect several shards. OmniLedger outperforms ELASTICO by implementing optimizations including (i) using ByzCoinX rather than PBFT to much more effectively proceed with the transactions inside of shards (i.e., To use better interaction styles like tree topology), (ii) Going to resolve interconnections only at transaction volume to reach greater block parallel processing, and (iii) incorporating a "trust-but-verify" verification structure for real-time confirmation of transactions. According to assessment findings, OmniLedger enables increased throughput linearly with the active validators, reaches a throughput in the thousands of transactions per second, and keeps latency under two seconds.

### 9.1.4.3 Randomized BFT

**HoneyBadgerBFT** [19]- (HBBFT) is an asynchronous BFT system that addresses the issue of atomic broadcast and hence consensus. It creates an adversarial network controller to delay communications and build an environment that finds weak synchronous BFT protocols like PBFT ineffective. Furthermore, HBBFT presents a unique atomic broadcast technique achieved by improving the common asynchronous subset (ACS) employing a new batching approach with threshold encrypted data. In tests, HBBFT outperforms PBFT in a poor synchronous network, processing thousands of transactions per second and scaling up to a thousand nodes.

**Algorand** [20]- Algorand is a revolutionary technique for scaling a blockchain system for millions of consumers. Algorand generates a presentive committee of arbitrarily and anonymously picked users for each block that indicates the extent. The committee members are actively involved in the byzantine consensus process since they are either verifiers or proposers, and others know about the committee's decision. Cryptographic sortition is, a technique that depends on the usage of valid randomized functions (VRFs) is used in the selection process. An adversary cannot attack the members of the committee since the selection procedure is kept confidential. These VRFs permit users to evaluate their own verification and if they have been elected as verifiers or proposers and create immutable evidence of the election's legitimacy. This procedure is best described as a private draw, with each participant receiving a ticket as proof. Multiple proposers emerge from the selection process, but only the individual online users also have the ticket with the shortest hash, and it becomes the original leader. The verifiers must agree with the leader's selected block. Moreover, messages are disseminated to all users via gossiping to become aware of the block agreement. As a result, Algorand can produce new blocks at the same frequency as messages are carried via the network, and it grows regardless of the network's total number of users. The risk-weighted election process distributes probabilities according to the number of stakes users have. Thus, as long as the attacker holds below 1/3 of the stakes, the system is secure with a high likelihood.

**Gosig** [22]- Gosig aspires to scale byzantine consensus on adversarial broad area networks. Cryptographic determination is being used to pick a leader randomly and secretly. Moreover, all members engage actively to achieve consensus in the signature-based voting system. It allows in making a secure system in an asynchronous situation. Gosig implements scalability enhancements at the implementation stage, for example, by grouping signatures into multi-signature forms to reduce communication costs. It also uses gossiping communication to perform the broadcast proposal and signature gathering stages effectively. Gosig has a lower performance cost than HoneyBadgerBFT, but it compromises verifiable liveness in the asynchronous environment. Additionally, it develops various techniques like fault detection and asynchronous signature validation to raise the likelihood of liveness on the greatest premise.

#### 9.1.4.4  Federated Consensus

**Stellar** [23]- The Stellar consensus mechanism gets adaptable trust and develops a framework for federated Byzantine consensus compared to byzcoin or tendermint. It does not use PoW or PoS, but it can also enable open membership because new nodes are not always trusted. In this mechanism, security relies on nodes to select trustworthy decisions, and there is no cash incentive to host a node and engage in a consensus mechanism. This might be an issue for network expansion and decentralization. The stellar network has about a hundred nodes and can handle thousands of transactions per second. On the other hand, Nodes are generally managed by the stellar foundation, or companies like IBM maintain the network's security.

## 9.2  Scalability Bottlenecks

**Data**- Distribution of commands to all duplicates. e.g., If a block includes 1MB of instructions, all bits must reach all validating duplicates. The apparent bottleneck is the network throughput of the system.

**Consensus**- Duplicates engage in a consensus mechanism after the orders come. e.g., If a consensus process requires two types of round-trips and the evaluating duplicates are located worldwide, the apparent latency barrier is caused by the light speed and the planet size.

**Execution**- The duplicates must execute the commands once the commands have been received and a consensus has been achieved on the overall ordering of such commands. The implementation algorithm is a process that takes the previous state and computes the new state and output using ordering commands. e.g., Whenever an execution needs a large number of cryptographic procedures, then the apparent bottleneck seems to be that duplicates must execute these cryptographic operations again.

### 9.2.1  Scaling the Data

#### 9.2.1.1  Better Network Solutions

The potential to grow bitcoin and other cryptocurrencies is highly dependent on the possibility of minimizing the time it takes for a successful block of orders to spread and reach all other miners. For example, forward error control and parallel processing codes are used by systems like falcon, fiber, and bloXroute to transmit blocks with decreased latency. Another option to increase data scalability is to use data through the network's dynamic capabilities and identify peers and access data.

### 9.2.1.2  Pushing Data to Layer Two

The bottleneck created by copying all the instructions can be alleviated by not copying them. Lightening, plasma, and other layer solutions try to limit data duplication by delegating some intermediary tasks to a tiny closed community and reporting periodic summaries to an original network. This method has a natural disadvantage: not duplicating all of the data results in a data availability issue. Each private group's security is dependent on having at least one truthful party who can respond quickly [6].

## 9.2.2  Scaling Consensus

### 9.2.2.1  Throughput Versus Latency Trade-Off

Some individuals use Transactions-Per-Second (TPS) as a metric for a consensus protocol's scalability. Scaling consensus requires addressing both latency and throughput. It's simple to increase consensus throughput (while lowering latency) by batch processing: achieve consensus mostly on a batch of a hash of all the data once a day rather than second data Apparently, the cost of consensus will be minimized only once a day, and consensus will not be a throughput bottleneck. Batch processing is a useful technique for reducing latency and increasing the throughput of consensus mechanisms, but it is not a solution for scaling consensus performance.

### 9.2.2.2  Performance Versus Security Trade-Off

Some believe that performing consensus on a limited collection of verifying copies will improve its performance. While reducing the number of verifying duplicates improves performance but compromises security. The most important thing that should be considered is increasing consensus performance while maintaining the level of validated copies. For enhancing the consensus performance without compromising the security of mechanism, various solutions are present, e.g., by lowering the number of rounds or altering the message difficulty from polynomial to linear [6].

### 9.2.2.3  Scale Versus Adaptive Trade-Off

The attacker can dynamically target the primary in consensus mechanisms based on the PBFT. A consensus mechanism's security is being defined by the attacker's size (which is defined by the total number of verifying copies) and by the attacker's adaptive ability. This mechanism dealing with adaptive attackers is frequently more expensive and difficult to scale. To scale byzantine consensus and defend it against adaptive attackers, algorand employs round-based cryptographic selection. The simulation findings for this approach are encouraging. An adaptable attacker can use Denial-of-Service attacks to prevent the system from developing. HoneyBadger proposes the first real asynchronous BFT method, which ensures liveness without making assumptions regarding time.

### 9.2.2.4  Avoiding a Total Ordering of All Commands

When all commands are dependent, then the only way to reach a consensus is to produce an overall order of commands Many workloads have commands independent of one another and may not conflict with each other. In certain circumstances, an order where A pay to B and for second-order, where C pay to D, in both cases they do not interact with each other, therefore there is no need to bottleneck the process and spend valuable consensus attempts to effectively provide a rank to these two commands. In epaxos, the non-byzantine paradigm was used. Various consensus mechanisms like DAG-based and Avalanche mechanisms enhance consensus throughput by enabling non-interfering commands to be recorded simultaneously.

### 9.2.2.5  Sharding

Sharding is the concept of splitting the state and the collection of validating copies at a higher level. Each shard is in charge of the state, while the entire validating duplicate population is in charge of consensus. There must also be some cross-shard system in place. The Ethereum "Sharding FAQ" is an excellent resource for learning further about sharding. Data, consensus, and execution bottlenecks may all be parallelized via sharding. The ability to parallelize data and execution is being expected on a workload with minimal disputation. Sharding is basically compared to a security tradeoff and performance from the viewpoint of consensus: rather than employing all verifying copies to protect one state machine, sharding produces many state machines. Among each valid copy, only one of them is secure. While having numerous shards (when conflict is minimal) might definitely enhance performance, it can also degrade security because fewer validating copies protect each shard. For examples of systems that employ sharding strategies, consider omniledger and ethereum 2.0. It aims to integrate each shard's weaker security with a greater safety of the global chain. The premise is that each weaker security shard can frequently confirm itself with the greater protection of the global chain, similar to second layer solutions. This provides a latency and security tradeoff awaiting highly secure needs wait for global chain final approval regularly.

## 9.2.3 Scaling Execution

One of the core structural components of state machine replication is the division of execution and consensus. To notify the benefits of this division, a command is being duplicated and saved; it must be performed on all verifying copies in the classic SMR design. The overhead of executing commands is the main scalability constraint in many systems. A DoS attack on the state machine replication network is an issue of legitimate commands that cause the system to lose time during execution. Many systems use Domain-specific languages (DSLs) to mitigate these threats. Bitcoin employs the bitcoin script, which keeps the computing cost of each transaction to a minimum. Ethereum employs a gas method to reduce processing overhead and promote optimal usage.

### 9.2.3.1 Parallelizing Execution

Machine parallelization is one potential method for reducing the operational computation bottleneck. When the commands in the block are usually free of conflict, this strategy works. The fundamental goal is to identify techniques to imitate linear execution using a mechanism that uses parallelism in the optimistic disputation-free scenario but ensures security in the case of conflict.

### 9.2.3.2 Don't Execute, Verify Using Economic Incentives and Fraud Proofs (Optimistic Rollups)

The commands are saved as data in this solution, but the execution of verifying copies is not done. The verifying copies serve as a layer of data availability; rather than having copies run the commands, users can engage in a financially game in which they can become executors by placing bonds. A bond operator can save the execution result. Any bond reporting party can submit fraud evidence that the operator made a mistake in the execution process. The executor is clipped, and the reporting agent is partially paid if the fraud evidence is accurate. The reporting agent reclines fraud evidence, and then their bond is cut. The work of employing on-chain encouragement has given which is effectively confronting the operator.

### 9.2.3.3 Don't Execute, Verify Using Succinct Proofs (zk Rollups)

The commands are saved as data in this solution, but the verifying copies can not execute the data. The verifying copies serve as a data availability layer for the commands. It is feasible to use brief non-interactive evidence instead of games and fraud evidence to check execution. These cryptographic algorithms enable companies to create very brief evidence with strong cryptographic correctness and completeness that can be confirmed quickly. Only one entity should be responsible for the execution and evidence creation. When one small piece of evidence is created, the execution engine's verifying copies only needs to verify the small evidence rather than execute the longer transactions again. To enhance the scalability of bitcoin, first, securely combine new and innovative substitute transaction mechanisms and enlarge the bitcoin's unreliable nature to other sorts of transactions. Using concise evidence has a significant benefit: once the evidence is prepared, the validation cost is minimal. The drawback is that producing evidence of command execution consumes significantly more power than just running the commands (both in terms of memory and CPU) Another issue is that these techniques add a significant amount of complexity to the system. Furthermore, concise proofs are available in a variety of flavors, some of which need nontrivial trusted setup processes. Rollups aim to eliminate the execution bottleneck while leaving the data bottleneck intact.

### 9.2.3.4 Executing (Validating) Requires State or Can It Be Stateless?

Another difficulty with execution is that it needs not just the command but also the present state. A complete node in ethereum takes several gigabytes, whereas an archive takes a few terabytes [18]. A complete node in bitcoin incorporates several terabytes. Getting a huge state raises the entrance barrier and reduces the speed where a new node may begin verifying commands. This is especially difficult in a sharding architecture, as participants must be shuffled around to minimize the strength of an adaptive attacker. If each shard's state is huge, the time it takes to sync with it becomes a huge obstacle to the speeds at which participants may be randomized among shards. As a result, lowering the quantity of data required for verification enables secure sharing and increased scalability. One solution is for the command to provide cryptographic evidence of all the required data for approval. Stateless clients are a concept that can be realized using Merkle tree evidence. Stateless validation is becoming more efficient due to the new cryptographic approaches.

## *9.2.4  RPoC Reputation Based Scalable Consensus Algorithm*

The current consensus mechanism optimizes PBFT from many angles; it does enhance consensus efficiency to some level when compared to regular PBFT.

Moreover, It addresses the problems of classical PBFT since they are not relied on the stability of nodes to filter consensus nodes. To address the aforementioned issues, we present an efficient consensus mechanism (EBRC) based on byzantine reputation, which is able to support large-scale and dynamic systems. The key challenge for a reputation-based blockchain consensus is to find a way to achieve the following [24]

- Create a distributed reputation score scheme for the miners based on a trust threshold
- Monitor the miners actions
- React to observed malicious actions, including provisioning resources to inspect and withdraw the authorization of the miner;
- Maintain a list of authorized miners that can change in time and organize a rotation mining scheme
- Perform network access control and self-organization for the number of authorized miners according to the size of the network.

The protocol uses reputation information and timestamps to maintain node reliability and improve blockchain security through a deposit penalty mechanism [14]. The node's activity has a strong influence on its reputation score. It may be used as a motivator to act quite enough to become consensus nodes. The EBRC's VRF arbitrary election process assures randomization and fairness. When it comes to choosing nodes in a blockchain. Meantime, evaluating EBRC replaces the traditional PBFT has significantly enhanced the consensus network's scalability and performance and lower costs, and the difficulty of communication among nodes. EBRC also has a dynamic join and exit protocol (DJEP), which PBFT does not have. Blockchain systems can swiftly adjust to changes in network conditions and obtain consensus using this method. Experimental results with nearly around 40 nodes were performed using a blockchain prototype using the EBRC mechanism. The presented EBRC method outperforms current techniques such as PBFT in terms of consensus throughput, efficiency, and network communication overhead.

## 9.2.5 EBRC Framework

The EBRC consensus mechanism includes reputation development and reputation assessment, DJEP, EBRC consensus mechanism, VRF random election process. Initially, the reputation assessment algorithm examines the reputation and reputation development rate of nodes in the block for node-set of all nodes to ensure better-quality gathering of consensus nodes. Second, the VRF function chooses nodes at random with the elected leadership to verify that the election is random and fair. Third, the scalability and validity of the consensus are ensured by the EBRC consensus mechanism. Finally, the DJEP mechanism completes the dynamic join and

exit of nodes in the system; the EBRC reputation-based consensus mechanism efficiently increases node reliability and system security while increasing the blockchain
network's scalability.

## 9.3 Summary

This chapter highlights the various permissionless and permissioned scalable consensus algorithms. This algorithm helps to increase the ability to serialize a more
number of the transaction at each time, which improves the transaction throughput
in terms of scalability. It also solves the consensus finality issue, in which forks
are eliminated by the various consensus and improves the system's performance.
This chapter also highlighted various PBFT-based consensus mechanisms, in which
they provide better performance in terms of scalability, but when considering several
nodes, then it is limited in various PBFT-based mechanisms. To solve this issue,
the helix and reputation-based scalable consensus mechanisms are being explored to
improve the performance in terms of nodes and transactions. It also provides an efficient solution for consensus in dynamic systems. It uses a trust-based threshold and
security-based mechanisms to enhance the system's security. Thus, the consensus
mechanisms make a system more robust and reliable.

## 9.4 Question Answers

1. Which of the following consensus mechanism achieves higher scalability in terms
of throughput?

(a) Algorand
(b) Byzcoin
(c) HoneyLedger BFT
(d) Stellar

2. Which of the following consensus mechanism are considered in permission less?

(a) Byzcoin
(b) PoW
(c) Bitcoin-NG
(d) All of the above

3. Which types of consensus protocols are scalable in terms of node and reliability

(a) Helix based consensus mechanism
(b) Reputation based consensus mechanism
(c) Both A and B
(d) All of the above

4. Which are the challenges of consensus scalability?

(a) Consensus Finality
(b) Transaction Throughput
(c) Both A and B
(d) None of the above

### 9.4.1 Short Questions

(1) List out popular scalable consensus mechanism in terms of number of nodes
(2) List out the types of scalable consensus mechanism in terms of transaction throughput
(3) What are the main issues of PoW based consensus mechanisms?
(4) How PBFT is used to achieve the higher consensus scalability?

### 9.4.2 Long Questions

(1) How blockchain is help to scale the byzantine consensus?
(2) How RPoC consensus mechanism ensures the security of the system?
(3) Describe the Working process of Bitcoin-NG.
(4) How nodes, users and committee interact in helix consensus mechanism?

## References

1. Yuan X et al (2021) Efficient Byzantine consensus mechanism based on reputation in IoT blockchain. Wirel Commun Mobile Comput 2021
2. Berger C, Reiser HP (2018) Scaling Byzantine consensus: a broad analysis. In: Proceedings of the 2nd workshop on scalable and resilient infrastructures for distributed ledgers
3. Yakira D et al (2021) Helix: a fair blockchain consensus protocol resistant to ordering manipulation. IEEE Trans Netw Service Manage 18(2):1584–1597
4. IIT kharagpur. https://nptel.ac.in/courses/106/105/106105184/ [Accessed: 26-Apr-2018]
5. https://www.geeksforgeeks.org/consensus-algorithms-in-blockchain/
6. Abraham I. https://decentralizedthoughts.github.io/2019-12-06-dce-the-three-scalability-bottlenecks-of-state-machine-replication/ [Accessed : 06-Dec-2019]
7. coindesk.com. Santander: Blockchain Tech Can Save Banks Dollar 20 Billion a Year. [Online]. https://www.coindesk.com/santander-blockchain-tech-can-save-banks-20-billion-a-year. [Accessed: 16-Apr-2020]
8. ibm.com. Blockchain basics: introduction to distributed ledgers. [Online]. https://developer.ibm.com/technologies/blockchain/tutorials/cl-blockchain-basics-intro-bluemix-trs/. [Accessed: 17-Apr-2020]
9. tradeix.com. 6 essential blockchain technology concepts you need to know. [Online]. https://tradeix.com/essential-blockchain-technology-concepts/. [Accessed: 18-Apr-2020]

10. leewayhertz.com. What are the key concepts of blockchain development?. [Online]. https://www.leewayhertz.com/blockchain-development-key-concepts/. [Accessed: 18-Apr-2020]
11. 101blockchains.com. Benefits of blockchain technology. [Online]. https://101blockchains.com/benefits-of-blockchain-technology/. [Accessed: 18-Apr-2020]
12. ibm.com. Introducing trusted computing base components. [Online]. https://www.ibm.com/docs/en/zos/2.2.0?topic=function-introducing-trusted-computing-base-components
13. epfl.com. ByzCoin - an innovative solution. [Online]. https://actu.epfl.ch/news/byzcoin-an-innovative-solution
14. Hathaliya J, Sharma P, Tanwar S, Gupta R (2019) Blockchain-based remote patient monitoring in healthcare 4.0. In: 2019 IEEE 9th international conference on advanced computing (IACC), pp 87–91. https://doi.org/10.1109/IACC48062.2019.8971593
15. tedermint. Building the most powerful tools for distributed networks. [Online]. https://tendermint.com/
16. Luu L et al (2016) A secure sharding protocol for open blockchains. In: Proceedings of the 2016 ACM SIGSAC conference on computer and communications security
17. Ovenden J. OmniLedger: a secure, scale-out decentralised ledger via sharding. [Online]. https://medium.com/primalbase/omniledger-a-secure-scale-out-decentralised-ledger-via-sharding-7a71adb9ec3b
18. Akram SV, Malik PK, Singh R, Anita G, Tanwar S (2020) Adoption of blockchain technology in various realms: opportunities and challenges. Secur Priv 3:e109. https://doi.org/10.1002/spy2.109
19. POA Network. POA network: how honey badger BFT consensus work. [Online] https://medium.com/poa-network/poa-network-how-honey-badger-bft-consensus-works-4b16c0f1ff94
20. Algorand. The future of finance. [Online] https://www.algorand.com/
21. Kumari A et al (2020) Blockchain and AI amalgamation for energy cloud management: challenges, solutions, and future directions. J Parallel Distrib Comput 143:148–166
22. Li P et al (2018) Gosig: scalable byzantine consensus on adversarial wide area network for blockchains. arXiv preprint arXiv:1802.01315
23. Stellar. Stellar consensus protocol [Online] https://www.stellar.org/papers/stellar-consensus-protocol?locale=en
24. Gupta R et al (2020) Blockchain-based security attack resilience schemes for autonomous vehicles in industry 4.0: a systematic review. Comput Electr Eng 86:106717

# Chapter 10
# Building Trust in Blockchain Network Using Collective Signing

**Abstract** Collective signing is used to protect the authorities and clients from malicious users. It uses the consigning protocol (CoSi) that gives affirmation of the authentication. CoSi protocol ensures the validity of the authoritative disclosure. Nowadays, bitcoin can be a game-changer in the digital world. However, completing any bitcoin transaction still requires around 10 min for the transaction to be committed. A protocol based on byzantine consensus, a ByzCoin, is introduced to solve this problem. The byzcoin maximizes the scalable collective signing and executing the transactions in a few seconds compared to the bitcoin transactions of committing. The byzcoin maintains the public memberships of bitcoin by dynamically creating a new group in such a way that it establishes group based who has high power for calculating hash and which displays the successful block mines. Byzcoin removes problems like double spending and self-mining by generating collectively signed transaction blocks in a few seconds of executing the transaction. This chapter discusses the fundamentals regarding collective signing. However, it is utilized in byzcoin, the problems faced in deploying byzcoin, and how BLS multisignature plays a crucial role.

**Keywords** Collective signing · Byzcoin · BLS signature

## 10.1  Introduction

Collective signing aims to provide terms and conditions, where the end-user is prepared to accept those. It is a method introduced to overcome the message signing problem by multiple signer [1]. Generally, people get confused between collective signing and digital signature [2]. When various stakeholders sign one document and send it to the recipient with their agreement, it is known as collective signing. However, the document signs digitally in the digital signature and certifies the trust between parties. It offers security and reduces paperwork that saves time, money, and space. Figure 10.1 shows the collective signing approach, where multiple stockholders sign a single document and certify that document in the presence of different

S. Tanwar, *Blockchain Technology*, Studies in Autonomic, Data-driven and Industrial Computing, https://doi.org/10.1007/978-981-19-1488-1_10

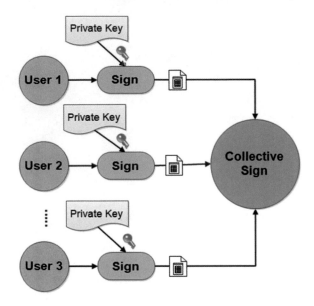

**Fig. 10.1** Collective signing

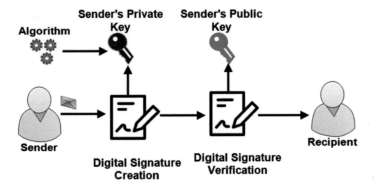

**Fig. 10.2** Digital signature

witnesses. Whereas, Fig. 10.2 shows the working of digital signature. It represents the creation and the verification of digital signature by sender's private and public key, respectively.

Collective signing is essential in numerous areas like the voting system, industry, banking system, organization, and corporate environment. There are multiple authorities involved in the design, such as time-stamping and logging authority, lotteries, digital notaries, and naming authority, in which there are issues of transparency, security, and privacy [3]. There is a need to protect these authorities from malicious users, attackers, and spy agencies. We use a blockchain-based approach in collective signing that provides transparency of data and security of the information. In today's world, cryptocurrency requires security and performance enhancement [4, 5].

Bitcoin is a decentralized public cryptocurrency and an alternative to the traditional currency managed by the bank authorities. Bitcoin is built upon the peer's networks [6]. Some miners take the transaction, solve a proof of work and add the transactions to the current block or the new block to form a public ledger. The original paper of bitcoin indicates that the blockchain transactions are secure and not reversible [6]. But later, it was found that there are security loopholes like double spending, transaction reversibility, and strategic mining at tracks. Another problem with bitcoin is it does not provide strong consistency. So it can also achieve other benefits as all miners should strongly agree upon the validity of blocks and not waste any resources. The client does not have to wait for a more extended period to validate whether his transaction has been committed or not. Finally, strong security should be there when the blocks are added to the chain, and they should remain immutable [7].

To overcome the aforementioned issues, there is a requirement to have a protocol that satisfies all the needs of blockchain consensus, such as byzantine fault tolerance (BFT) system, strong consistency, and scalability. Over time, byzcoin has been introduced, which works on the principle of a practical byzantine fault tolerance algorithm (PBFT) and proof of work (PoW) [8]. The byzcoin provides a solution for various issues like open membership, scalability of cryptocurrency, proof of work block conflicts, and transaction commitment rate [7]. The use of the BFT earlier variant in private blockchain limits scalability. A combination of collective signing and PBFT is used to overcome the scalability issues. Collective signing reduces both the cost of PBFT and the light cost for the client to verify the commitment of the transaction. Some assumptions against the PBFT are that it works in the closed group of replicas and employs proof of work that protects against the Sybil attack. Thus, byzcoin solved this problem dynamically by creating a group of consensus from the recently mined blocks, sharing some shares to the recently mined blocks, and voting power for the hash they generate using their resources. Byzcoin uses proof of concept though it is vulnerable to the DoS attack that any node in byzantine can perform [9]. Byzcoin provides security against the only attack that manages less than three consensus group shares. It can overcome some issues, which offers strong agreement. However, there are some problems in byzcoin like performance failure, reliability, and high failure due to the improper use of collective signing. In the following sections, we will learn more detail about the byzcoin and collective signing.

## 10.2 Byzcoin: Securely Scaling Blockchains

In this section, we will discuss the architecture of byzcoin and CoSi protocol for collective signing [10].

## 10.2.1  Byzcoin Architecture

There are four requirements of the blockchain consensus protocol, which can be solved by byzcoin. Figure 10.3 shows these requirements, and is discussed in detail as follows:

- The system should work even if there are any malicious users.
- System should guarantee strong consistency across the replicas, which means whenever we create multiple copies of the blockchain, every miner should be able to maintain that copies.
- The system should perform well if there is an increase in the number of transactions by providing the required throughput.
- The last requirement is the system should be scalable in such a way that when the network size increases, then it does not affect the performance of the system [9].

Byzcoin solved the above mentioned requirements, thereby providing the byzantine fault tolerance system that delivers a strong guarantee. Byzcoin is a proof of work (pow) and PBFT-based system. It is designed for the public or un-trust network where the transaction gets delayed, duplicated, or reordered. It also provides scalability with increasing workloads and network size. The byzcoin system is a collection of N block miners to generate key pairs. Every node has a limited amount of hash power which can correlate hashes to the most number of block headers that the node can perform per second [11]. A subset of miners will be handled by the malicious attacker who is considered faulty at any arbitrary time. The byzantine miners can behave arbitrarily and divert from the protocol to attack the system. The total hash power required in this process is less than one-fourth of the hash power of the system, which is less at any moment of the time.

A group of PBFT replicas referred to as trustees have been resolved and agreed that most trustees are faulty. In PBFT, at any moment, these trustees can be the leaders who propose the transactions and maintain the consensus process. Their trustees collect the transactions from the clients, add them to the block, and claim that there will be only one blockchain that it can never rollback. On the other side, the Sybil attack can easily break or destroy the membership protocol. So, to solve this problem, proof of work mining is used. In this, the miner who has the proper resources can become a member of the consensus group. This mechanism manages the power balance within the consensus group of the BFT over a given time of stationary sliding share window. Figure 10.4 shows the byzcoin design that contains micro blocks. A micro block is

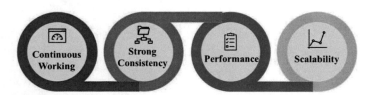

**Fig. 10.3** Requirements of blockchain consensus

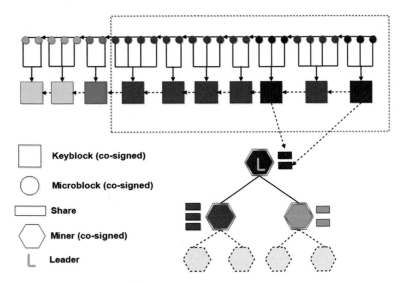

**Fig. 10.4** Byzcoin design

a simple block that the consensus group produces every few seconds. Each micro block has collections of the transaction, collective signature, hashes of the previous blocks, and key blocks. The key blocks state which leader of the consensus group window generates the signature of the micro block [11].

### 10.2.2 Collective Signing—CoSi

The whole byzantine network is based on collective signing (CoSi). The goal of CoSi is that there are groups of people, and few people in some groups behave maliciously. If we get a signature from all people from the group and believe that 50% of people in that group are valid people, we will get a document having signatures from the valid nodes [7]. In this way, we can prove the validity of the system.

The collective signing can help the system get signed by the many signers instead of the one signer. Due to this, we will know the validity and the integrity of the micro blocks. CoSi is a method that protects the authorities and their clients from undetected exploits [13]. It ensures that every authoritative statement is validated and openly logged by a group of people who work as a witness before any client accepts it. Figure 10.5 represents the CoSi architecture. Here, the leader manages the witness in a tree structure. It is a scalable way of combining signatures that come from the successor. There are three rounds of PBFT, i.e., pre-prepare, prepare, and commit, which can be simulated using two rounds of CoSi protocol [13]. Figure 10.6 shows the two rounds of the CoSi protocol. The first phase is the announcement, where the leader announces sharing to all the witnesses. The second phase is commitment, where

**Fig. 10.5** CoSi architecture [12]

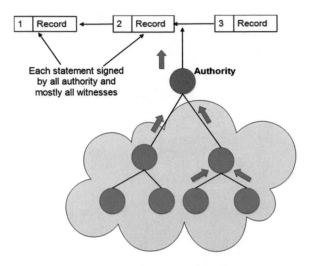

**Fig. 10.6** CoSi architecture phases

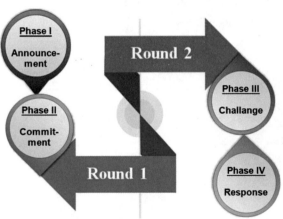

leader collects the aggregate commitment from their successor and computes their value. The first and second phase makes one round of CoSi protocol that implements the pre-prepare and prepare phase of PBFT. The third phase of the CoSi protocol is the successor sending collective challenges, and the leader computes the challenges to all the nodes. In the last phase, all the nodes send the collective response to the leader. Here, the third and fourth phases implement the PBFT commit phase.

CoSi uses the Schnorr signature, which believes the discrete logarithmic problem is challenging to solve for a group of prime order [12]. Schnorr signature follows a four-step process, i.e., key generation, signing procedure, verification procedure, and proof. The first step is a key generation, which follows that basic digital signature concept. Here, the node is signed by the private key, whereas the sign validates by the public key available to other users. In the signing procedure, the leader aggregates the sign of a particular response. After that, the sign generates by the private key aggregates the verification key at the time of verification. Here, every node verifies the sign and message for the verification. Finally, the proof of the sign is collected.

### 10.2.3 Byzcoin: Securely Scaling Blockchain

There is no secret that bitcoin is in the current scenario facing severe scalability problems. Its consistently rising prevalence has been joined by a regularly expanding volume of exchanges (tx), pushing the cryptographic money to the brink of collapse. Sometimes every part of the system is overloaded, which thwarts regular operation. As of 3rd March 2016, bitcoin reached its 1MB-per-block transaction processing cap. It resulted in a backlog where up to 30,000 transactions were waiting for many hours. In some cases, it requires many days to process the transaction [14]. ByzCoin offers a solution to bitcoin's scalability issues. It shows how it can safely process the hundreds of transactions per second of decentralized miners.

## 10.3 BLS Multisignature

The BLS multisignature came into existence to address the schnorr signature issues that require two communications round trips. With BLS, multisignature, we do not need two round trips; a single communication round trip is sufficient [15]. BLS signatures are of less size in length. BLS signature is 160 bits, while the digital signature is 320 bits long. BLS signature is based upon the gap of the Diffie Hellman algorithm. Further, the BLS signature schemes work in prime order and support the generation of the signature, crucial generation threshold, and combination of signatures. The systems use bilinear pairing and hash functions. Bilinear pairing is used for verification of the signature over the elliptic curves, while the hash function is a one-way hash function [15]. The various steps involved in the BLS multisignature scheme are as follows:

- **Key Generation**- The secret key of the signer $u \in U$ is a random element $x \in z * p$, users public key $u = g^{xi}$ . Here U is a group of signers who take part in signing statements digitally.
- **Signing**- A one-way function hash H is used, which generates an output of a random element in group G & n is the statement is used for signing. Therefore, signer $u \in L$ computes $H = H(n)$ and will return $\sigma = H^{xi}$
- **Aggregation**- The person or the system who provides multisignature will finally collect all $\sigma's$ generated by $u_i$ and will computer aggregate functions as $\sigma = \sum_i^n$ and will return n, L, $\sigma$ .
- **Verification**- When verifier provides g, n, L and $\sigma$ the system collects all ui by L and computes $v = \sum_i^n v_i$ , $h = H(n)$ and verifies $\delta(g, \sigma) = \delta(v, h)$.

By using the aforementioned steps, anyone can sign the statement and verify their signature. If the signer generates a signature $\sigma_i$ and the signature is correct, then verifying that signature can be done using the following equations of bilinear pairing.

$$\delta(g, \sigma) = \delta(g, \sum_{i=0}^{n} \sigma i) = \delta(g, \sum_{i=0}^{n} hx_i) = \delta(g, h) \sum_{i=0}^{n} xi \qquad (10.1)$$

$$\delta(v, h) = \delta(\sum_{i=0}^{n} vi, h) = \delta(\sum_{i=0}^{n} gx_i, h) = \delta(g, h) \sum_{i=0}^{n} xi \qquad (10.2)$$

If $\delta(g, \sigma) = \delta(v, h)$ then each individual signature $\sigma_i$ which are generated are correct. If we compared it with schnorr multisignature, then schnorr supports the multisignature, but incorporating the signature can only occur when signing happens. Thus, it needs multiple round protocols between the signers. While In BLS, combining the signature can take place openly via basic multiplication even though the signature is generated and signers are not there.

### 10.3.1  Applications of Using BLS Multisignature

The author provides an important application which is related to the bitcoin and described as follow:

- **Multi sign addresses**- When we use BLS multisignature in bitcoin, anyone can combine multiple signatures into a single signature. Hence, it reduces the size of the transaction [15].
- **Multi-input transactions**- When there are multiple inputs for the transaction, one can combine all the signatures into that particular transaction. The other is to build a system to compute all signatures over the same message using BLS multisignature concept.

## 10.4  Deployment Challenges

While developing the functionalities for the Byzcoin, which is built upon the bitcoin. It comes with a few challenges listed as follows:

- Creating the application so that it is compatible with the old version of the bitcoin until new code or application of miners supports the Byzantine consensus [7].
- Once the miners support the new byzantine consensus, an initial group of consensus needs to be developed, then replace the new consensus mechanism.

- Need to handle PBFT deadlock event which occurs very rarely because many miners may vanish in a short time and there will be no 2/3 majority available in the present group of consensus.

To solve the first and second issues mentioned above, we can use the already running Nakamoto consensus as a tool for bootstrapping. From the external point of view, we can say that the bitcoin would run as it is in a standard way until bootstrapping is not completed. Some of the things get shifted from the miner's viewpoint. However, every miner places the symmetric key, and the information like IP address and port number into the block miner are created. While recognising the symmetric key of the miner helps to claim their share's which is blocked until the miner finds the vital information of the proof of work [7]. IP address and port number are needed so that members of the consensus group locate each other, thus creating a communication tree. When the number of distributed shares hits the peak point or the maximum share window size, all the miners in the group of consensus shift to the byzcoin. The leader is selected when the last miner joins the group and the members in the group sign the critical block of the leader. The leader then generated the fresh microblock of the subchain from his key block and started developing and submitting micro blocks to the procedure of co-signing consensus. To solve the third issue, we can use the Nakamoto consensus as a backup. When the miners find too long a lack of input from the PBFT consensus group, they will return to their key blocks to make and commit the transactions [6]. Due to this, the framework will revert to its pre byzcoin agreement mechanism successfully. When some amount of the shares reach some value and are distributed, the miners will restart byzcoin consensuses. Another alternative is to use the bitcoining mechanism as a backup which provides the same performance as byzcoin, but some security features of byzcoin will not be available [7].

## 10.5 Use Case of Collective Signing

In the previous sections, we discussed the collective signing. Now, in this section, we are going to discuss one case study on a collective signing. One of the use cases is the combat environment, where there is a requirement to pass the correct message to the relevant person. We consider the case where collective signing plays a vital role in giving authentic information.

### 10.5.1 Determine the Combat Fake Content Using Collective Signing

There is an explosion of social media in the present era, and people suffer from fake digital content. They are finding it difficult to identify authentic and fake news. To

**Fig. 10.7**  Requirement of combat fake content determine system

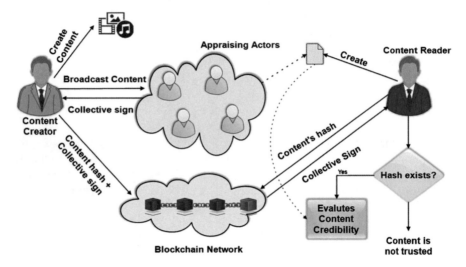

**Fig. 10.8**  TRUSTD model

resolve this issue, Jaroucheh et al. [16] proposed a blockchain and collective signature enabled system, named, TRUSTD. They have used a schnorr signing mechanism with the collective signing approach as mentioned in Fig. 10.7. The authors of [16] proposed TRUSTD system, where the user receives the correct content and gets the credit for the same. The author provided a decentralized and open-access platform that is accessible to everyone. In a combat environment, a decentralized ledger and collective signing approach enables security and trust between the different entities and identify the fake content. Figure 10.8 shows the TRUSTD approach, where they had specified two types of user such as content creator and reader. In their proposed model, the content creator creates content and broadcasts it to the appraising actor. The collective signature generates using the CoSi algorithm from the appraising actor to the content creator. After receiving the collective signature, the content creator sends the content hash with a collective sign to the blockchain network.

Next, the content reader gets the collective sign from the blockchain network. The trust policy is created between the content reader and the appraising actor. To assess the content creditability, the content reader authenticates the hash of the content. Rather than combat environment, this model is applicable in different areas such as academic publishing, where the various stockholders give their review.

## 10.6 Summary

This chapter shows the fall of the bitcoin value and at the same time how byzcoin is gaining momentum having a strong guarantee for the various issues. Byzcoin works on the principle of the practical byzantine fault tolerance. This chapter also discussed the byzcoin architecture and collective signing of BLS multisignature works in the multi-party environment to verify the digital signature. As well as introduced idea regarding the BLS multisignature is helpful in a multi-input transaction and multisign addresses.

## 10.7 Question and Answer

### 10.7.1 MCQ Questions

1. A collective agreement shall—

(a) be in writing and signed by the parties to the agreement;
(b) contain the date on which it is to become effective;
(c) contain procedures for the avoidance and settlement of disputes arising out of the interpretation, application and administration of the agreement, which may include a reference to conciliation or arbitration;
(d) provide for such other matters as may be agreed between the parties.

2. Which is value in the key block used in micro block generation?

(a) Nonce value
(b) Public key
(c) Coinbase transaction
(d) Unix time

3. How is the micro block fork handled in Bitcoin-NG?

(a) Micro block fork transactions are added in the next key block
(b) Micro blocks are added only after waiting for the propagation time of the network
(c) Micro blocks are merged with the previous micro blocks of the corresponding key block
(d) None of the above

4. Collective signing is used in the following protocol:

(a)  Bitcoin PoW
(b)  Bitcoin-NG
(c)  Byzcoin
(d)  PBFT

5. How many private values are used by each signer in Schnorr Multisignature?

(a)  One
(b)  Two
(c)  N (the number of signers)
(d)  log N

6. The topology used in CoSi for sharing the information is:

(a)  Chain-based
(b)  Ring-based
(c)  Tree-based
(d)  Star-based

7. How does the previous block reference in each key block of Bitcoin-NG?

(a)  Previous key block
(b)  Previous microblock
(c)  Both the previous key block and microblock
(d)  Either previous key block or microblock

8. In case of collective signing, if the system allows at most f false witnesses, then what is the minimum number of true cosignature required on the statement for ensuring the honesty in the system?

(a)  f
(b)  2f
(c)  f + 1
(d)  2f + 1

9. Which of the following strategies are used to manage the scalability of the Byzantine Agreement in the Algorand network?

(a)  All the nodes are added as a part of the random committee
(b)  Selecting a specific number of nodes to form the random committee
(c)  Make a single node to decide on the Byzantine Agreement
(d)  None of the above

10. Byzcoin is a—

(a)  Proof of work based-system
(b)  PBFT-based system
(c)  Combination of proof of work and PBFT-based systems
(d)  None of the above

## 10.7.2 Fill in the Blanks

1. CoSi protocol uses ....... signature?
2. Scale up the CoSi protocol further using ..... signature?
3. BLS signature uses ..... for verification?
4. ..... is a communication complexity for PBFT?
5. PBFT can be implemented over two subsequent CoSi rounds ...... and ..... ?
6. PBFT support ..... consistency ? 7. POW scale well to increase ..... where as PBFT scale well to increase ..... ?
8. In collective signing, the leader manages the witness in ..... structure.

## 10.7.3 Short Questions

1. What is the CoSi protocol?
2. How collective signing will change the short coming of Bitcoin-NG?
3. Discuss CoSi based on Schnorr signature.
4. What are the advantages of BLS?
5. How CoSi is incorporated in case of Byzcoin?
6. What is difference between strong and week consistency in blockchain?

## 10.7.4 Long Questions

1. In what ways do Blockchains use a public witness?
2. How does signing a transaction work?
3. How does ethereum transaction signing work?
4. Explain how BLS signatures is better than Schnorr? Justify your answer.
5. How to enhance the bitcoin security and performance with strong consistency via collective signing?
6. Why collective signing is important? Explain with example.
7. Briefly describe the CoSi architecture.

## References

1. Rjaško M, Stanek M (2012) Attacking m&m collective signature scheme. Fundam Inform 114(3–4):319–323
2. Mehta P, Gupta R, Tanwar S (2020) Blockchain envisioned uav networks: challenges, solutions, and comparisons. Comput Commun 151(02 2020). https://doi.org/10.1016/j.comcom.2020.01.023

3. Syta E, Tamas I, Visher D, Wolinsky DI, Ford B (2015) Certificate cothority: towards trust-worthy collective CAs. Hot Top Priv Enhancing Technol (HotPETs) 7 (2015)
4. Jay P, Kalariya V, Parmar P, Tanwar S, Kumar N, Alazab M (2020) Stochastic neural networks for cryptocurrency price prediction. IEEE Access 8:82804–82818. https://doi.org/10.1109/ACCESS.2020.2990659
5. Patel MM, Tanwar S, Gupta R, Kumar N (2020) A deep learning-based cryptocurrency price prediction scheme for financial institutions. J Inf Secur Appl 55:102583. https://doi.org/10.1016/j.jisa.2020.102583, https://www.sciencedirect.com/science/article/pii/S2214212620307535
6. Nakamoto S (2008) Bitcoin: a peer-to-peer electronic cash system. Decentralized Bus Rev 21260
7. Kogias EK, Jovanovic P, Gailly N, Khoffi I, Gasser L, Ford B (2016) Enhancing bitcoin security and performance with strong consistency via collective signing. In: 25th {usenix} security symposium ({usenix} security 16), pp 279–296
8. Bodkhe U, Mehta D, Tanwar S, Bhattacharya P, Singh PK, Hong WC (2020) A survey on decentralized consensus mechanisms for cyber physical systems. IEEE Access 8:54371–54401. https://doi.org/10.1109/ACCESS.2020.2981415
9. Bano S, Al-Bassam M, Danezis G (2017) The road to scalable blockchain designs. USENIX; Login: Mag 42(4):31–36
10. Kokoris Kogias E, Jovanovic PS, Gailly N, Khoffi I, Gasser L, Ford BA (2016) Bitcoin meets collective signing. In: 37th IEEE symposium on security and privacy. no. POST_TALK
11. Alangot B, Suresh M, Raj AS, Pathinarupothi RK, Achuthan K (2018) Reliable collective cosigning to scale blockchain with strong consistency. In: Proceedings of the workshop on decentralized IoT security and standards, pp 1–6
12. Blockchain architecture design and use cases (2022) https://nptel.ac.in/courses/106/105/106105184/. Accessed 23 Jan 2022
13. bt Abd Halim NS, Rahman MA, Azad S, Kabir MN (2017) Blockchain security hole: issues and solutions. In: International conference of reliable information and communication technology. Springer, pp 739–746
14. Jovanovic P (2016) ByzCoin: securely scaling blockchains. Hacking, Distributed, August
15. Boneh D, Drijvers M, Neven G (2018) BLS multisignatures with public-key aggregation
16. Jaroucheh Z, Alissa M, Buchanan WJ, Liu X (2020) TRUSTD: combat fake content using blockchain and collective signature technologies. In: 2020 IEEE 44th annual computers, software, and applications conference (COMPSAC). IEEE, pp 1235–1240

# Chapter 11
# Adoption of Blockchain in Enterprise Computing

**Abstract** Nowadays, enterprises require more security because it integrates with national and geographical boundaries. Reaching the security level in the traditional way of noting the progress of the business is not a feasible solution. Many of the enterprises use IT technologies to improve profit. To achieve security in the business cycle, blockchain provides a prominent solution. As well as, blockchain can help to identify faulty ends. Blockchain also provides facilities to select public or private blockchains as per requirement. In this chapter, we have defined the enterprise computing and the use of the blockchain in enterprise computing. Various actors and components of blockchain solutions are also discussed in detail. The scenario of the blockchain in the enterprise is explained with a digi locker, and supply chain management case study. The possibilities of adapting blockchain in today's buzzing enterprises are addressed as well.

## 11.1 Introduction

Businesses work on large-scale developments and operate in departments. The afore-mentioned kind of organization or enterprise has different departments for different applications. Keeping track of every activity arising in the business is tedious work to handle. In this case, a single person or simple paperwork can not be a feasible solution. Likewise, if an organization chooses a team to handle the processes and data during the development, confusion may arise. An of adoption software tool to handle the organization's data and the processes can be an expedient way to solve the problem. Enterprise software is an acquisition of computer software that fulfills organization-specific needs. Figure 11.1 shows the services provided by enterprise software. On the same note, enterprise computing is designed to provide information technology (IT) tools for organizations. According to the business research guide, "Enterprise computing is a term that refers to a myriad of information technology (IT) tools that businesses use for efficient production operations and back-office support" [1]. Organizations are dependent on IT tools that provide various functions to know the growth of the business, Alternative solutions if any problem occurs, databases to keep records and many other features. Common IT tools are [1],

© The Author(s), under exclusive license to Springer Nature Singapore Pte Ltd. 2022     287
S. Tanwar, *Blockchain Technology*, Studies in Autonomic, Data-driven and Industrial Computing, https://doi.org/10.1007/978-981-19-1488-1_11

**Fig. 11.1**  Services provided by enterprise software

- Enterprise Resource Planning Systems (ERP)—Helps in gathering, collecting, retrieving and analyzing information from various business departments for business professionals.
- Customer relationship management systems (CRM)—Collects information from different sources like transactions between parties, social media interactions and other feedback of customers and helps market research analysts, business managers and sales managers to manage pricing of the products, suggest modifications of products or packaging and identify new markets for existing products.
- Integrated Supply Chain Management Systems (SCM)—Supply chain involves various functions like collecting materials, producing goods, warehousing and distributing them for sale to end-user. These processes are very challenging. So, SCM helps logistics managers transporting goods to centers at correct timing.
- Product Lifecycle Management Systems (PLM)—Previously business leaders and engineers were storing data locally related to the product design but now, with the

help of PLM the layout, development, examination, generation, sustenance, and destruction of products that are generated by organizations can be easily accessed and modified with proper credentials.

### *11.1.1   Enterprise Computing*

Enterprise computing security is a crucial finding of an enterprise computing-based application. Security is the main concern that affects the overall performance of an enterprise computing-based application displayed in Fig. 11.2. An essential filtering requirement for application deployment is security sign-off. This chapter explores many facets of web security in enterprise computing. For a better understanding, the ideas, best practices, and security recommendations are supplemented with use cases and instances. The aspects in this chapter are the result of implementations and experience gained from a variety of real-world enterprise computing-based projects [2]. Enterprise security refers to the methods, strategies, and procedures used to protect information and IT resources are used to prevent illegal access and threats that might compromise the systems' integrity, confidentiality, and availability. Enterprise security expands the protection of information in movement across the linked network, servers, and end-users, built on the conventional cybersecurity assumption of securing assets at the local level. It includes all of the people, technology and procedures to keep digital assets safe. Because it covers the whole enterprise, this

**Fig. 11.2** Enterprise computing

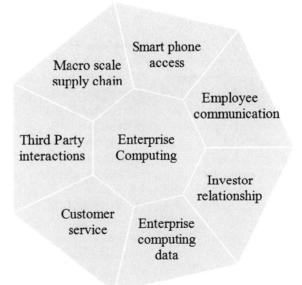

security focuses even more on the legal and cultural requirements of securing assets that belong to a company's users.

## 11.1.2  Security Flow in Enterprise Computing

Nowadays, the top technological trend is enterprise security. Security is required in every area of an industry's data architecture. Managing security has become more complicated as businesses' networks continue to grow and evolve. The security operations budget for many enterprises is more diminutive than 1% of their overall budget. Firewalls are the most common security device purchased when securing a corporate network. Firewalls should not be utilized as the sole way of network protection, but when implemented properly, they can reduce the risk of network security vulnerabilities and loss of data. The firewall may provide a variety of security applications in a single source. The integration of technologies like spam filtering, anti-viruses, URL filtering, and virtual private networks (VPNs) [3]. Enterprise computing, referred to as business-related information technology, is crucial to managing daily updates and operations. It is not feasible for everyone to manage and see the growth of business and make a decision based on it. Sometimes it is challenging to provide security in specific operations of business [4]. Due to the lack of privacy, data modification is possible. To address this issues, blockchain can provide end-to-end visibility and transparency in the business, which helps to increase the growth of the business. It provides security to the system so that customers can trust and invest in the company.

## 11.1.3  Motivation

Enterprise computing connects with multiple companies at various locations. All companies are connected with different banks. Initially, the two companies work together and transfer money through the bank. In this process, the company can request the bank for the transfer. If the bank successfully verifies the company and allows it to transfer the money, it will forward it to the receiver's bank. These interactions of the bank in the current enterprise computing-based system become expensive and inefficient. Thus, it is very difficult to track and control the assets in an enterprise system. The promising solution to this issue is a blockchain-based enterprise computing system. In this system, each user or participant can securely share their knowledge and information. It provides visibility of the entire system to each participant. Thus, the participant can see the growth of a specific company and invest based on growth. Blockchain-based enterprise computing system makes an enterprise system more reliable, immutable, and transparent. It also improves the security and traceability of the enterprise computing-based system, which helps in the overall growth of the business.

## 11.2 Blockchain Integration in Enterprise Computing

Nowadays, multiple organizations apply IT terms like artificial intelligence, Big data, and cloud computing to the betterment of the business. The improvement of business facilities along with the security of the organization's data is essential. For that, blockchain is a feasible solution. It is a decentralized system that works without a central authority [5]. Blockchain is a crucial technology whose demand is increasing in the future because it enhances the security of the system. There is no role of central authority in the Blockchain system, as it works in a decentralized manner. The Blockchain network is dependent on all the users present in its network. Every transaction performed by any user in the network will be recorded to every node in the distributed ledger. So, if any malicious activity occurs in the network, it can be easily identified. Every newly added block is connected to its previous block with the previous block's hash value. So, more security can be added to the network.

Figure 11.3 shows the blockchain-based enterprise computing solution. Here, nodes are called miners. To add a block in the blockchain, validation of the block must be done using miners. Validation is done by the miners with the use of consensus mechanisms. Permissionless, permissioned and consortium blockchain are the types of blockchain networks. All three types of blockchain use different consensus mechanisms. Plenty of consensus mechanisms have been presented until now, like Proof of Work (PoW), Proof of Burn (PoB), Proof of Stake (PoS), and Proof-of-Elapsed-Time (PoET) for permissionless blockchains, RAFT, and PAXOS for permissioned blockchains, etc. PoW is mostly used in public or permissionless blockchains. Trust in the blockchain is increasing day by day, and many businesses use blockchain to secure organizations' data. A smart contract is a self-executing contract that helps businesses automate business workflows [6]. Blockchain for enterprises can help in building new financial technologies, maintaining track of physical assets or voting rights, tracking ownership, digital assets, and locating the source of a document [7]. According to The FinTech 2.0 Paper, "Blockchain technologies could diminish banks' infrastructural costs by $15-20 billion a year by 2022" [8] and showed that "distributed ledger technology (DLT) would save banks some money by removing central authorities and evading slow and expensive payment platforms" [28].

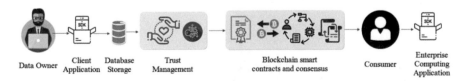

| Data Owner | Client Application | Database Storage | Trust Management | Blockchain smart contracts and consensus | Consumer | Enterprise Computing Application |

**Fig. 11.3** Blockchain-based enterprise computing

## 11.3 Problem: How It Is Difficult to Track Asset Transfers in a Business Network

Anything capable of being owned or can be controlled to produce value is termed as an asset. They are mainly classified into two types that is tangible and intangible. Tangible includes car, house, and intangible includes patents, music, etc. Today's business needs are demanding the transferring assets on national and international levels between the associates like producers, consumers, suppliers, partners, market makers/enablers, and other stakeholders [9]. Every associated party can control their assets in terms of granting privileges, access control, ownership, and modification. Assets can be transferred from one party to another, and ownership changes accordingly. This transfer is a transaction in terms of business networks [9]. The transactions between parties are recorded in the business ledger. The business ledger keeps track of economic activities occurring in the business. Businesses also include authorities like banks. Agreements and contracts of the third party like the banks are also recorded in the business ledger. Businesses use multiple ledgers to keep records.

Tampering of business ledgers is the biggest problem in business networks. Managing the ledgers are tedious and can be costly also. Any type of failure in the ledger can cause unreliable transactions and reduce trust in the business. Concurrent transactions also make the ledgers faulty. The above-stated problems can be solved if the business ledgers adopt blockchain technology. There is no need to rely on a third party for validating transactions. Member nodes in a blockchain network can use a consensus protocol to validate ledger content, digital signatures, and cryptographic hashes that are used to ensure the integrity of transactions. A shared network technology allows any participants in the network to see system records. Participants agree by the consensus on the updates of records. "No third party is involved in this. Every record in a distributed ledger has a timestamp and a unique cryptographic signature, thus making all ledgers an immutable history of transactions in the network" [9].

## 11.4 Key Concepts and Benefits of Blockchain for Business

### 11.4.1 Key Concepts

The key concepts of blockchain and different consensus algorithm are explained as follows:

- *Smart Contracts*—A Smart contract is a legal document that is script or code developed by a developer and deployed into a blockchain [10]. It is also used to automatically update the shipment details of the product. This facility eliminates the need for time and saves time. The smart contract is a digital code or program that automates the execution of business logic and agreement [11].

- *Mining*—Mining is the process of adding the block in the blockchain. Miners are users who utilize computational services to mine for blocks [12]. Every miner is competing to solve the block hash. In block hash difficulty target is set. The difficulty target is leading zeros in the hash. Miners have to find the appropriate nonce that can fulfill the Difficulty requirement. The above explanation is for mining in PoW. Different consensus algorithm does the process differently.
- *Proof of Work*—In proof of work, before adding any blocking network, it proposes a challenge for the miners who want to add the block into the chain. Miners have to solve that challenge and the first miner that solves the block will be able to add the current block into the chain of blocks. Challenge is to solve the hash value as per the given difficulty target. Here, a difficulty target is the number of leading zeros in the hash value. For solving challenge, miners require resources like computing resources and I/O resources. To solve any particular challenge resources are used and according to that rewards are given to the participated miners. "The concept behind this technique is to solve complicated mathematical problems and provide a solution. It requires a lot of computational power to solve a mathematical problem, proof of work has certain limitations. More a network grows, more the power is required" [12].
- *Proof of Elapsed Time (PoET)*—It is one of the safest consensus algorithms which is created for permissioned blockchain. Here mining rights or voting principles are decided by permissioned network. As the system demand identification of miners, a consensus algorithm uses a secure login to ensure identification. PoET determines the winner as a just means only [12].
- *Practical Byzantine Fault Tolerance (PBFT)*—This algorithm assumes that there are certain possible failures in the network and all autonomous nodes might not work correctly at certain times. PBFT is created for an asynchronous system [12]. All nodes are rearranged in a particular order. From all nodes, a single node act as primary and other acts as a backup plan. All the nodes perform functions and communicate with each other.

## 11.4.2 Benefits

Benefits or Advantages of blockchain are as follows:

- *Better Transparency*—With the help of blockchain, an organization can go for a complete decentralized system where there is no need for a central authority, which improves the transparency of the system. Peers take part in carrying out a transaction and validate it [13]. Further, to provide validation, a consensus algorithm is used. After a transaction is validated, each node stores a copy of the transaction record. That's how blockchain handles transparency [14].
- *Enhanced Security*—"Blockchain provides better security compared to other platforms. Each transaction is encrypted and has a proper link to the old transaction using a hash method" [14]. Security is enhanced as each node keeps a copy of the

transaction [15]. So if an attacker wants to make changes in a transaction, it will not happen because other nodes will reject his request to modify the transaction. "Blockchain network is immutable, which means once a transaction is written can not be reverted. Blockchain is formed by a complicated string of mathematical numbers and is impossible to be altered once created" [16]. The various consensus algorithm is used here to validate entry; it removes the risk of duplicate entry.

- *Reduced Cost*—As in blockchain, there is no need for a third party or middle man, the cost of blockchain can be reduced. Therefore less interaction needed for validation of the transaction.
- *True Traceability*—Companies with the help of blockchain can create a chain that processes with both the vendors and suppliers. "In this chain, it is hard to trace items that can lead to loss of goods, theft, etc. Blockchain enables every party to trace the goods and ensure that it is not being replaced or misused during the supply chain process" [14].
- *Improved Speed and Highly Efficient*—Blockchain solves the time-consuming issues because it processes which automatically maximize efficiency. "It minimizes human-based errors with the help of automation. The digital ledger makes everything possible by providing a single place to store transactions. The streamlining and automation of processes also mean that everything becomes highly efficient and fast" [14]. Blockchain allows verification without being dependent on the third party.

## 11.5    Degree of Centralization: Permission-Less Versus Permissioned Blockchains

A permissioned blockchain needs prior approval like you are required to have permission before accessing that blockchain, and permissionless blockchain lets anyone participate in it like bitcoin [17]. So in a permissionless environment, anyone can run a node, access a wallet, and transact in the blockchain.

### 11.5.1   Permissionless Blockchain

Permissionless or public blockchain is open to the public and requires some consensus algorithm to alter data. Here anyone can participate in read/write or audit transactions without permission. Each user in the network is anonymous to each other; a copy of the ledger is distributed across the globe, which makes it more reliable in terms of security. It is open to the public so anyone can download it and run it on the local machine, send a transaction, validate the transaction, and track it at any time on the blockchain. The decision-making process is done by consensus algorithms like PoW or PoS [18]. Digital Assets, decentralized, and transparency are the main characteristics of permissionless blockchains [17]. If a public blockchain has less number of nodes, it can be more vulnerable to attack. Businesses usually do not have a large number of nodes. So, it could turn risky to use the permissionless blockchain environment in organizations. As well as applications like controlled data reversibility, data privacy, transaction volume scalability, system responsiveness, and ease of protocol updatability required in business networks are not covered by the public blockchains [6]. So, most organizations do not favor using permissionless blockchains and prefer private blockchains.

### 11.5.2  Permissioned Blockchain

This type of blockchain is closed to the public and requires authorization to access it. These blockchains are generally used in corporations known as the private blockchain. In many cases, there is no need for a consensus mechanism to add blocks to the chain [30]. Instead, it depends on validators that it has preselected to endorse the transactions. Such blockchain is developed by an individual or private organization where everyone cannot audit/write/read the transaction. Mainly, write/audit permission is kept centralized, and read permission may be restricted to a limited number of people. The use of permissioned blockchains is favored by corporations or organizations with particular needs [19]. Semi-private blockchains are also a new term in the blockchain network. It covers some of the aspects of private as well as public blockchains. Differences of both blockchains are concerns of application-level, not architectural aspects in semiprivate blockchains [6].

### 11.5.3  Similarities of Permissioned and Permissionless Blockchains

Some common similarities between permissioned and permissionless blockchain are as follows:

- Both of these are distributed ledgers
- Both of these blockchains are immutable means that the stored which is stored cannot be modified or altered without having sufficient power.
- Both of these use consensus algorithm [17].

## 11.6  Actors in BC Solution

Like any application development, a business needs a designer, developer, network admin, and a customer who interacts with the developed application. In the blockchain, with these main actors or participants, who act as an examiner, the certificate authorities issue certificates for obtaining the data sources to obtain the data. Each participant has a specified key role in developing and maintaining the blockchain [20].

- Process Proprietors—They are specialists to effectively execute business processes. Their primary objective is to discover business challenges and assist in finding their corresponding solutions. In addition, the proprietors are essential partakers in specifying different use cases [20].
- Regulatory authority—They are the authentic authority that has wider access in the deployed area, and it issues authorizations by providing licenses to the participants. Furthermore, they specify various access control mechanisms and issue certificates to all the associated blockchain members. They guarantee that only the authorized person can access the relevant information from the blockchain network.
- Blockchain designers—They are the developers who design and implement blockchain into the enterprise to which they should know essential parameters such as their peers, security, chain code, consensus, ledgers, and application. Moreover, they assist in using consensus algorithms, defining practical standards and best practices [20].
- Solution providers—The designer decides the infrastructure of the blockchain network. Once it is determined, the solution provider of the blockchain ensures persuasive management, technical support, setup configuration, and maintenance. Further, they also deliver security solutions and maintain smart contracts [20].
- Network operators—The network operator of blockchain emphasizes creating a business network, configuring it, providing access control, and monitoring the business network [20]. They mainly concentrate on the regular operation of the blockchain network.
- Blockchain developer—Blockchain developers focus on developing, such as creating smart contracts, applying cryptographic solutions, etc. They should have knowledge of code in the high-level programming language, and he needs to have some basic understanding of the blockchain. These developers develop smart contracts and also test smart contracts. They use APIs in code to interact with the blockchain network [20].
- Blockchain end-user—These are end-users who can consume blockchain smart contracts with the help of web applications. This will allow the end user to execute transactions against smart contracts, and the underlying blockchain remains transparent to them.
- Auditors—The intermediaries consist of different organizations and parties with complexities in their business, leading to disputes. Therefore, an authentic regulatory body should have specific rights to confront disputes. Nowadays, these central authorities are replaced by authorized regulatory bodies encompassing smart contracts that manage the legal documents. The smart contract auditors verify the smart contract, its interface, and the application that are utilizing the smart contracts; this helps in tackling severe vulnerabilities hindering the performance of the business.

## 11.7 Components of BC Solution

The blockchain network has multiple participants as shown in the above section. Participants communicate with or to components of the blockchain. According to the [21] components in a blockchain solution [29],

- Ledger—Ledger is distributed among all the participants. Each change in the network will be recorded in the ledger which makes blockchain more transparent and trustworthy.
- Smart Contract—Smart contract is an agreement between all the participants. It is a self-executed code which does not require any human interaction. It is a software being executed on a ledger, that defines rules for transactions and modification of assets.
- Peer network—Network of nodes that are validating each transaction in the business network. Each peer holds ledger so that they know the complete transaction flow.
- Membership—As it is permissioned blockchain, it must require authentication, authorization, and management of identities. The mentioned services can be fulfilled by the membership component.
- Events—Events that occurred in the business blockchain are notified to the participants.
- System management—It grants the ability to formulate, edit, and observe blockchain components.
- Wallet—It manages the credentials of the user as well as cryptocurrency available with the user. Every node owns the wallet.
- System Integration—Used to connect the blockchain bi-directionally with the outer systems.

## 11.8 How Application of Selected Example Interact with Ledger

Many business applications adopt the blockchain for security and interact with a ledger differently. For example, supply chain management.

### 11.8.1 Case Study: Supply Chain Management

Supply chain management requires transparency from collecting raw materials to creating an end product and sending them to customers. Figure 11.4 [22] shows multiple actors participation in the system. All members participate in the supply chain, such as raw material producers, manufacturers, wholesalers, retailers, and customers. All the transactions performed by any participants must be recorded in

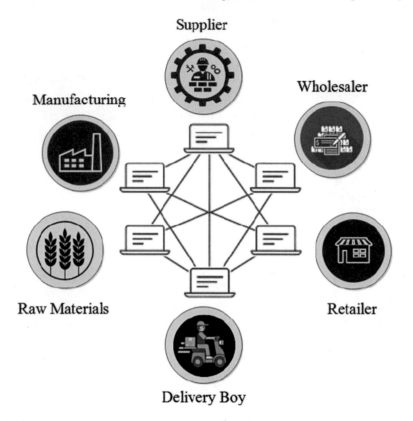

**Fig. 11.4** Supply chain management with blockchain [22]

the ledger. When any transaction takes place, the participated nodes do validation and verification. The participants who initiated transaction is directly associated with the smart contract agreements. Smart contracts are self-executable, and every participant in the blockchain is in the agreement. After the validation and verification, the transaction must be added to a ledger. Each transaction has a few parameters like the cryptographic hash of the transaction, sender address, receiver address, Merkle root, the timestamp, transaction currency, and some other information [22]. This information is aggregated and added to the ledger. Once the transaction is added to the ledger, it can not be altered or changed. To change the transaction, participants have to regenerate the transaction and repeat the entire process. Here, transactions are related to transferring money and tracking assets. For example in [22], Fig. 11.5 shows a process of delivery. In this process, a delivery boy who transmits the goods from a manufacturer to the wholesaler also participates in the transactions. Once the goods are transferred to the wholesaler, the delivery boy initiates the transaction and confirms that the goods are delivered. As a result, he obtains the transaction hash, and that transaction is recorded in the ledger. This implies the security of assets

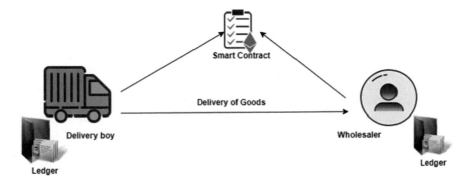

**Fig. 11.5** Process of delivery

**Table 11.1** Comparison of enterprise computing with traditional and blockchain-based approach

| No | Parameter | Traditional approach | Blockchain approach |
|----|-----------|----------------------|---------------------|
| 1 | Ledger | Business ledger (can be tempered) | Blockchain in ledger (temper proof) |
| 2 | Security | Low | High |
| 3 | Environment | Permissioned | Permissioned |
| 4 | Data availability | Available with few participants | Available with every participant |
| 5 | Transaction authority | Third part like banks | No central authority data is validated by ever node |
| 6 | System | Centralized | Distributed, Decentralized |
| 7 | Asset tracking | Less efficient | 100% efficient |
| 8 | Execution | By authority | Self executable with the help of smart contract |
| 9 | Accuracy | Low | High |
| 10 | Trust of the participants | Low | High |

and builds a trustful environment. Here, Table 11.1 shows the benefits of using the blockchain approach over the traditional approach.

## 11.9 Enterprise Computing Categories

The categories define the planning and structure of the business. For instance, how a business can run the flow and how it manages the relationship with its customer. Such categories are presented as follows:

## 11.9.1  Enterprise Resource Planning

With the advent of higher, lower cost, and much more adaptable computer systems, Enterprise resource planning has a chance to execute infrastructure that handles and organizes numerous routine or complicated operations for the facility. Enterprise resource planning (ERP) is a business system software that aids a firm or organization in successfully organizing, planning, tracking, and using assets (man, machine, material, and money) throughout the project. An ERP software is a generic software application that replaces stand-alone programs with a unified system. For many years, this network of components has been used in a variety of corporate settings and has formed the backbone of how businesses handle all activities. Figure 11.6 shows the structure of ERP system. There are several software suppliers on the market, including SAP, Microsoft, and Oracle, among others. ERP unifies a company's many functions into a single system. Earlier to ERP, if the Accounting department needed to know how often and what kind of inventory was on hand, they had to travel to that department and ask for the information, assuming it was accessible. Due to the extremely

**Fig. 11.6**  Structure of enterprise resource planning

controlled permission processes to get the information, accuracy frequently got in the way of acquiring awareness. If the information were available at the centralized system, there still could be some delay in terms of the data's timeliness and even correctness. With today's ERP, insight into activity outside of a single department is accessible immediately and is correct within a day or less. When making judgments in an environment that might change by the minute, providing such visibility is critical [23].

## 11.9.2  Enterprise Planning System

Large firms use enterprise planning to consolidate and synchronize planning across various departments, including finance, marketing, sales, HR, information technology, and operations. Enterprise planning aims to give planners a comprehensive view of the company's financial performance. Enterprise planning breaks down the operational barriers and combined operational and financial planning procedures. Enterprise planning solutions combine data from several departments into a single source that is accurately managed to make proper decisions. In enterprise planning, finance and operations enhance cooperation and improve corporate performance. In terms of technology, enterprise planning solutions combine non-financial and financial data in a single system and allow users to communicate across departments. This enhances the information flow. Finance has more visibility into the present, and future financial results, as well as how those drivers influence operations, due to an enterprise planning system that unifies planning processes throughout a company [24]. Figure 11.7 shows the structure of the enterprise planning system. Finance and operations are brought together through enterprise planning. Using enterprise planning, planners can identify the cause-and-effect link between financial and operational data. This offers planners a better understanding of resource requirements, leading to better company performance. Cross-functional cooperation is being aided by enterprise planning. Plans are generally isolated from each other when businesses are planned by department. Every department plans with other departments when a unified planning approach is implemented across the business. As a result of this understanding, operational plans that address financial outcomes and more effective quality management are developed. Performance is illuminated by enterprise planning. Many businesses restrict planning by confining methods to financial and financial alone. Organizations may, however, coordinate their whole business by using an enterprise planning approach. Once all planning operations are being integrated across the firm, planners from all departments can set aims and estimates budgets and plans that reflect all aspects of the business. Decision-making is aided by enterprise planning. By empowering important information across divisions, enterprise planning tools improve planners' analyses. Planners can identify the primary reason for performance by shifting and sorting corporate data according to numerous factors, taking action, or changing direction.

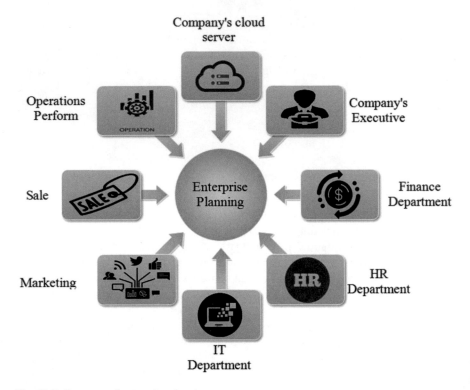

**Fig. 11.7**   Structure of enterprise planning

## 11.9.3   Customer Relationship Management

Customer relationship management (CRM) is an approach to monitor the relationship and interaction between the company's customers. The objective is straightforward, that is, to improve commercial connections. It aids businesses in keeping in contact with customers, streamlining procedures, and maximizing profits. A CRM solution focuses on the company's relationships with specific people such as service users, suppliers, customers, colleagues throughout the relationship's lifecycle, which is an excellent opportunity for new customers to earn their business and offer support and extra features. Organizing client and prospect information in an effective way to help to create strong connections with them and expand their business quicker. It also helps you to locate new customers, gain their business, and keep them satisfied. CRM systems begin by gathering information from a company's website, phone, email, social media, as well as other sources and networks. The CRM tool organizes data into a comprehensive record of persons and businesses, allowing users to easily understand their relationships over time. Today's CRM systems are more open, and they may link with your favorite business tools like accounting, billing, surveys, and document signing, allowing data to flow both ways and give a complete 360-degree

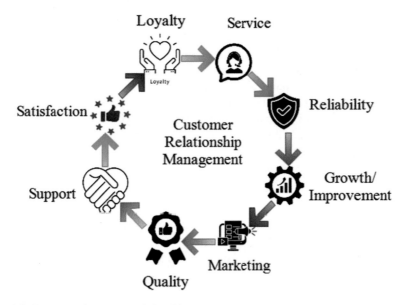

**Fig. 11.8** Structure of customer relationship management

perspective view to the customer. A new CRM generation goes much further, such as it incorporates intelligence that simplifies administrative processes like service case routing and data entry which remarks to concentrate on responsibilities. Automatically produced insights assist in better understanding their clients, even forecasting what they will think and act accordingly so the company can prepare the appropriate outreach [25]. Figure 11.8 shows the structure of customer relationship management. Following are the observations of the CRM when integrating it with any business,

- Add to company's bottom line improvements—Integrating a CRM platform was shown to obtain real results, which includes direct financial benefits. CRM software has an established track record of enhancing.
- Leads should be identified and categorized—A CRM technology can help identify and add new prospects and precisely categorize them. Sales can highlight the prospects that will complete transactions by concentrating on the right leads, and marketing can discover leads that need nurture and prepare them to become quality leads by focusing on the correct aspect. Sales and marketing can focus their efforts and resources on the correct clients if they have stored extensive and accurate information about them.
- Increase the number of referrals from current customers—Cross-selling and increased sales opportunities become apparent once a company understands their clients and offers the opportunity to acquire additional business from present customers. The customer can be satisfied with the better service provided by the company. These customers can bring trust and invest more in the future.

- Provide better customer service—Nowadays, customers want prompt, customized service at all hours of the day and night. A CRM system can assist their customers and provide better quality service than their customer's expectations. The agents can instantly see what products customers have bought and keep track of every interaction to provide clients with the information they want swiftly.
- Products and services must be improved—A strong CRM system will collect data from a wide range of sources inside and outside the company. This provides the company with unmatched insight into how their consumers feel and what they're saying about their company. It enables them to enhance what they provide, identify gaps, and discover problems early.

## 11.10    Case Study on Enterprise Computing

In the enterprise computing case study, we identified the supply chain management and digilocker for a business; in both case study, we identified the roles of the stackholders and defined a smart contract to execute their role and verify the product and documents to improve the better performance of any enterprise computing system. By looking at some of the real-world uses of blockchain in enterprise and other applications of blockchain that help revolutionize industries. Blockchain-based projects are increasing rapidly with time and some use cases such as healthcare, Energy, land registry system, Protecting digital Identity, etc., are given below [26, 27].

### *11.10.1    Case Study of Diamond Supply Chain*

Blockchain technology is a promising solution in identifying an item's legitimacy and tracking its route from its origins to its final destination. The value of high-end items such as jewels, artifacts, and fine wine is largely determined by their provenance. The blockchain is used to verify the provenance, making the system more valuable. Everledger is a well-known enterprise that gives digital transparency and profitable solutions for blockchain-based applications. Everledger now has data sets from over 1.6 million diamonds in its database. In reality, we must trust the retailers to validate the things we're purchasing, and retailers must expect their suppliers to do the same. Sadly, this belief is not practical in real markets as it is vulnerable to theft, prone to many errors, and lacks effective management. To tackle the trust issues in the retailer market, they must adopt blockchain technology that emphasizes the trust you provide on the goods. Unfortunately, various countries have big markets in terms of manufacturing goods, and their supply chain still has not incorporated blockchain in their business operations. In particular, in the diamond market, there are a lot of scandals of corruption and thievery, which makes the system incapable of tracing the origin country of a diamond. In addition, several challenges hinder the smooth operation of the diamond market, which is difficult to identify, such as

**Fig. 11.9** Diamond industry supply chain

record manipulation, forged documentation, exploit the mineral labels, and double financing. The reality that the diamond market has to face is that all intermediaries, from manufacturers to diamond cutters, laborers to the bank, and insurance have to mutually adopt blockchain-based solutions, which make the documentation and regular operation of the diamond market transparent to all the members (as shown in Fig. 11.9) associated with it.

## 11.10.2 Putting Diamonds on the Blockchain

A company named Everledger was founded in 2015 with the goal of adding clarity to the diamond market. Everledger was established in 2015 with an aim to provide cutting-edge solutions for the blockchain application. Recently, it has joined hands with IBM to incorporate an open-source program, i.e., hyperledger, to assemble blockchain. They developed many tools which are used in the manufacturing of the diamond, such as scanning, simulation, and cutting the gems. The manufacturing devices can be aligned with the blockchain to accurately operate and store the data in the immutable blocks of the blockchain. The protocol that Everledger utilizes prompts the user to provide necessary details of each production phase, such as the timestamp of the operation user information that handles the particular process in the production unit. Furthermore, the retailers can insert information such as store location and warranty details about the jewelry that holds the diamond. To access this information, a customer has to provide their credentials in an appropriate interface. The organization has a massive database with many diamond records, which improves the market rates and controls the supply chain.

Human activity poorly influences the electronic control program as there might be an adversary manipulating the business, manufacturing, and production data. A blockchain needs to overcome this vulnerability by using digital signatures in such instances. For instance, when diamonds are discovered in the mines, their location in the form of the transaction can be stored inside the blockchain. This ensures security and trust in the diamond market, where rogue diamond businesses cannot take place. Not anyone can access and write on the stored transaction; if someone tries to do that, it automatically gets detected by the blockchain miners and is considered a fraudulent business. In addition, it provides a mechanism by which we can securely trace the provenance of the diamond and other related products by employing the digital signature and machine-to-blockchain application. Consequently, it conveys value to the diamond market, its commodity, and all the entities associated with the supply chain.

### 11.10.2.1   Implementation

The process of the diamond supply chain contains various entities such as manufacturer, diamond retailer, diamond miner, jeweler, and customer. In this process, raw data of the diamond is mined in the smart contract of a diamond miner. Then, the mining procedure loads the diamond for transportation and moves it to the manufacturer. Next, the smart contract of the manufacturer collects the diamond, and if diamond details verify with the data miner, the manufacturer is payable. Next, the jeweler starts designing jewelry for the customer. Finally, after verifying the diamond, the required members of the supply chain are payable in their specific wallet address. Figure 11.10 shows the verification process of the members connected in the diamond supply chain process. In this smart contract, each and every member is verified and included in the supply chain process. If any unauthorized user was found, then withdraws that member from the supply chain process. Figure 11.11 shows the function details of diamond supply chain process. In this process, every function is associated with verification, product information, members information, purchase details store into the blockchain. In this, jeweler can be include after the successful verification of jeweler details. Figure 11.12 shows the mining process of diamond. In this mining process, the diamond unique product code, product notes, diamond miner name, and mined diamond information collect and store into the system. Figure 11.13 shows the process of diamond purchase. In this, the diamond

**Fig. 11.10**   Verification process of diamond supply chain members

**Fig. 11.11**   Functional details of diamond supply chain process

**Fig. 11.12**   Data collection for mining procedure of diamond

and its retailer information is collect and verify. If it is successful then the diamond is ready to purchase for a customer. Figure 11.14 shows the deployment process of all members connected in the supply chain process. After the product verification at each handover, the requested members are payable in their wallet address. The transaction details stored in the block of blockchain.

**Fig. 11.13** Data collection for diamond purchase

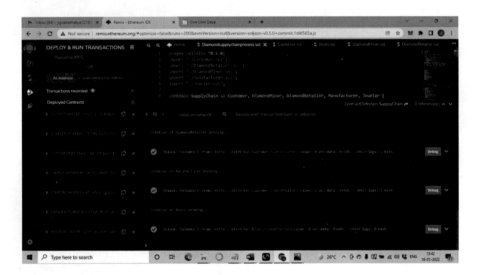

**Fig. 11.14** Deployment process of diamond supply chain roles

### 11.10.3   Case Study of Digilocker

DigiLocker is an online digitizing service provided by the Indian government Ministry of Electronics and Information Technology (MeitY) as a part of its Digital India project. Every Aadhaar user (unique identification number given to the Indian citizen) is given a cloud account in which they can obtain legitimate documents/credentials in digital format from the original holders of these certificates, such as a driver's

Peoples

Digitally document          Documents access
collection                        online

Data collector          DigiLocker          Data Requestor

**Fig. 11.15**  Architecture of digilocker

license, car registration, or academic mark sheet. It also gives each customer 1 GB of
storage space to upload scanned images of old documents. To use DigiLocker, users
must have an Aadhaar number. In order to use this service, a user has to provide his
Aadhaar number, followed by the one-time password (OTP) issued to his Aadhaar-
registered phone number in the DigiLocker web interface. Figure 11.15 shows the
architecture of the digilocker. Digital India, the Indian government's flagship program aiming at converting India into a digitally transformed knowledge and society
economy, includes DigiLocker as a significant component. DigiLocker is aligned
with Digital India's aim of giving citizens a shared personal space on a public cloud
and making all papers and certifications accessible through this cloud. DigiLocker is
a platform for issuing and verifying documents and certifications in a digital format,
withdrawing the need for physical documents. When Indian individuals join up for
a DigiLocker account, they receive a cloud-based storage space connected to their
Aadhaar number. Organizations that have signed up for DigiLocker can send citizens
electronic versions of papers and certifications (such as driver's licenses, voter ID
cards, and school certificates). Citizens can also use their accounts to save scanned
copies of their old documents. The eSign feature may be used to electronically sign
these older documents. The following are the advantages of digilocker:

- People may access and exchange their digital information at any time and from
  any location. This is effective and save time.
- It lowers government agencies' administrative costs by reducing their usage of
  paperwork.
- Because the documents are provided immediately by the registered issuers, DigiLocker makes it easy to verify their legitimacy.

- The eSign function may be used to digitally sign documents that have been self-uploaded (which is same as the process of self-confirmation).

### 11.10.3.1   Implementation

For a digilocker case study, we define three types of role in a smart contract such as data collector, data requester, and users. The smart contract of each role individually perform their task such as data collector can verify the documents and add the details into a digilocker portal, while data requester can access the document and modify the document as per their requirement, and user can give their documents to secure their document from the risk. Figure 11.16 shows the recorded transaction details and deployment of smart contract details. Figure 11.17 shows the registration process of document. In this, the data collector can obtain the document information from data owner and stored into the digilocker portal. Include and withdraw the document details from this portal as per owner's requirement. Figure 11.18 shows the history of specific document such as owner name, reference URL of current owner and next owner's reference URL address collect and store into digilocker portal. Figure 11.19 shows the successful modification process of owner. In this process, the previous owner's information verify from the portal and assign the next owner to specific document.

## *11.10.4   Transportation Services*

There are many other existing application available where blockchain has been accurately integrated as shown in Fig. 11.20. In transportation services like Uber and Ola,

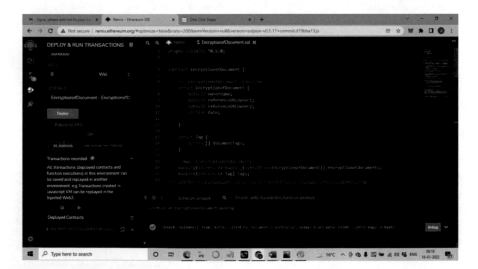

**Fig. 11.16** Deployment of digilocker smart contract

**Fig. 11.17**  Registration process of document

**Fig. 11.18**  Data collection of current and next owner details

blockchain can be used. The Trust of the customer can be achieved by removing central authority from the system. Additionally, the blockchain can also achieve location security where customers or riders will validate the location. The smart contract can help set rules of distance covering as well as monitory transactions. The smart contract is self-executable so that the customer can trust the services provided by the transportation services.

**Fig. 11.19**  Successful modification process of owner

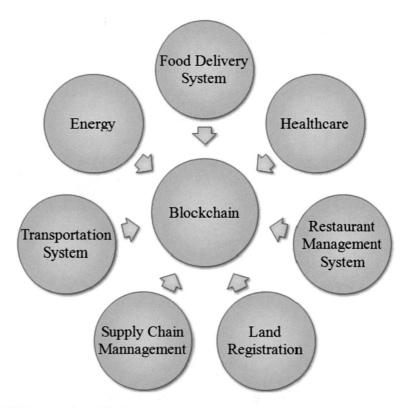

**Fig. 11.20**  Integrating with existing systems: possibilities

### 11.10.5  *Food Delivery Services*

Blockchain can be useful in food delivery services for efficient delivery of the system. For example, the cancelation of an order must return the currency, but this feature is lacking in the traditional system. So with the help of the blockchain, smart contract rules can be set. Participants in the system are customers, delivery persons, restaurants, and application owners. Blockchain ensures efficient delivery and valid transactions without a central authority.

### 11.10.6  *Restaurant Management*

Restaurants have processes like collecting raw materials from different sources such as payments of electricity bills, utensils requirements, collecting bills from the customers, etc. It is challenging to handle restaurant businesses with the help of a centralized business ledger that can be tempered. Blockchain can be helpful by providing an immutable ledger. Self-executable smart contracts can help the business collect

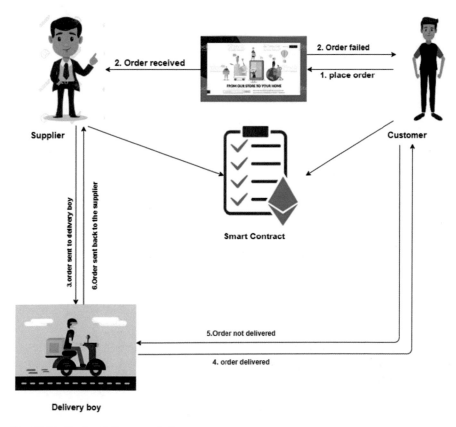

**Fig. 11.21**  Product delivery work flow

the user's bills as illustrated in Fig. 11.21. Some restaurants run on partnerships where multiple participants have invested their money in the business. Blockchain's distributed, and decentralized approach can be helpful to build trust among the participants.

## 11.11   Problems/Implementation: E-commerce Product Delivery System

### 11.11.1   Description

Here, the supplier enters the platform to sell his product and customer register himself to buy products and transaction happens. Here, the role of blockchain comes when customer makes transaction with the supplier, presented in Algorithm 1.

---

**Algorithm 1** Product delivery system

---

**Input:**
**Output:**
  **while** ($Customer \in blockchain$) **do**
    **if** ($Order\_item \in available\_stock$) **then**
      Customer has selected item available in the list.
    **else**
      Item out of stock
    **end if**
    **if** ($Customer\_balance > Order\_amount$) **then**
      Customer will choose payment method
      $Customer\_balance = Customer\_balance - Order\_amount$
      $Supplier\_balance = Supplier\_balance + Order\_amount$
      Order Confirmed.
    **end if**
    Order sent for delivery.
    **if** ($Delivery\_state = Successful$) **then**
      Delivery successful.
    **else**
      $Customer\_balance = Customer\_balance + Order\_amount$
      $Supplier\_balance = Supplier\_balance - Order\_amount$
    **end if**
  **end while**

---

### 11.11.2   Benefits of Blockchain in the System

- Blockchain eliminates the requirement of third party authorities like banks.
- Every transaction is recorded into the immutable ledger and a copy of the ledger is with all the participants so that the system is less likely to be attacked.

- Self-executable smart contracts will help in defining rules and the system will become less prone to errors.
- Blockchain builds the trust of the customers, thereby increasing the profit of the business.

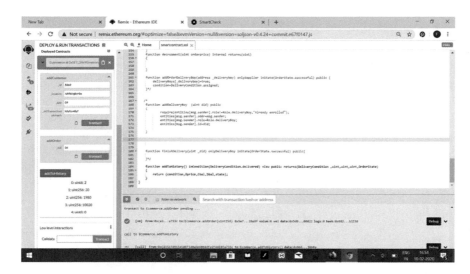

**Fig. 11.22**   Product delivery smart contract

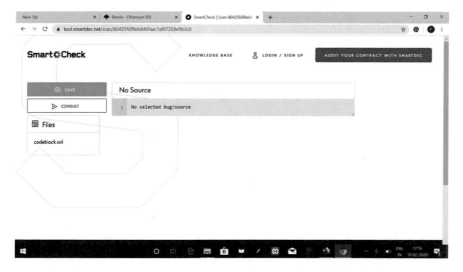

**Fig. 11.23**   Implementation

### 11.11.3 Implementation

Figures 11.22 and 11.23 shows the smart contract implementation of e-commerce product delivery system.

## 11.12 Summary

In this chapter, most of the aspects related to enterprise computing are covered to the best of our knowledge. It shows that using blockchain in business can help to achieve a more trustful environment and keep the data secure. Any change in the business network will be added to the ledger and validated by each participant. As ledgers are immutable in blockchain so, asset tracing will become easy in blockchain solutions for enterprise computing. A supply chain management case study shows that asset tracking gets easy because each participant has to validate the asset location in the form of the transaction. Blockchain provides greater transparency, security, efficiency, and tracking capabilities in business networks. So, it builds trust among different parties of the organization and increases the organization's profit.

## 11.13 Practice Questions

### 11.13.1 Multiple Choice Questions

1. Which of the following service monitors compliance of operations with governmental similar as a regulation?

(a) Governance
(b) Data based services
(c) Security based services
(d) None of the above

2. How blockchain will help in business enterprise computing application?

(a) Improves visibility and transparency
(b) Secure business operation
(c) Efficient and cost saving
(d) All of the above

3. Which are the important aspects for creating a enterprise computing application?

(a) Customer Satisfaction
(b) Resource Planning
(c) Both a and b
(d) All of the above

4. Which of the following function is performed by service bus?

(a) Message translation
(b) Auditing
(c) Routing
(d) None of the above

### 11.13.2  Short Questions

(1) List out popular platforms for designing a blockchain-based applications
(2) List out the types of enterprise computing?
(3) What are the main parameters are considered during the enterprise application generation process?
(4) How blockchain and enterprise computing integrate with web application?

### 11.13.3  Long Questions

(1) How blockchain work as a enterprise computing solution in IBM?
(2) How smart contracts are being used to ensure the security of enterprise computing-based applications?
(3) How smart contracts can be used to track asset management?
(4) How blockchain helps to reduce the cost in enterprise computing?

## References

1. businessresearchguide.com, 'What is enterprise computing?' (2020). https://www.businessresearchguide.com/faq/what-is-enterprise-computing/. Accessed 15 Apr 2020
2. https://www.bmc.com/blogs/enterprise-security/
3. ShivaKumar SK (2015). https://www.sciencedirect.com/topics/computer-science/security-enterprise. Accessed 2015
4. @incollectionAGGARWAL2021345, title = Chapter Seventeen - Blockchain for enterprise working model., editor = Shubhani Aggarwal and Neeraj Kumar and Pethuru Raj, series = Advances in computers, publisher = Elsevier, volume = 121, pages = 345–354, year = 2021, booktitle = The blockchain technology for secure and smart applications across industry verticals, issn = 0065-2458, doi = https://doi.org/10.1016/bs.adcom.2020.08.017, url = https://www.sciencedirect.com/science/article/pii/S0065245820300723, author = Shubhani Aggarwal and Neeraj Kumar, keywords = Blockchain for enterprise, Benefits and advantages of blockchain in business

5. Akram SV, Malik P, Singh R, Anita G, Tanwar S (2020) Adoption of blockchain technology in various realms: opportunities and challenges. Secur Priv 3:e109. https://doi.org/10.1002/spy2. 109
6. Hamida EB, Brousmiche K-L, Levard H, Thea E (2017) Blockchain for enterprise: overview, opportunities and challenges
7. espeoblockchain.com, 'Leveraging blockchain for enterprise applications' (2020). https:// espeoblockchain.com/blog/blockchain-for-enterprise-applications/. Accessed 16 Apr 2020
8. coindesk.com, 'Santander: blockchain tech can save banks dollar 20 billion a year' (2020). https://www.coindesk.com/santander-blockchain-tech-can-save-banks-20-billion-a-year. Accessed 16 Apr 2020
9. ibm.com, 'Blockchain basics: introduction to distributed ledgers' (2020). https://developer.ibm.com/technologies/blockchain/tutorials/cl-blockchain-basics-intro-bluemix-trs/. Accessed 17 Apr 2020
10. Gupta R, Kumari A, Tanwar S, Kumar N (2020) Blockchain-envisioned softwarized multi-swarming UAVs to tackle COVID-19 situations. IEEE Netw. https://doi.org/10.1109/MNET.011.2000439
11. tradeix.com, '6 essential blockchain technology concepts you need to know' (2020). https://tradeix.com/essential-blockchain-technology-concepts/. Accessed 18 Apr 2020
12. leewayhertz.com, 'What are the key concepts of blockchain development?' (2020). https://www.leewayhertz.com/blockchain-development-key-concepts/. Accessed 18 Apr 2020
13. Shah K, Chadotra S, Tanwar S et al (2022) Blockchain for IoV in 6G environment: review solutions and challenges. Cluster Comput. https://doi.org/10.1007/s10586-021-03492-0
14. 101blockchains.com, 'Benefits of blockchain technology' (2020). https://101blockchains.com/benefits-of-blockchain-technology/. Accessed 18 Apr 2020
15. fool.com, '5 big advantages of blockchain, and 1 reason to be very worried' (2020). https://www.fool.com/investing/2017/12/11/5-big-advantages-of-blockchain-and-1-reason-to-be.aspx. Accessed 18 Apr 2020
16. forbes.com, 'The benefits of applying blockchain technology in any industry' (2020). https://www.forbes.com/sites/ilkerkoksal/2019/10/23/the-benefits-of-applying-blockchain-technology-in-any-industry/5c48871c49a5. Accessed 18 Apr 2020
17. blockchain-council.org, 'Permissioned and permissionless blockchains: a comprehensive guide' (2020). https://www.blockchain-council.org/blockchain/permissioned-and-permissionless-blockchains-a-comprehensive-guide/. Accessed 18 Apr 2020
18. learningactors.com, 'Private blockchain vs public blockchain' (2020). https://learningactors.com/private-blockchain-vs-public-blockchain/. Accessed 19 Apr 2020
19. sungardas.com, 'Blockchain series - part 3: permissioned vs. permissionless blockchains' (2020). https://www.sungardas.com/en/blog/cto-labs-blog/blockchain-series---part-3permissioned-vs.-permissionless-blockchains/. Accessed 19 Apr 2020
20. packtpub.com, 'Exploring blockchain and BaaS' (2020). https://subscription.packtpub.com/book/data/9781789804164/1/ch01lvl1sec10/blockchain-actors. Accessed 19 Apr 2020
21. medium.com, 'Blockchain simplified notes' (2020). https://medium.com/moatcoin/part-5-blockchain-simplified-notes-nptel-9eadc5a4e5f8. Accessed 19 Apr 2020
22. Litke A, Anagnostopoulos D, Varvarigou T (2019) Blockchains for supply chain management: architectural elements and challenges towards a global scale deployment. Logistics 3(1):5
23. https://www.mbaknol.com/management-information-systems/integration-of-blockchain-and-enterprise-resource-planning-systems/
24. https://www.wolterskluwer.com/
25. https://www.salesforce.com/
26. techgenix.com, 'Blockchain in the enterprise' (2020). http://techgenix.com/blockchain-in-the-enterprise/. Accessed 19 Apr 2020
27. wikipedia.org, 'Enterprise/software' (2020). https://en.wikipedia.org/wiki/Enterprise_software. Accessed 15 Apr 2020
28. leanix.net, 'Blockchain in enterprise' (2020). https://www.leanix.net/en/blog/blockchain-in-the-enterprise. Accessed 15 Apr 2020

29. ibm.com, 'Top five blockchain benefits transforming your industry' (2020). https://www.ibm.com/blogs/blockchain/2018/02/top-five-blockchain-benefits-transforming-your-industry/. Accessed 18 Apr 2020
30. 101blockchains.com, 'Introduction to permissioned blockchain' (2020). https://101blockchains.com/permissioned-blockchain/. Accessed 19 Apr 2020

# Chapter 12
# Blockchain for Supply Chain Management

**Abstract** Nowadays, supply chain management is a prime concern to transfer products/medicines at a time without any interference. The management of the movement of products and services is referred to as supply chain management, and it encompasses all procedures that turn raw materials into items. It includes effectively optimizing a company's supply-side processes to increase consumer growth and achieve a competitive edge in the market. This process is shared with multiple parties; if a party is trusted, it results in a reliable supply chain. The product can be tracked using sensors in the supply chain management, but it sometimes does not provide reliable tracking. Blockchain technology addresses this issue and provides transparency to the supply chain data so that users facilitate with end-to-end visibility, which helps to locate the contaminated source quickly and generates trust in the supply chain. The use of blockchain in the supply chain has the potential to increase transparency and traceability while also lowering administrative expenses. A blockchain supply chain can assist participants in managing the supply chain by recording price, date, location, quality, certification, and other sensitive information. The accessibility of this data inside blockchain can improve material supply chain traceability, reduce counterfeit and drop market losses, enhancement of compliance and visibility, expand manufacturing of contracts, and potentially boost an organization's performance. Blockchain uses a smart contract for the verification of stakeholders and to verify product quality. In this process, all required stakeholders are payable in their wallet address after successfully verifying the product's quality. A blockchain-based supply chain makes a supply chain process more reliable, transparent, and easily locate and track the product.

## 12.1 Introduction

The physical supply chain's involvement in modern life is becoming more complicated. Consumer needs are easily conveyed, and the delivery of products and services is monitored to ensure consistency in the supply chain. When the supply chain requires better trust between stakeholders number of factors have to be considered. Figure 12.1 shows the end-to-end process of the supply chain. If participants of

**Fig. 12.1** Supply Chain

the supply chain process do not trust each other, it is necessary to create trust among them to increase the supply chain's reliability. Another part of the supply chain is the ability to quickly detect the source of contamination. Moreover, to increase the supply chain's traceability in this chapter, we first analyze the supply chain's specifications and functionality with the goal of incorporating blockchain technology. The use of blockchain technology is producing a significant shift in supply chain management. This technology obtained identification for its bitcoin link and its promise to create a secure and open transaction distributed ledger.

Blockchain is a platform for distributed ledgers that enables individuals and businesses to share open transactions [4]. Clarity and traceability are features of this technology that increase confidence in each and every exchange of data, products, and financial resources. Governments and organizations have been studying the deployment and implementation of this technology in several disciplines, ranging from economic, social, and legal sectors to manufacturing, network engineering, and supply chain. To understand the significance of blockchain, it is necessary to first understand its current condition and its implementations. The potential value of this technology in the supply chain offers an effective manufacturing solution. Nowadays, the processing of cardboard boxes is used to explore how this technology might be applied in a worldwide supply chain network. Customers are looking for uniformity across the supply chain; therefore, there is a need for an advancement in a supply chain [1].

The distributed shared ledger reduces the need for mediators and has the potential to transform supply chain management adroitly [5]. This gives stakeholders better control and management as per their needs, whether ocean and inland carriers, terminal owners, freight borders, service providers, or financial authorities. The stakeholders connected in the supply chain process gain access to a blockchain-based, fully integrated system that marks all alterations and monitors the exchange of hands without the need for human intervention. Blockchain efficiently and accurately stores each transaction of the supply chain process. As the chain progresses, the approach ensures that no product item appears twice in the same place. Blockchain is expected to play a critical role in cost reductions, performance improvement, and fraud prevention in the following ways [13],

– Blockchain generates a trust and provides transparency in procurement by the use of permanently kept records to verify everyone which engaging in a smart contract. As a result, each side confirms the other's trustworthiness [7]. It delivers greater analysis, which means that such analysis is processed more efficiently to improve the performance of the supply chain process.

– In order to make transactions easier and efficient, blockchain adds transparency to a dynamic supply chain [22]. It combines digital, physical, and political data to reveal economic leakage sources ranging from daily failures to fraud and terrorism. It also aids in the identification of novel techniques to deal with the dynamic supply chain.
– In a blockchain, retailers and buyers frequently mention that they are required to verify and authorize the products and assure the product's quality. The entire invoice cycle will become increasingly inadequate because the product's price cannot be changed. When a purchase order is characterized as a stone, the blockchain creates a tamper-proof solution for a physical object [18].
– Blockchain uses a decentralized ledger to ease payments at retail banks, especially international payments that have large costs and take many days to complete [8]. It reduces the resources used by banks to verify client IDs.
– Blockchain provides an opportunity to discover and improve ethical standards through manufacturing chains. The current supply chain management system is unable to assure the employment regulations, environmental conditions, and impacted raw data. Only a blockchain allows the user to manage everything in one place. [2].

## 12.2 Supply Chain Fraud and Addressing

Nowadays, due to a lack of trust among stakeholders in the supply chain, there is a possibility of fraud in amount transfer without verification of product quality. Stakeholders can claim that they are not payable. The contract between the two companies is being managed by a third-party owner, which is not a trustworthy source. In this process, various top companies are adding new technologies to aid and compete with it. Figure 12.2 shows various frauds that occurred in the supply chain. The following are examples of warning signs of supply chain such as waste, fraud, and violence:

– The auctioning procedure is neither reliable nor self-contained.
– Lack of sufficient consistency like invoices of third-party holders
– Efficient execution, and arrangements with third-party holders.
– The use of sole-sourced third-party transactions eliminates the need for a precise explanation to be established as price arrangements with no specified expenditure descriptions or other associated terms.

The supply chain has become increasingly complicated. It is a process of transfer between a producer, retailer, consumer, and seller. It takes much time to send the product from one place to another within a time limit. In this process, legal agreements enable banks and attorneys to provide facilities, which contributes to the time and expenses for the process of delivery. It is also impossible to trace products and materials back to their original producers, making it tough to correct flaws. Multiple handovers and stakeholders exist in this process, which increases the possibility of fraud

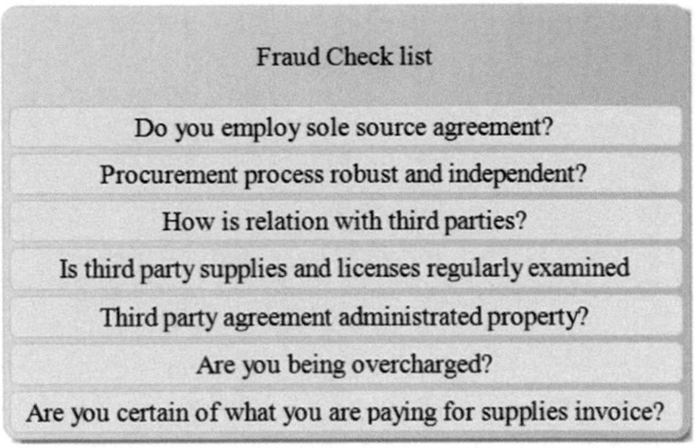

**Fig. 12.2**  Supply chain
Fraud addressing

due to a lack of transparency, end-to-end visibility, and trust between the stakeholders. It affected the growth of supply chains from well functioning. Rather than direct
communication with one other, customers and suppliers communicate through a centralized third-party entity. Simple operations have become more complex, resulting
in multi-step processes. The impact of fraud on the supply chain might strain the business performance; consequences could include financial loss as a result of the scam
and costs associated with legal actions and inspection criteria. Long-term implications might consist of the actual damage, company loss, and a potential loss of market
share. Blockchain, which is getting momentum among CEOs, can bring success and
provide a product in a modern style, reducing supply chain misuse, theft, and waste.

Blockchain platform uses a public environment, an immutable record that is resistant to modification, and transactions may be trusted. Authenticated users can store,
show, and share the digital file in a monitoring environment. It aids in the promo-
tion of honesty, transaction transparency, and integrity, all of which are important
aspects of business relationships that may be extended to financial transactions and
identity preservation in the supply chain. It improves transactional accountability
and efficiency in the supply chain, encouraging businesses to utilize blockchain for
the supply chain for fraud prevention, misuse, and corruption through third-party
cooperation monitoring and transaction implementation [20]. Blockchain is a popular option to lead in the direction of financial fraud since it is loaded with advanced
technology [14] and identification. It is used to handle ownership and asset records
for shares, smart contracts, land, and other entities, which makes the fraud-resistant
system. It's not to be neglected, but technology is also a fraud accomplice. Low
performance is the result of ineffective information management software that isn't
working properly. Because the individual has the permission to submit fraudulent
data, the blockchain could only discover the inputs and facts to classify legitimate
data and identify the invalid information. In the physical environment, the individual
relies on trustworthy data sources and permissions. If you modify the procedure, the
individual will be able to take advantage of it.

## 12.3   Why Blockchain is Needed in Supply Chain Management

This section describes the different scenarios of the supply chain and identifies the key terms to check the provenance of supply chain management. The different scenarios are present as follows:

– **Reliable Supply Chains:** If the supply chain system involves a few known and trusted sources (Organization and third parties) for end-to-end visibility and traceability of the product supply chain, then there is no requirement for a blockchain-based supply chain management system.
– **Real-time traceability:** If the supply chain provides real-time traceability and checks the product quality at each stage. If any product is contaminated due to any threat, if it can be easily identified by the system then there is no requirement for a blockchain-based system.

In supply chain management, if any product is contaminated, it takes more time to track its source. For faster identification of the contaminated source, blockchain provides end-to-end traceability and visibility of the system. Blockchain facilitates automation when the end-user checks product quality. If it is correct, it automatically initiates the transaction and transfers the required amount to the supplier in the bitcoin wallet. Blockchain gives transparency to the system, which results in a trusted supply chain eliminating the third parties. Figure 12.3 shows the need for blockchain in supply chain management. There are various scenarios where the blockchain is an essential requirement to make a system more reliable, transparent, and traceable. When the product is sensitive and requires privacy for that specific product, blockchain is used to provide privacy to the data. Any unauthorized person cannot know the information about the product. Only legitimate users can be verified using the smart contract and added as a stakeholder in the supply chain process. After the quality check of the product, all legitimate stakeholders are payable.

## 12.4   Supply Chain Traceability and Transparency Using Blockchain

In production systems, the supply chain consists of various processes, individuals, skills, physical resources, contractual contracts, and transactions that make getting a product from a supplier to a customer easier. Creating an aggregated description that includes all bought products made within a large supply chain network isn't easy. This information is usually processed across several locations and made available to other companies. In structures, consumers now have minimal access to the information. In some instances, a portion of the data is being considered a retailer product. By leveraging durable digital archival, centralized storage, and monitored user accesses, the blockchain network potentially addresses challenges of confidentiality and traceability within the industrial supply chain. A transparent logistics chain

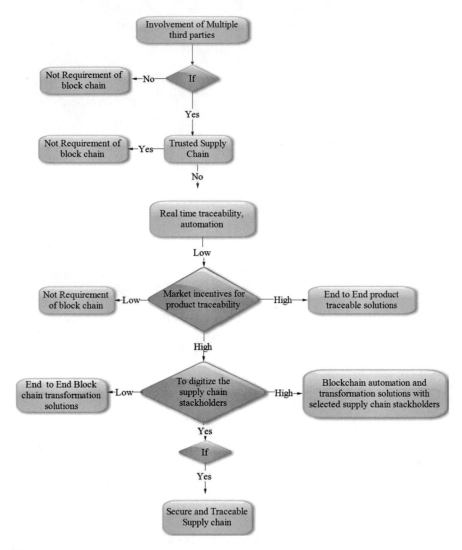

**Fig. 12.3** Need of Blockchain in supply chain management

that uses blockchain to gather, store, and manage important commodity information throughout the life cycle of each product. This central block successfully offers reliable, public transaction details for each particular product and specific product data.

### 12.4.1   Blockchain Ready Manufacturing Supply Chain

Figure 12.4 shows the summary of the potential use of blockchain-based manufacturing supply chain [6]. Blockchain-based network infrastructure is being proposed

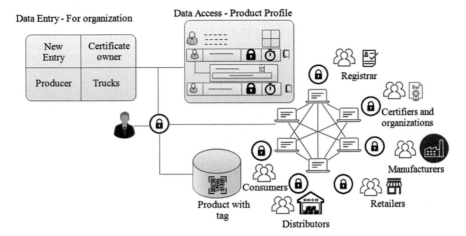

**Fig. 12.4**  Data access model of blockchain

to capture, store, and manage a product's basic commodity information during its life cycle. It offers each commodity with a sustainable public distribution directory as well as detailed product information. Several actors hold a product along its development cycle, including producers, marketers, retailers, dealers, and end-users. These stakeholders are critical to the scheme's success, logging into the public blockchain to record critical product information and its current condition.

Each object receives an identity tag, which might be in the form of a barcode, like QR code and RFID. This tag represents only one electronic cryptographic identification linking a physical system component to its virtual identity. This visual identifier is considered as part of the user's digital identity in an on-device application. Actors can have their own digital channel profiles, which are being created using registration. This profile includes information such as a definition, location, certifications, and affiliation with a company. An actor signed a contract in order to link the contractual profile to such actor's identification. The actors of the system register with the system through such a registrar, which is a certification service that offers expert knowledge and a distinct identity to the actors. On such registration, each actor receives a public-private cryptographic key pair. The public key recognizes and confirms the actor inside the system, as well as the time of interaction with it, and the private key verifies the actor. Characters can only interact with the network after cryptographically confirming themselves with their secret key. This allows actors to sign each commodity remotely as it is exchanged or equipped with the supply chain. This system consists of various types of actors defined as follows,

- Registrar gives unique identifies to the network users.
- Organizations create connective systems based on fair trade.
- Certifiers issue certificates to participants, and allow them to participate in the network.
- Suppliers, producers, retailers, marketers, dealers, and waste management.

– Consumer's basic product related details—like purchase goods, and in some circumstances, the blockchain can verified the access of commodity knowledge.

Every member in this program gains access to a unique blockchain system consumer
experience. The actors' software architecture is being customized to a company's
digital presence. A group of trustworthy people developed the application software,
and it can access and perform on various networks from registered companies and
organizations. Consumers must have a customized consumer experience model that
enables the visibility of associated participants' information. This software operates on a blockchain, which enables the execution of programmable code, such as
the ethereum smart contract, and subsequently the storage of data in the ethereum
blockchain. [6]. The system's records are stored in a database and only accessed
through a legitimate verification process. This information is only shared with those
directly involved in the supply chain process. The level of data access given to the
participants fluctuates based on their function in the supply chain. The program rules
are being written as smart contract codes and stored on the blockchain. These rules
specify how network members should interact with the model and how data should
be shared within the network. Since the stakeholders cannot change the laws, the
process is being assumed to be immutable, transparent, and reliable. If the rules are
written on the blockchain, they work precisely as specified and will not be changed.
Certifications and certified systems can be included in this model, with network
accountants and registrars visiting the warehouses and manufacturers to ensure that
the certified system needs are addressed. A standard organization must digitally
approve this after it has been examined by suppliers to validate its legitimacy, the
actor's identification, and its items. All of the participants are being evaluated by the
certifiers in order to verify their identities. The certifiers must disclose the names
of all network members by way of the registrar. It improves the accountability of a
program item while maintaining data privacy and security.

## 12.5   Supply Chain Visibility What? How? And Reality

The transparency principle benefits business partners such as manufacturers, suppliers, and customers [15]. The decision-maker is being associated with the classification of various supply chain ambiguity styles, including cost, range, price, time,
and commodity configuration. Managing the difficulties via knowledge exchange
and increasing awareness among supply chain partners. Supply chain transparency
is being implemented, and it is only feasible because of the worldwide interchange of
commodity information. It's only a matter of personal choices and dedication, which
was not the case when it comes to transmitting "official data" for people. The most
significant key obstacle for all aspects of the supply chain, from the manufacturer to
distributor, suppliers, reseller, and even the end consumer, is a lack of field of view.
Lack of information and lack of influence contributes to a range of issues, including stock limitations, price fluctuations, and faster time to market. It improves the

productivity of the supply chain system by providing data that is freely accessible to all stakeholders, including customers, sellers, manufacturers, and programmers.

It plays a variety of important responsibilities in businesses. Visibility is a sort of information exchange. The supply chain network aids technology in encouraging and quickly reacting to the shift that allows selected customers to take charge of altering commodity needs or shifting supply. Incorporating the network from both the internal and external through information sharing may improve the supply chain's success. On the other hand, visibility becomes important for manufacturing logistics operations; it produces a suitable method of transferring information among supply chain managers, allowing them to be experienced and sensitive to all exchange indicators. It enables manufacturers, producers, transporters, distributors, and consumers to understand the supply chain and the location of their items.

### 12.5.1  End-to-End Visibility

Dealing with climate change demands a web-based, automated, and electronic infrastructure that provides the required visibility all across the supply chain. An environment that offers a complete ordering network among producers, dealers, and manufacturers, with stock prices and home delivery capabilities, enhances greater efficiency at the initial step. Distributors can better manage their warehouses, stock rates, and employees, and the product can be tracked at any time, regardless of the supply chain process. Distributors are permitted to make clear receipts during the selling, offer accurate orders to clients, and have details about the manufacturer's vulnerability. Retailers have total control, in which they have to compensate for what they have purchased. Payables and investors can be included in the system for improved financial oversight. The circle can also be restricted to give incentive services and logins while maintaining maximum visibility. Blockchain can provide a facility to suppliers, marketers, and dealers to use smart insights to test technology across key performance indicators and provide relevant real-time feedback. The structured files could be taken at any time, and periodic notifications can be set up when the exceptions are met. This allows for real-time management of the business. Data from the digital supply chain might be utilized for predictive analytic to allow for more flexible and adaptable decision-making across the enterprise.

## 12.6  Research Areas of Adopting Blockchain in Supply Chain

Improved supply chains naturally decrease overall delivery costs while maximizing consumer experience and loyalty. These hypotheses are mostly supported by evidence in the form of studies. Researchers are working to find more suitable methods in this sector, which might lead to more safe and productive supply chains [16]. It includes

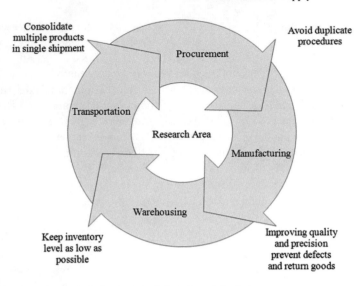

**Fig. 12.5**  Research areas of adopting blockchain in supply chain

several basic strategies to enhance the trust and transparency of the system [11, 17]. Researchers working on supply chain models are confronted with difficulties that are important to this domain. Customized prototypes may be constructed and changed to fit the expectations of different supply chains if differences arise. Otherwise, these model-building tasks would be difficult.

Figure 12.5 shows the various research areas of adopting blockchain in supply chain management. It was created with the goal of improving service quality by eliminating the product manufacturing process. Effective manufacturing was created to reduce manufacturing costs while maintaining product quality and customer service standards. Although such technologies have been widely acknowledged and accepted, their use in the supply chain has been limited, consumer service, cost-cutting, market exchange, self-service exchange, and increased productivity for a company's supply chain all produces maximum customer value, leading to positive customers' expectations and ultimately increased customer loyalty. Consumers are more dedicated to firms that are part of a connected supply chain aids in businesses that are part of more traditional networks, which is crucial to address this issue. Risk assessment is a necessary requirement to reduce supply chain disruptions, and uncertainties [17].

Entities, procedures, and operations are all supply chain components. A supply chain with various companies collaborating involves numerous activities and defines a job for each member, as well as design features in a smart contract that helps to validate each member's wallet addresses. Experts from all parts of the supply chain explore such issues in depth. Increased oil prices, significant global change threats, restricted accessibility, sub-assets, carbon footprints, and international businesses become critical considerations for companies, governments, and consumers. These challenges are essential macro-societal research topics. Identifying core topics like activities, costs, and electorate parts in supply chain concepts illustrates places where supply chain work is being done, while far more is likely that could advance awareness and hypothesis growth rates in the supply chain.

## 12.7 Steps to Be Taken Care While Preparing Case Study

This section highlights key points for designing any blockchain use cases in the area of the supply chain. The case study gives an idea about a real-life application, how blockchain is being used in supply chain management. In any case study, the first step is to identify the stakeholders, then after role specification of each stakeholder designed in terms of a smart contract. Afterward, the functions define the interaction of roles in a main smart contract. In the end, the execution process of the smart contract is presented. This smart contract verifies the wallet addresses of each stakeholder and is stored in it. After the product validation, all stakeholders are payable in their bitcoin wallet address. The following steps help to make any case study,

- Identify the business problem and business network with different stakeholders, assets, participants, and transactions.
- Identify the different solutions to solve the problem
- Ensures the trust of the system when it involves multiple participants.
- Address a solution to mitigate the risk factor of the system.
- Identify the individual roles of the connected participants (Organization, industry) in the network.
- Identify the Key users of the network and their roles in the network.
- Identify the process of interacting with the other participants in the network.
- Assigned the Job roles for each participant of the network.
- Design an access control role to restrict the participants for join the network.
- Permit only verified users can participate in the network to preserve privacy.
- Identify the involvement of various assets and its associated information of the network.
- Identify the assets, participants, owner, and transactions associated with each stakeholder.
- Design a contract among the participants, if certain conditions meet successfully, then automatically transfer the amount to the required stakeholder.
- Identify the need for legacy system integration in the network.
- Visualize the workflow which is executed by the connected participants of the network.
- Identify the efficiency, reliability, scalability, provenance of the system.
- Analysis of cost estimation of the system.

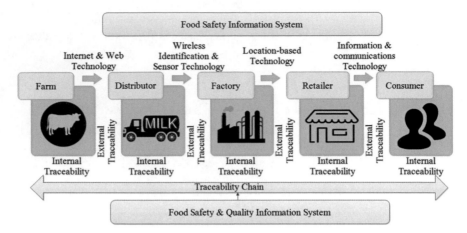

**Fig. 12.6**  Food Supply Chain

## 12.7.1  Food Supply Chain

Blockchain can monitor food market growth in the agrifood business in food supply chain management. The capacity to trace food items throughout their complete life-cycle, from their origins to every point of contact with the client, readily improves quality, integrity, and well-being. Consumers may also use a QR code search to monitor their food from "farmers to consumers." Blockchain holds the possibility of bringing about a dramatic change in the future, but it is not without its challenges.

Figure 12.6 shows the food supply chain process, in which blockchain helps to improve traceability in the field of agriculture. This makes it easier for marketers to track poisonous items back to their sources to determine if they should be supplied globally or not. It prevents infection and increases the cost of retrieving a medicine due to disease. The many ways for tracking the merchandise include QR codes, bar codes, and radio frequency detection. It offers a useful traceability mechanism that can help with food supply chain control, easy product identification, food security, and quality control. There are several flaws in the current food supply system, including a lack of responsibility. Because each organization's data is managed as separate repositories inside its architecture.

Blockchain Technology (BCT) has emerged as a viable alternative that uses hundreds of devices to validate transactions in a huge, decentralized ledger. It cannot be used as a transparent network thus, all connections with the supply chain's enterprises should be monitored at all times. Every action is recorded in blocks that are verified, sorted in a linear manner, and cannot be modified or deleted after being recorded in the blockchain. It may also comprise end-to-end traceability in the form of commodity data, such as manufacturing information, expiration dates, batch numbers, and environmental conditions [3]. The fundamental benefit of a blockchain-based food supply chain is that it is transparent. As a result, it is possible to infer the reliability, protection, and requirement of identifying and monitoring the items can be improved. It assists clients in gathering the high-quality information required to make better-informed decisions. It also gives a high level of confidence regarding the results' security. With the use of blockchain, products can be promptly identified, respected, and tested in a supply chain process.

**Implementation** As the key solution of blockchain, the smart contract has been designed to eliminate the use of third-party holders. In the supply chain smart contract, each stakeholder participant assigns a role in the supply chain process. For that, it can create different functions and design access control rights for specific stakeholders. Every stakeholder can manage their bitcoin wallet for the transaction. The smart contract is being designed for a farmer, milk deliver, transporter, distributor, manufacturer, retailer, and consumer in the food supply chain. Each stakeholder can define their functions, verify from the other stakeholders, and include and withdraw particular stakeholders. Step-by-step implementation of smart contract is as follows, Figure 12.7 shows the verification process of all stakeholders. In this, verify all stakeholders bitcoin wallet address and if it is valid then include the wallet address as per the requested stakeholder after each handover. Figure 12.8 shows the selling process of milk and milk id and its result. This milk id is verified by each stakeholders and validates the product of milk. Figure 12.9 shows the food supply chain process smart contract, where all functions are being designed to track the product at each handover and initiate the transaction process with the required stakeholders. The various functions related to including and withdrawing the particular stakeholder after each handover from the food supply chain. Figure 12.10 shows the verification and including process of farmer bitcoin address into the blockchain. Figure 12.11 shows

**Fig. 12.7** Verification process of each stakeholder

**Fig. 12.8** Selling process of milk

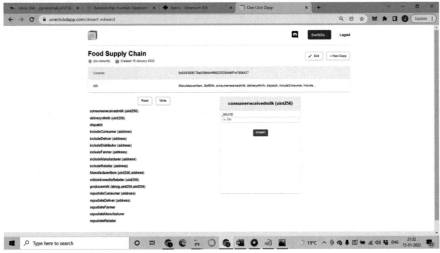

**Fig. 12.9** food supply chain process to track the product

the process of milk producer and how it can be verified at manufacturing process. Here milk information is stored in this contract. Figure 12.12 shows the recorded deployed contract of each stakeholder. The various role to run a food supply chain is present in Fig. 12.12.

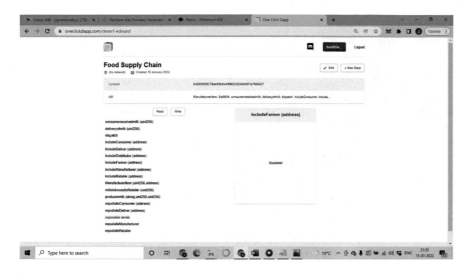

**Fig. 12.10**   Verification and insertion process of stakeholders in system

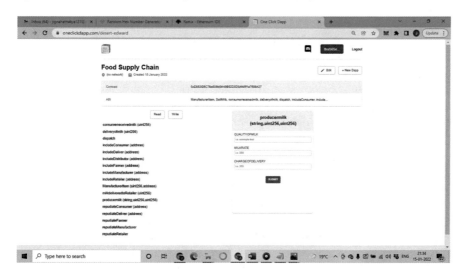

**Fig. 12.11**   Include milk producer information

## *12.7.2   Case Study of Walmart Industry*

Several organizations, including Walmart, Sainsbury, USFood, Shippo, and others, have adopted digital supply chains and used blockchain to promote transparency, resilience, and trust.

**Fig. 12.12** Deploy and record transaction of each stakeholders

**Wallmart industry:** Walmart Canada has announced the world's initial large-scale production blockchain technology for any corporate usage, and digital supply chain looks deeper into the supermarkets.

**Application:** In collaboration with DLT Laboratories, the business also launched a new blockchain-based shipping and payment network. The system uses distributed ledger technology to track deliveries, verify transactions, and automate payments among Canada Walmart and its suppliers, who supply merchandise to over 400 retail locations around the country each year.

**Benefits:**

1. **Data collection and quality:** This case study's central archive aims to promote trust and transparency by sharing information and automate activities and take measurement to save human-based activities, and enhances the accuracy.
2. **System performance:** efficient use of standard techniques, such as faster responses, efficient recording and monitoring, and early discovery of flaws.
3. **Processing time is reduced:** The real-time convergence of both operating rules and transfers to create consolidated invoices which reduce payment response time and allow faster payments [19].
4. **Disputes are eliminated:** Both parties will now be able to manage the dynamic delivery, invoice, payment, and arbitration processes with ease.
5. **Increasing cost:** Costs are growing as a result of increased productivity, and resulted in higher business operations and operational expenditures for both parties.
6. **Enhanced finance and planning:** Now reliable and real-time data is being used for improved analytics and predictive analyses.

**Fig. 12.13** Walmart industry supply chain

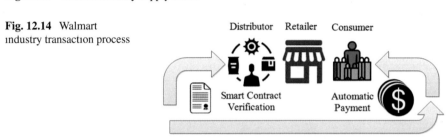

**Fig. 12.14** Walmart industry transaction process

Data placed on a reliable public ledger may be distributed, persistent, and highly auditable due to blockchain technology. Since the industry has discussed the usage of blockchain technology, this is the first real deployment on a large scale that directly highlights the key benefits of blockchain. The internet of Things (IoT) plays a prominent role in supply chains; a tremendous quantity of data is being generated, which must be processed and integrated. DLT Laboratories and this supply chain network have formed a partnership to improve transportation and transactions of payment data using a DL-based asset management system. The new transportation and transaction network gathers and synchronizes all supply chain and logistical data in real-time using blockchain [12]. Figure 12.13 shows the billing and payments, and transaction details. The method also maximizes the different measures required to support real-time payments, invoices, and transactions [19]. Simultaneously, it integrates seamlessly with each company's legacy infrastructure, and businesses are permitted to modify their present activities without learning or investing in new technology. Figure 12.14 A smart contract defines the roles of manufacturer and consumer, in which they are allowed to check the product quality and if verified successfully, then the required stakeholders are payable.

**Implementation**

1. **Define digital assets:** Every item transferred between distributors with name, QR code, description, and form User access regulations are set to keep track of the data at on each handover. Both the buyer and the seller come up with a basic agreement, develop a smart contract, and make a decision.

2. **Define participants:** Specific identities and information, such as balance, name, and services, are used to identify each participant. The producer, dealer, and administrator are the three categories of people involved in this scenario. The manager gets full access to all device details, and it enables real-time monitoring of all network resources. The actions must be taken at the POS (Point of Sale).

3. **Define transactions:** For easy product authentication and monitoring, each commodity is assigned a unique QR code. When a QR code is being scanned, every item information is displayed as function arguments, including ID, type, description, timestamp, delivery date, and shipment date. It aids in the execution of a payment-independent solution. Every payment transaction is added to a chain that contains information about the products sold. Before a transaction takes place, the authentication processes and laws are examined. To switch on a smart contract in real-time.

4. **User interface implementation:** A web framework is introduced, and a smartphone QR scanner device is used to support it.

The blockchain software application that has been approved is an interface that displays two essential items, i.e., all system transactions, and the other is all regulations and laws that all parties recognize as smart contracts. Smart contracts accept repayment if producers and retailers have reached a prior agreement. The two parties predefine the rules and regulations set by smart contracts within the network. Users have a tendency to keep their security keys on their business card ID. It allows everyone in the system to communicate through a computer without registering. In the blockchain, each user has a separate wallet with a unique public key address.

Figure 12.15 shows the verification process of each stakeholders. In this process, every authorized participant and their wallet addresses. if it is successful, then, it includes the wallet address after each handover. Next, Fig. 12.16 shows the supply chain process at each handover. In this process, various functions are designed to validate the product and validate the producer, distributor, retailer, and customer. Figure 12.17 shows the data insertion and withdrawn process of stakeholders in the blockchain system. If functions execute successfully then it gives a result as a success. Figure 12.18 shows the pick product process, which stores a value of producer and product information such as unique producer product notes, producer name.

### *12.7.3 Case Study of Drug Logistics*

Blockchain-based drug logistics is designed with the current supply chain management process to keep a record of drug data in an immutable chain. It begins by approving the procurement of drugs to the suppliers, getting the drugs to the warehouse, ensuring the quality of drugs with QR code testing, modifying the drug standards, and generating alerts to other involved stack holders at each level of the supply chain management process. In this process, quarterly basis indent volume along with the hospital, apparent to the similar warehouse, and supplying these drugs to hospital

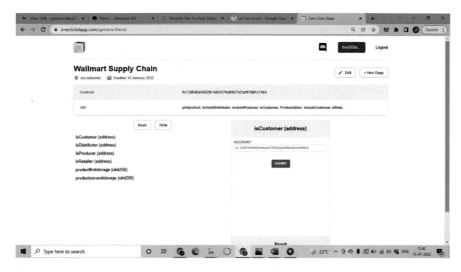

**Fig. 12.15**  Verification process of stakeholders of Walmart supply chain

**Fig. 12.16**  Functions of Walmart supply chain

and individual pharmacy shop, where it reaches the patient/consumer. Various characteristics are added to the blockchain at each phase of the drug logistics life cycle. The drug logistic life cycle is present in Fig. 12.19

Next, it shows the process of acquisition and scheduling where each supplier receives a purchase order (PO). It specifies the quantity of each drug and warehouse to be delivered and the delivery date. The scheduling is revealed to the public so the supplier may make plans to deliver the drugs to the appropriate warehouse. The

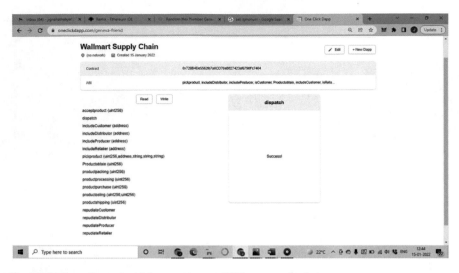

**Fig. 12.17** Insertion and withdrawn process of Walmart supply chain

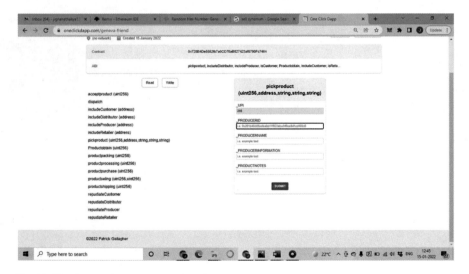

**Fig. 12.18** Pick product process with various parameters

blockchain is intended to hold the details of the PO number, PO date, notification of award (NOA) number, drug name, supplier information, amount, warehousing details, and quantity. Further, the pharmacy chemist gets drugs by following the purchase order's batch information. The quantity of drugs received in person is compared and verified to the quantity listed on the invoice. Requesting the data of suppliers and the pharmaceuticals to be delivered to the relevant warehouses from blockchain might reveal the pre-conditions before getting the same. The blockchain

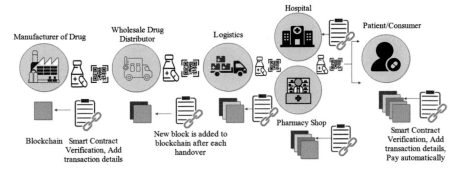

**Fig. 12.19** Drug logistics

stores information such as supplier information, warehousing details, drug name, PO No., date of receipt, amount, batch number, batch Qty., invoice date., invoice No., Exp. date, and Mfg. date. At this point, the smart contract code recorded in the chain may be used to initiate payment to the supplier.

Figure 12.19 shows the drug logistics process using blockchain and smart contracts. Each batch of the drug is subjected to a quality assurance, which involves choosing samples from three warehouses at random. The drug will be sent to the quality checking section from each of these warehouses. The QC department chooses one sample from each of the three warehouses where the drug was tested. By producing a QR code, one sample is transmitted to laboratories for testing. The chosen batch has become ready to be sent to the hospitals and pharmacy shop. If any of the drugs are being contaminated and do not ensure the standard quality of it, the entire batch is frozen in warehouses and hospitals, preventing further distribution. The blockchain stores the following information, drug name, batch number, extracted Qty., warehouse, QR code, and outcome [9]

If any batch does not have a standard amount of drug, the notification is sent to the approval authority. The batch of drugs must be frozen and should notify all involved parties. Hospitals' Quarterly Indent—hospital to warehouse mappings has been completed [during primary data generation], the hospital's quarterly invoice is only assigned to the mapped warehouse. The hospital's in-charge approves the pharmacist's request. The warehouse validates the invoice with or without modification in the requested standard amount of drugs by the hospitals based on the remaining stock at a warehouse. The warehouse approves the drugs, which are then delivered to the hospitals. When the amount exceeds a limit, the indent can be lifted. Blockchain stores the warehousing details, Indent no., year, month, standard drug amount.

Furthermore, the hospital accepts drugs delivered by the warehouse. The drugs are distributed towards the warehouse on a daily/weekly basis by the hospital's pharmacy shop. Blockchain stores the information of warehouse, inbound and outbound No., Date, Indent, Indent No., Mfg Date, drug batch No., Exp Date, Qty, amount, and supplier so that the hospitals only receive the authentic drug batches. It also stores information such as Institute, Issue No., Drug Batch No., Mfg and Exp Date, Quantity, and issued date. A patient can verify the expiry, manufacturer information, and drug batch standard before purchasing any drug. The goal of incorporating blockchain Technology (BCT) into a drug logistics process is to ensure that patients receive authentic drugs. To accomplish this, various organizations must use the blockchain to check the quality and expiration of drugs, which makes a system more reliable. The warehouses might also check the batch information to verify that no fraudulent batches are sent out. Consequently, all stakeholders can determine the total number of drugs in various hospitals and warehouses to verify the drug's validity.

**Implementation** The drug logistics smart contract comprises various roles for distributor, manufacturer, retailer, warehouse, pharmacy shop, and patient. Each role contains the information of their wallet address, access rights of a specific user, drug batches information, collector, and shipper address. It makes a supply chain process more reliable and transparent. In the process of drug logistics, the drug smart contract contains information about a drug, drug batches, expiry, information, and the temperature that should be maintained for a specific drug. This smart contract is being connected with other smart contracts and gives the information to the required participant in the system. During logistics, first, the specification of a drug is being verified with a smart drug contract, and if valid, that drug batches forward to another handover. If any drug is being contaminated during the process, it can be easily tracked using these smart contracts.

Figure 12.20 shows the deployment process of drug logistics. This smart contract connects all the participants and provides visibility of drug logistics at each stage of the process. After the verification of all the participants, only authorized participants to allowed to access the drug information.

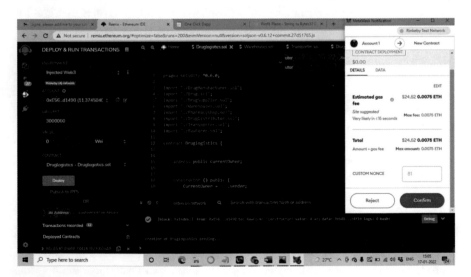

**Fig. 12.20** Deployment process of drug logistics smart contract

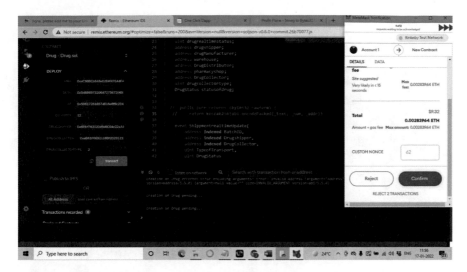

**Fig. 12.21** Deployment process of drug smart contract

Figure 12.21 shows the deployment process of the drug smart contract. In this process, drug manufacturer address, drug description, quantity of a drug, drug shipper, collector, and drug collector type add in the form of address after a transaction is being initiated during the time of deployment. In this deployment process the metamask is open and if the transaction is valid then confirm the transaction. The transaction is being recorded into the new block.

Figure 12.22 shows the deployment process of a warehouse. In this smart contract, the drug batches are being selected from the various drugs and packed and forward to the pharmacy shop after the validation of drug quality and verification of the pharmacy shop. It contains a batch of the drug, drug sender(manufacturer), drug shipper, and drug collector address. After adding the details, the transaction is initiated and the amount is being transferred into the required stakeholders. Next, Fig. 12.23 shows the deployment process of pharmacy shop smart contract. It contains the drug information, and source of the drug, where it comes from the drug sender, drug shipper's address. It also maintains a stock of drugs to notify the stakeholders to know the information of drug batches. The pharmacy shop records a transaction with metamask wallet and stores transaction details into the new block. Figure 12.24 shows raw form smart contract information. It contains the addresses of drug suppliers, description of drug, unique QR number of drug, quantity, drug shipper, drug collector, and drugid. After the successful verification of the drug logistics process, all accounts are payable as required. Finally, Fig. 12.25 shows the information of deployed smart contracts and their addresses with recorded transaction details.

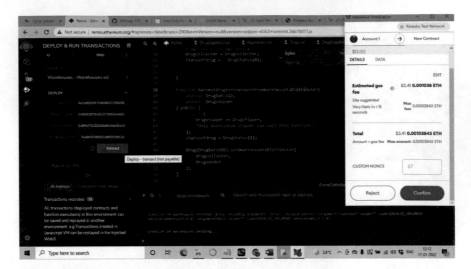

**Fig. 12.22** Deployment of warehouse smart contract

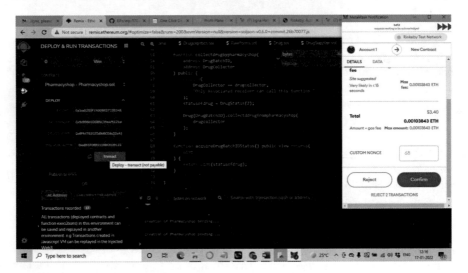

**Fig. 12.23** Deployment process of pharmacy shop

### 12.7.4  Case Study on Human Organ Supply Chain

The human organ supply chain and governance is a complex procedure that traverses the entire life cycle, from pre-assessment of organ placement through the supply chain's travels and significant analysis of donors. Many healthcare institutions and hospitals are not able to maintain the wide collection of data used in silo operation due to the limited scope of accessibility and interoperability of the healthcare data.

**Fig. 12.24**  Deployment process of raw form smart contract

**Fig. 12.25**  Deployed smart contracts and their addresses

Lack of trust and accessibility of data makes the process more difficult for hospitals and healthcare organizations. They prefer to spend their resources and energy on the decision-making for the patient's health, such as evaluating organ donor suitability and the importance of organ match associated with patient risks of death. Additional issues might arise as a result of probable organ mix-ups, DNA contamination during organ transplantation, unethical organ supply, and audit record visibility connected to these activities. The topic of how to build a significant single source of information,

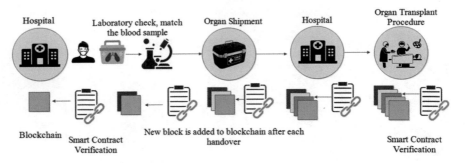

**Fig. 12.26**   Human Organ Supply Chain

blockchain, may offer prominent solutions. Due to its immutability, security, and traceability qualities, blockchain has become a more desired technology in the healthcare field while also giving the guarantee of transparency and auditing record. Blockchain appears to be an ideal fit for managing the organ placement/procurement in the supply chain and an audit monitoring tool for analyzing data in any pre/post-operation. The integration of a cyber security framework and supply chain process helps to improve the privacy of the data in the healthcare industry. In this integration, all people who signed up to use the blockchain for supply chain logistics would follow the ethics and requirements and require transparency from those granted access.

Figure 12.26 shows the human organ supply chain process using blockchain and smart contracts. The potential for developing a blockchain framework is to handle the organ transplantation lifecycle to improve efficiency, ethics, and transparency. It also ensures the management of the organ supply chain efficiently. This provides additional protection against illegal activity contamination and protects the rules prohibiting entry. It guarantees that organizations wishing to join the blockchain-based organ supply chain assure compliance with regulatory standards and are regularly audited.

The first step is to design a cyber security framework to prevent the organ supply chain from any malicious intent. This sort of blockchain continues to act in a distributed manner, with software services running on various stakeholders and not relying on a central authority. The consensus method is an identical process to agree on the same data; this study uses the PBFT consensus method. It can offer 1000 transactions per second if it operates with less than 100 nodes with a minimal payload size. The health records in blockchain only manage data text, although it could also handle many rogue nodes. Blockchain-based supply chain achieves high throughput, efficient auditability, and transparency of immutable data by restricting the number of participants in the blockchain. The system must provide data access management and anonymization of health data to ensure compliance with data privacy rules.

The framework is used to design a blockchain-based system that tracks and audits the supply chain using smart contracts to assist healthcare organizations in tracking and auditing organ transplantation. The audit system should have numerous capabilities, such as the ability to transform audit log data into a blockchain-based

**Fig. 12.27**  Deployment process of human organ donor smart contract

distributed ledger that can be shared across all participants of the network. To ensure record validity, it should also incorporate data integrity. This should restrict rogue nodes from modifying transaction timestamps to ensure the security of the system. The access control restricts the users, and only specific users grant access to join the network. This should enable auditability and transparency of documents and a tamper-proof audit log, integrity, evidence of compliance, and time stamp verification for the transaction.

Figure 12.26 shows the blockchain-based human organ supply chain management. In this study, a smart contract is designed to preserve the privacy of the data, and it can also easily locate the contamination source using the traceable solution. At each handover, the data is visualized and tracked in real-time using blockchain. After completing the quality check process, the smart contract is being executed to check user validity. If the smart contract satisfies the quality and required parameter conditions, the block is verified, validated, and added to the existing blocks of the blockchain network. This helps in preventing the behaviors that have encouraged both criminal activities and urgency from people who want to sell their organs fraudulently and donors who want to procure them. It can also aid in the post-donor study of failures and the tracking and tracing of the sources (if the organ was not a perfect match and was contaminated, etc.).

**Implementation** In the human organ smart contract, initially, the hospital found the list of organ donors and matched all the requirements with the patient. If all requirements are satisfied, it is forwarded to the patient. First, information of organ and organ donors is being stored. Afterward, this information is being given to the required stakeholders of the system. In the next step, organ details and the required

**Fig. 12.28**  Deployment process of human organ supply chain smart contract

temperature for the organ's preservation are specified in the human organ smart contract. During the organ transportation, it must preserve the required temperature and successfully reach the patient. During the supply chain process, if it was contaminated due to environmental conditions or hinders with security threats, it can easily locate the source where it is damaged or manipulated. The smart contract can verify all the stakeholders and their wallet addresses added into the block.

Figure 12.27 shows the deployment process of a human organ smart contract, in which we first obtain the information from an organ donor, QR code of human organ, the status of the organ, and input the information to the human organ supply chain.

Figure 12.28 shows the deployment process of the human organ supply chain. In this contract, various functions are used to collect the organ from the hospital. It moves using an organ transporter to a laboratory. After the verification of the organ, it is picked from the laboratory and reached to the specific patient. After the completion of the process, the required stakeholders are payable. Then, Fig. 12.29 shows the information of organ transportation function, in which organ can forward from laboratory to organ transporter and after that send that organ to a specific hospital, and then reached to the required patient. Figure 12.30 shows the information of the received human organ. It contains the address of various stakeholders who is present in the supply chain process. In this, we add the wallet address of each stakeholder to initiate the transaction after verification of the organ. After this, transaction details are being stored in the new block in the blockchain. Finally, Fig. 12.31 shows the information of deployed contracts and recorded transaction summary.

**Fig. 12.29**   Information of organ transportation function

**Fig. 12.30**   Information of received human organ at hospital/patient

Fig. 12.31  Deployed smart contracts and recorded transaction details

## 12.8  Summary

Blockchain technology is a game-changing innovation that has the potential to transform many present centralized institutions into safe, open, accessible, and inclusive networks to motivate the customers to take a benefit of this technology. This chapter examined the essential elements of blockchain technology in several case studies. The goal of blockchain-based supply chain management is to demonstrate future advantages and encourage people to work as suppliers. The various case studies analyze several aspects of the blockchain-based supply chain. The potential customer and manufacturer benefits, as well as the impact on the environment, are expected to offer adequate motivation for change. In this chapter, we have addressed the various issues of existing supply chain. It then highlight various frauds, and give a solution to overcome the issue. Afterwards, we describe the integration of blockchain in the supply chain system, and their benefits such as immutability, reliability, and many more. How the system makes a more reliable, and how all assets, entities are payable after checking the product quality. This chapter also explore various case studies to address the real word problem of supply chain management and how it solves using the blockchain. In the blockchain, we define a role of stakeholders using a smart contract. This smart contract helps to automate the transaction process between the various participants connected in the blockchain-based supply chain process. Initially, all stakeholders are verified and added into the system for a supply chain traceability. All stakeholders can provide their wallet address to next member in the supply chain process using smart contract. The transactions are being initiated after the completion process of product and stakeholder verification. At the end, all stakeholders are payable in their bitcoin wallet address. If any product is contaminated during any handover then it easily locate using blockchain traceability. Thus, blockchain-based supply chain system makes a system more reliable, transparent, easily traceable, and provide a robust solution to existing supply chain system. The blockchain-based benefits in the supply chain process are described as follows:

– Automatic process of transfer amount to the supplier
– Achieve a traceability of supply chain process
– RFID code provides the identity of the product and it helps to track the product.
– Digital currencies are paid in the bitcoin wallet address of all stakeholders

## 12.9   Question Answers

1. Which of the operations are used by manufacturers to achieve agreement requests?

(a)  Durable acquisition
(b)  End-to-End Visibility of supply chain.
(c)  Planning of supply chain
(d)  Risk management

2. How blockchain will help in business supply chain management?

(a)  Improves visibility and transparency
(b)  Secure trading of supply chain data
(c)  Smart contract Management
(d)  All of the above

3. The supply chain system aims to apply socially responsible for executions encourages positive brand awareness for a business

(a)  True
(b)  False

4. Which of the following are important in creating a supply chain management strategy?

(a)  Ensuring the customer is satisfied
(b)  Using as many resources available as effectively as possible
(c)  Implementing technology and management software that best fits your business needs
(d)  All of the above

5. Which of the following actions helps companies avoid or manage potential supply chain risks?

(a)  Moving away from sources in earthquake-prone areas
(b)  Dual sourcing
(c)  Sourcing raw materials or goods from a more politically stable region
(d)  All of the above

6. The blockchain can address the errors in the current supply chain by

(a) Automatically tracking the information and verifying that using the smart contract
(b) Only verifying the information using the smart contract
(c) Only tracking the information
(d) Making the data available to auditors to manually verify and correct errors

### 12.9.1   Short Questions

(1) Why blockchain is required for supply chain management?
(2) How is any product traceability is ensure using blockchain?
(3) What are the main concerns to be considered in supply chain management?
(4) How product validity can be checked using the smart contract of the supply chain process?

### 12.9.2   Long Questions

(1) How Blockchain is being leveraged as an emerging technology start-ups to help organizations track their supply chain in COVID 19 outbreak. Use diagrammatic representation for your viewpoints.
(2) How smart contracts are being used to ensure the security of organs, vaccines, and medicine?
(3) How smart contracts can be used for validation of the stakeholders present in the supply chain process?
(4) is it safe to give all information about the product to all members which are stakeholders of the supply chain process?

## References

1. Francisco K, Swanson D (2018) The supply chain has no clothes: technology adoption of blockchain for supply chain transparency
2. Hallikas J, Dahlberg T (2017) Digital supply chain transformation toward blockchain integration
3. Abeyratne SA, Monfared RP, Blockchain ready manufacturing supply chain for distributed ledger
4. Gupta R, Kumari A, Tanwar S, Kumar N (2020) Blockchain-envisioned softwarized multi-swarming UAVs to tackle COVID-19 situations. IEEE Netw. https://doi.org/10.1109/MNET.011.2000439

5. Mehta D, Tanwar S, Bodkhe U, Shukla A, Kumar N (2021) Blockchain-based royalty contract transactions scheme for Industry 4.0 supply-chain management. Inf Proc Manage 58(4):102586. ISSN 0306-4573. https://doi.org/10.1016/j.ipm.2021.102586
6. Seebacher S, Schuritz R (2017) Blockchain technology as an enabler of service systems: a structured literature review. In: Proceedings of the international conference on exploring services science, Rome, Italy, pp 12–23
7. Kumari A, Gupta R, Tanwar S (2021) Amalgamation of blockchain and IoT for smart cities underlying 6G communication: a comprehensive review. Comput Commun 172:102–118. ISSN 0140-3664. https://doi.org/10.1016/j.comcom.2021.03.005
8. Rajesh G, Sudeep T, Neeraj K (2021) Blockchain and 5G integrated softwarized UAV network management: architecture, solutions, and challenges. Phys Commun 101355. ISSN 1874-4907. https://doi.org/10.1016/j.phycom.2021.101355
9. Hofmann E, Rusch M (2017) Industry 4.0 and the current status as well as future prospects on logistics. Comput Ind 89:23–34
10. Earley K (2020) Supply chain transparency: forging better relationships with suppliers available online: https://www.theguardian.com/sustainable-business/supply-chain-transparency-relationships-suppliers(Accessed on 2 April 2020)
11. Stock JR, Boyer SL, Harmon T (2009) Research opportunities in supply chain management. Academy of Marketing Science
12. https://www.logisticsbureau.com/how-blockchain-can-transform-the-supply-chain/. Last Accessed 4 Oct 2017
13. https://www.mhlnews.com/global-supply-chain/article/22050821/who-is-causing-supply-chain-fraud. Last Accessed 4 Oct 2017
14. https://www.scmr.com/article/supply-chain-crime-can-be-addressed-by-blockchain-strategy-says-deloitte-st. Last Accessed 4 Oct 2017
15. https://www.businesswire.com/news/home/20180508006009/en/Top-Reasons-Supply-Chain-Visibility-Important-Quantzig. Last Accessed 4 Oct 2017
16. https://www.researchgate.net/publication/266614041-Research-Opportunities-in-Supply-Chain-Management. Last Accessed 4 Oct 2017
17. https://www.zlc.edu.es/research/core-research-areas/. Last Accessed 4 Oct 2017
18. https://www.nap.edu/read/6369/chapter/15. Last Accessed 4 Oct 2017
19. https://www.supplychaindigital.com/topics. Last Accessed 4 Oct 2017
20. https://www.linkedin.com/pulse/supply-chain-fraud-common-examples-risk-management-best-pei-li-wong. Last Accessed 4 Oct 2017
21. https://www.supplychaindigital.com/technology/comment-blockchain-solving-supply-chain-management-challenges. Last Accessed 4 Oct 2017
22. Kaid D, Agad Ellizaber M (2019) Applying blockchain to automate installments payment between supply chain parties. Academy of Marketing Science

# Chapter 13
# Blockchain for Government Services

**Abstract** The government sector is suffering from bribe issues, third-party mafias, and many more nowadays, which hinder the overall functions of government services. Blockchain technology is a viable solution that efficiently manages government services to overcome these issues. It is an emerging technology that works on the distributed ledger principle. Elimination of third-party, data transparency, high security, and privacy are the significant characteristics of blockchain technology. Blockchain can be applied for various functionalities in the government sector, including tax payments, land registration, funds tracking, and many more. In tax payments, the primary focus is towards the traceability of taxes and blockchain's immutable ledger stores all traces of historical data transactions, i.e., the record of each and everything. In the land registration scenario, the aim is to eliminate the third party and include transparency using cryptographic hash functions. Andhra Pradesh (a state of India) has taken the initiative and decided to incorporate blockchain technology in its land registration system. Further, this chapter includes various use case scenarios for an in-depth discussion with smart contracts, which explains the technicality of the blockchain-based system.

**Keywords** Blockchain · Tax payments · Land registry records · Smart contracts

## 13.1 Introduction

In today's era, everything is becoming smart right from a small wristwatch to the big giant machines and from a small home to a smart city. Augmented reality (AR), virtual reality (VR), and artificial intelligence (AI) play a vital role in the aforementioned revolution. Consider a scenario of smart city which is administered by government bodies for its proper and judicial functioning. A government is a regularity committee or group of people governing an organized community, often a state [1]. The smart city of a state governs various activities such as vehicle management (administered by the state transportation office), currency fraud (administered by the Reserve Bank of India), and many more. All smart devices in a smart city communicate with each other in the internet of things (IoT) environment over the

© The Author(s), under exclusive license to Springer Nature Singapore Pte Ltd. 2022    355
S. Tanwar, *Blockchain Technology*, Studies in Autonomic, Data-driven and Industrial Computing, https://doi.org/10.1007/978-981-19-1488-1_13

open wireless communication channel. All processes like storing, processing, and retrieval of information are performed in a digitized way.

The government officials from various state government sectors have their citizens' personal information, such as personal account number, Aadhar number, bank account number, and property information. Citizens can access their data online anytime, anywhere via the Internet, which is an open channel and susceptible to various security threats, i.e., under the eyes of malicious users/hackers. A malicious user is always trying to find the bugs and possible breakthroughs in the government system to access the citizen's personal information. Security of citizens' information is a prime concern for the government, as "everything in the digital world is hackable and have security breakthroughs". We can not guarantee here a 100% unbreakable system, but we try to maximize the data security. Blockchain technology is a viable solution in providing security to the citizen's data while in communication. The conceptual description on blockchain technology can be found in Chap. 1.

### 13.1.1  Blockchain Technology: The Data Security Viewpoint

Blockchain is a shared ledger that stores data in a distributed, trusted and immutable manner. It is a chain of blocks that are hashed together, i.e., the next block's hash depends on the current block's hash. The data stored in the blockchain is secured using public-private keys. The sender encrypts the information with the general public key of the receiver and the receiver decrypts it with their private key. It has the concept of digital or smart contracts that execute upon encountering some condition (related to policies, regulations, and financial transactions) [2]. Smart contracts are written in specific languages such as solidity, Go, javascript, SQL, and java [3]. Data will be stored into the blockchain only if it satisfies the smart contract conditions [4]. It also eliminates the need for centralized third-party systems to maintain trust between the participating blockchain network members [5]. The reliability and integrity of information stored in the blockchain can be achieved using consensus algorithms. Every node has to perform a certain complex task to add their blocks to the blockchain called mining.

Based on the nature of operations, the blockchain is categorized into public, private, and consortium blockchain. A brief description of the type of blockchain is as follows.

- *Public Blockchain*: It is a permissionless and perfectly distributed ledger that anyone can join the network. The data in a public blockchain cannot be altered once written into it. Examples of public blockchains are Ethereum and bitcoin.
- *Private Blockchain*: It is a permissioned blockchain and is being controlled by a single organization. All the participating members of the blockchain are known to each other. An unauthorized user cannot join the private blockchain network. An example of a private blockchain is hyperledger fabric.

- *Consortium Blockchain*: It is also a kind of permissioned blockchain under the control of multiple organizations.


### 13.1.2 Advantages of Blockchain in Government

A blockchain is required where the sharing of data with multiple parties is in place. As discussed above, establishing trust between parties without any centralized trusted third party is a challenging task that can be easily done through the blockchain. The government is the only system with the sensitive data of all citizens of the country, so there is a need to secure the data [6]. Blockchain helps a lot in this and the following are the benefits of the convergence of blockchain in government systems

- As part of daily routine, the government generates certain data such as import-export data, expenses done by the government, stock exchange data, etc., which need to be stored and retrieved without any third-party intervention. Anyone can see what is happening but cannot modify it.
- The government holds certain assets like public properties, land records, and government buildings that need a secure and reliable data storage and must be accessible by the concerned citizen of the country. Blockchain technology helps achieve the said goal of data access with assured security and reliability.
- The government also maintains detailed information about every citizen like what income he holds, how many properties he has, how much tax he pays, insurance policies, etc. This data should be open so that a person can fill his information independently with complete trust.
- Tracking individual citizens is essential for any government. If a person completes his graduation, he must upload their degree to the employment portal to prove that he graduated. Blockchain shows the trueness of the particular degree uploaded in it after validation and verification.
- Nowadays, during the covid-19 scenario, blockchain offers edges in the field of vaccination. Information storage concerning vaccination benefits the medical field, universities, insurance, and schools. Government officials can validate the vaccination data and medical status early.


## 13.2 Theft of Government Data

The "Quebec government" has acknowledged a cyberattack that could influence about three lakh sixty thousand teachers working in Canada [7]. Using stolen credentials, attackers compromised the system and got access to their databases, containing information about the teachers. Determinating how many were affected is quite difficult. Later, a savings plan supplier, which ties to the "Government of New Zealand", announced a security issue that around twenty-six thousand people were affected [8]. Third-party unauthorized access was made, which lasted in the range of months. No clue was found what data is compromised and whose information is compromised raised a big question.

Consumers of the water distributor in "Colorado" became victims of a string of attacks on transaction applications for "the Click2Gov municipality" [9]. The attacker modified the piece of computer code and tried to bypass the security standards. Later, the vpnMentor cybersecurity research team, led by Noam Rotem and Ran Locar, has uncovered a leaking S3 Bucket with 36,077 files of visible data on an Amazon server belonging to JailCore [10]. The auction of alleged government information containing the private data of some 92 million Brazilian citizens [11]. Sberbank, Russia's largest banking and monetary services organization, is investigating a suspected information leak impacting a minimum of 200 customers [12]. The biometric authentication discussion was reignited another time in the week, as reports surfaced of an alleged breach at a security firm that handles sensitive information on behalf of UK law enforcement, among different organizations [13].

In May 2017, WannaCry ransomware cryptoworm carried out a global cyberattack named "The WannaCry ransomware". It specifically targets computers running on MS Windows OS, encrypting data and bid for the ransom cryptocurrency like bitcoin [16]. It propagated through an exploit named EternalBlue, which the United States National Security Agency (NSA) flourished for older Windows systems. More than 200,000 computers across 150 countries were affected with total damages in the range of hundreds of millions to billions of dollars. Air India also suffered from a data breach/cyber attack in the year 2021 [17]. As per their report, cyberattack affects around 4.5 million customers' information. Attackers theft the ticket booking details, credit card information, and passport regarding data of the customer from 26th August 2011 to 20th February 2021. However, CVV and CVC numbers were not sorted in targeted the database server. Table 13.1 shows the summary of some cyberattacks that happened to date. It describes the region where theft was admitted, the year of the attack, and detailed descriptions.

## 13.3  Traditional Way of Processing Tax Payments

In India, there are mainly two types of taxes such as direct and indirect taxes [18]. The direct tax is one which can we pay directly from our income to the government, whereas in indirect tax, some other party collects the tax from us and pay to the government such as the purchase of goods or take any service

- *Direct Tax*: It is classified as both income tax and corporate tax. Income tax is the tax that each salaried employee has to pay on their total income and the percentage of the amount of tax is fixed. Whereas corporate tax is the tax that the person has to pay against the benefits earned from the business settled under the government laws [19].

**Table 13.1**  List of government data thefts

| Index | Theft admitted to | Year | Description |
|---|---|---|---|
| 1 | Government of Quebec, Canada | Feb 2020 | The "Quebec government" has acknowledged a cyberattack that could influence about three lakh sixty thousand teachers working in the region of Canada [7] |
| 2 | Generate, New Zealand | Feb 2020 | A savings plan supplier with ties to the "Government of New Zealand", announced a security issue that affected about twenty-six thousand people [8] |
| 3 | Aurora Water, US | Jan 2020 | Consumers of the water distributor in "Colorado" were new victims of a string of attacks on transaction applications for "the Click2Gov municipality" [9] |
| 4 | Jailcore, US | Jan 2020 | Scientists found that Jailcore, a US jail care facility, was revealing twenty thousand jail-related information [10] |
| 5 | Brazilian government | Oct 2019 | It has been stated that information of as many as ninety-two million Brazilian citizens is on sale on the dark web [11] |
| 6 | Sberbank, Russia | Oct 2019 | "Russian police" have filed charges against an unidentified current "Sberbank worker who reportedly admitted to trading consumers' credit card details on the dark web [12] |
| 7 | Suprema, UK | Aug 2019 | Private biometric company Suprema, which supplies organizations including London's Metropolitan Police, exposed a database that included more than one million fingerprints, user names, passwords, and facial recognition data [13] |
| 8 | NRA, Bulgaria | Jul 2019 | "Bulgaria's tax authority was hacked in 2019, with the incident affecting more than five million people" [14] |
| 9 | City Power, South Africa | Feb 2019 | Ransomware hit the supply of power throughout "Johannesburg". With the state-owned City Power cutting off access to their consumer-facing facilities. More than twenty-five lakh people got affected by this incident [15] |

- *Indirect Tax*: The indirect tax depends on goods and services. It is one which we can pay for daily used items which are goods and services like restaurants, parks, etc. The government mainly uses this tax to generate revenue. So, we primarily focused on indirect taxes.

GST is a form of indirect tax incorporated in 2017 to replace all previous taxes. Table 13.2 shows the types of taxes before and after the GST. It has one major benefit, i.e., it has removed the tax on tax, where at each step in a supply chain, there is a tax, which makes the product costly (cascading effect of taxes). GST has three components: first is CGST that the central government collects on intra-state sale,

**Table 13.2** Taxes before and after GST

| Taxes before GST | Taxes after GST |
|---|---|
| "Central Excise Duty, Duties of Excise Additional Duties of Excise, Additional Duties of Customs, Special Additional Duty of Customs, State VAT, Central Sales Tax, Purchase Tax, Luxury Tax, Entertainment Tax [20]" | "CGST (Central Goods & Services Tax), SGST (State Goods & Services Tax), IGST (Integrated Goods & Services Tax) [20]" |

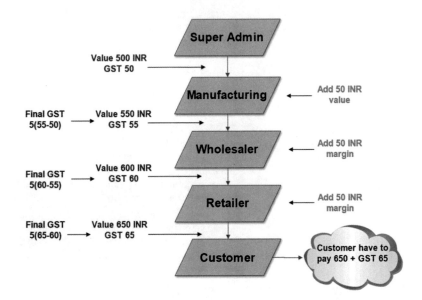

**Fig. 13.1**  GST in supply chain

and second is SGST that is possessed by the state government also for intra-state sale, the third is IGST which the central government collects for inter-state sale. Figure 13.1 shows the supply chain and tax payment at each stage for a particular item where GST Rate is 10%. Suppose the raw material has a value of 500 INR, then at the manufacturing process. In that case, when a manufacturer purchases the raw material, it has to pay 10% of the price as a tax that makes the value of raw material 550 INR (500 INR product value and 50 INR for GST tax). After that, when a manufacturer has processed that raw material to make items, it has added value worth 50 INR, so now product value is 550 INR. So, when a product goes out for sale in the market again, there is 10% of tax. So, now the manufacturer will not give the whole 55 INR for value 550 INR because the manufacturer has already paid a taxation amount of 50 INR on Raw material. It has to cut down the amount from this tax, so the manufacturer has to pay $55 - 50 = 5$ INR only.

After that, wholesalers also have to pay the tax at the time of purchase, which is 55 INR. So, at that point, the product value becomes 550 INR, and a profit margin of 50

INR added, so the product is now worth 600 INR, and at selling time, the wholesaler has to pay tax. Still, he already pays tax to the manufacturer, which is 55 INR, so now the wholesaler has to pay only $60 - 55 = 5$ INR. This process is repeated at each step of the supply chain. In the end, a customer gets a product at 650 INR and 10% of its tax 65 INR gets paid by the end consumer only. The amount of tax the government receives is $50 + 5 + 5 + 5 = 65$ INR. So, the burden of taxation from a manufacturing perspective and in the supply chain got decreased in comparison to pre-GST taxes [21].

So, to maintain the whole process of the GST government has issued a 15-digit common identification number to every taxpayer, i.e., Goods and Service Tax Identification Number (GSTIN), and there is a GST pilot portal to maintain all records that dealers update. So, this way, the whole GST payment process works [22]. Under GST, tax invoices are primary documents needed at every stage of the supply chain, from the purpose of purchase to the filing of returns [23].

## 13.4  BCT-Based Tax Implications

The government needs funding to provide good public services and function the nation that rises the economies. The present tax payment method is complex; the payers, who don't seem to be aware of the tax laws, cannot follow the tax exemption process according to their liabilities. Therefore the tax recovery executed is dead unfairly, time-consuming, and it additionally becomes inefficient. The assesses desire the tax return procedure to become easier and less labor-intensive, whereas taxation authorities want to collect and examine tax data digitally [24]. Figure 13.2 describes issues the tax authority faces like verification of transactions and tax returns due to individual and business tax fraud. So, there is a demand for technology that gives automation within the tax collection method, transparency in the network, authentication, and verification of returns and tax data. Blockchain becomes promising technology that revolutionizes tax management systems by providing transparency, speed, security protection on fraud, automation, and immutability of the information. Blockchain initially targeted indirect taxes such as value-added tax (VAT), payroll tax, and transfer valuation in the tax management approach. After that, its focus is on direct tax, such as income tax.

### 13.4.1  BCT for VAT Tax

VAT is an indirect consumption tax, where the price of the product is considered from the manufacturing to the selling price [25]. As it gives the highest benefits to the government budget, the tax administration admires an effective tax collection and obtains more income. VAT tax is applicable on the national and international levels. Hence, the monitoring and tracing of VAT transactions make problems for the tax authority.

**Fig. 13.2**  Reason to change tax collection process

Blockchain offers an effective and efficient VAT system using digital signatures and smart contracts to mitigate the above-mentioned issues. The author Setyowati et al. [26] proposed a blockchain-enabled VAT acceptance system for e-Invoice and compared it with the traditional VAT system. They discussed the dealer and treader position approach for the Indonesian VAT system and introduced a blockchain-based VAT system specifically for e-invoicing. China, Finland, Germany, Sweden, and the Netherlands are the countries that adopted this new emerging technology and started testing on the blockchain-based VAT system [27].

Based on the World Economic Forum report [28], 2015, "73% of respondents anticipate that by the year of 2023 government will start collecting tax via a blockchain technology". The joint production of Microsoft, Price water house Coopers and Belasting adviseurs N. V. the Netherlands and Vertex introduced VAT fraud avoidance approach using blockchain [29]. They addressed the trust issue in the VAT system, i.e., data inequality issue between tax governance agencies and taxpayers. Their solutions alleviate the fraud and construct transparent VAT infrastructure to formulate the "VAT Fraud Prevention Prototype" under "Microsoft Blockchain Essentials solution" and "Azure Blockchain Workbench". The proposed prototype uses 3 Ethereum Ledger nodes, produces an off-chain data repository, and produces results on Node.JS. Figure 13.3 shows there are three phases to implement the VAT Fraud Management system. In phase I, they introduced information interchange using blockchain between stockholders in multinational scenarios. Automated VAT calculation and return process, VAT payment between various countries, and modification in VAT accounts established using the smart contract in phase II. Phase III discussed the use of cryptocurrency in VAT transactions and collection procedures.

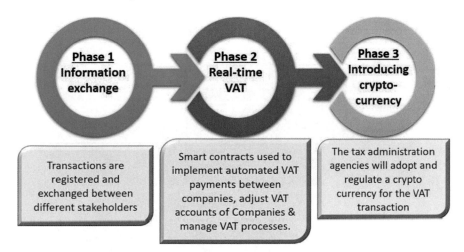

**Fig. 13.3** Blockchain technology in VAT Fraud Management

### 13.4.2 BCT for Payroll Tax

An employer remitted a certain amount of money to the government from the salary as a part of employee's pay known as payroll tax [24]. Payroll tax is directly deducted from the salary and compensated to the internal revenue service (IRS). Blockchain technology is applied in the payroll service to deal with real-time data. In the organization, blockchain-based payroll is used to transfer and calculate tax. Moreover, it has a decentralized ledger and transparency mechanism to provide security in the employee's salary. Employer fills the gross salary details in a blockchain-based system. Figure 13.4 shows steps on implication of blockchain in payroll system [30]. Here, a smart contract is used to validate that system and verify the social security of the data. Accordingly, the tax amount will display on the tax authority account and the net salary reflected in the employer account. A blockchain-based payroll system accelerates the tax calculation process and makes it cost-efficient. Multiple organizations adopted blockchain to calculate the payroll tax, such as J.P.Morgan. They developed a permissioned blockchain system using Ethereum protocol that correlates the amount of payment between commercial organizations, which is fully visible to governors, and private for the participants [31].

### 13.4.3 BCT for Transfer Pricing

Transfer pricing is considered among the different divisions of the industry that represents the change in prices of the goods and services affect between two divisions [25]. It applies to the national and international levels. The motive behind to use of

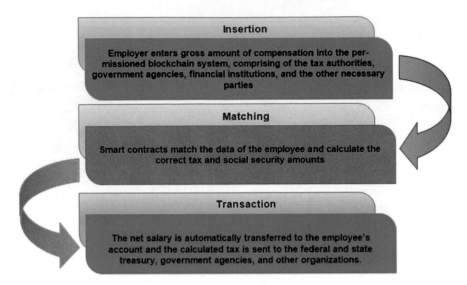

**Fig. 13.4**  Blockchain technology in payroll tax

transfer price is it overcome the tax-relevant burden from the core company. For example, the multinational company has two divisions such as M and N. If division M is in the higher tax country and division N is in the lower tax country, then division M sells its products to division N at the lower price. In this way, division M generates lower revenue by reducing the profit. On the other side, division N generates more profit using the costs of goods sold. Here, Blockchain brings transparency and reduces the risk factor of data falsification in the system. The digital ledger tracks the transaction, and the smart contract gives an easy and exhaustive solution to control the transfer pricing.

### 13.4.4  BCT for Income Tax

Blockchain enables a long-term vision by providing security between the income tax department and bank, authentication, real-time tracking of the information, and digital approval to the income tax authority [32]. Figure 13.5 shows how the income tax department will be able to resolve the tax avoidance problem and the queries quickly.

Nowadays, the tax applicability on the digital currency has become a vital issue. Several agencies provide consultancy on the tax issue on cryptocurrency. Deloitte is one of the leading organizations that provide consultancy, risk, and financial advisory, tax, and audit corresponding services [33]. They guide the taxpayers to identify the income tax, sales, and gross receipts taxes-based issues on digital property and provide Distributed ledger technology (DLT)-based solutions using blockchain. This

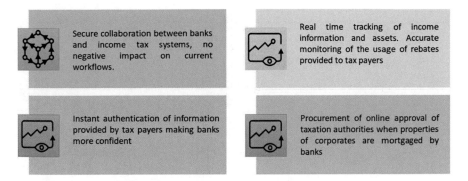

Secure collaboration between banks and income tax systems, no negative impact on current workflows.

Real time tracking of income information and assets. Accurate monitoring of the usage of rebates provided to tax payers

Instant authentication of information provided by tax payers making banks more confident

Procurement of online approval of taxation authorities when properties of corporates are mortgaged by banks

**Fig. 13.5**  Blockchain benefits in income tax

agency dispenses the bitcoin and tokens under digital assets. Rob Massey, Partner and Global Tax Leader (Deloitte Tax LLP) [34] said, "Building out a blockchain solution involves significant technical considerations in architecture and design. Having a tax lens at the table during the design phase enables tax efficiencies in compliance processes, and a greater ability to gather sensitive tax data for use in compliance, planning and support of a tax examination."

## 13.5  Case Study: GST With and Without BCT

As we discuss in Sect. 13.3, the process of GST involves multiple parties and all have to pay their taxes individually to the government. The government needs to check all transactions carefully, calculate the GST, and give tax relief to every individual.

### 13.5.1  GST Without Blockchain

In the current scenario (without blockchain technology), there are many issues by which the government is losing. Sometimes, the tax entry of a party (which is a part of the supply chain) is missing, as per the example in Sect. 13.3. A vendor, i.e., a raw material supplier, collects the goods tax from the manufacturer and does not pay it to the government. How can a government give relief to the manufacturer in their paid tax amount, i.e., 50 INR. So, in the next step, when a manufacturer sells an item to the wholesaler, it has to pay the full taxation amount because the vendor has not submitted the tax amount to the government. The above mentioned is a loophole in the existing system. The taxpayer needs to put additional effort to upload the invoices with accurate details regularly [35].

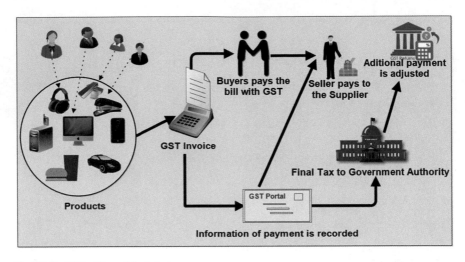

**Fig. 13.6** GST without blockchain

Another problem is that the government is not able to return or cut down the GST amount of an individual party until the government has not received the invoices at every step for their calculations. This delays the issuing of a refund to the parties involved in the supply chain process. Some small-scale retailers accept cash for the items purchased and do not generate an invoice to avoid taxes. So this way, they can make significant differences when they file GST forms. The current implementation of GST has been successful, but the IT module under GSTN is not capable of handling such a vast transition [36].

Let's suppose we want to buy a saree that has a different workflow at the time of its manufacturing. Generally, the production process is as the farmer produces cotton. After production, it comes to the vendor side. They require other stuff such as needles, colors, etc., to make a saree, then it comes to the shop or shopping mall from where we can purchase it. In this process, the government applies tax to the farmer who sells the cotton, cotton ball manufacturing industry, and other relevant companies. So, according to the GST process, the government collects tax from customers who will purchase this saree. After collecting tax, the government refunds the tax amount to production houses and other relevant entities. Figure 13.6 shows when the buyer pays the bill, the seller generates the GST invoice. The information will be reflected on the GST portal. At that movement, the government commission has to calculate how much tax will be refunded to the relevant entities of the productions and intermediaries. After that, the final tax amount will be distinguished between state and central government. Figure 13.6 shows the GST process without blockchain. The seller generates a GST invoice when the customer/buyer purchases any product. The details of that purchase are stored in the GST portal [37]. The seller pays the GST bill to the supplier. In the end, from the GST portal, the government authority adjusts the tax and return payment details.

## 13.5.2   GST With Blockchain

Blockchain is a real-time distributed system that stores transactions in a block. Blockchain smart contract defines a set of rules for each taxpayer, which is self-verifiable and self-enforceable before making an entry to the blockchain network. Blockchain is a plausible solution for the current GST system. Let's discuss how blockchain can be deployed in the existing taxation system, i.e., GST. Figure 13.1 shows when vendors sell an item to the manufacturer, then a transaction is generated and stored in a block of a blockchain. It is a decentralized and distributed ledger technology, where the transactions are secured with cryptography techniques that make it tamper-proof. Any kind of data manipulation can be easily tracked, as it changes the hash of preceding blocks in the blockchain network. A block transaction has details like vendor name, invoice, item quantity, name of the buyer, etc. The system calculates the GST and non-GST amounts as per the given transaction details. When a manufacturer pays the tax against the purchased item, the smart contract automatically calculates the tax amount and applies rebates accordingly before it gets credited to the government account via a blockchain network.

In the supply chain system, when a manufacturer sells an item to a wholesaler, the transaction is recorded into the blockchain. The smart contract calculates the GST accordingly as the manufacturer has already paid the 50 INR tax, which cuts down its tax amount. The manufacturer has to pay only 5 INR as a tax and also record it into the blockchain network as an immutable entry. So this way, the whole supply chain and its transactions are stored in blockchain, which keeps track of all the transactions made by parties in the supply chain system. It also maintains the trust between the participating blockchain network members, i.e., the supply chain hierarchy. This way, most of the goods and services are recorded to boost a nation's economy. Also, if the end retailer does not make an entry, then it's easy to detect. In this system, every user directly pays the tax to the government, which complexes the tax refund tracking. So, Fig. 13.7 shows how blockchain helps to provide a solution to the above issue of the GST taxation system.

## 13.6   Case Study: BCT for Taxation

The federal tax administration of Brazil enforced the tax management system using BCT known as "bCPF", which distributes taxpayer registry information among tax and federal, municipal, and state government regulatory organizations [38]. They aim to create a blockchain-enabled registry for the legal entities. Infosys provides a case study on the Indian income tax department to upgrade the tax process using blockchain technology [39]. Table 13.3 shows the current income tax system along with its key issues and provides a blockchain-based solution. Figure 13.8 shows the proposed blockchain-based tax process system. It has three layers such as user interface, third-party applications, and blockchain layer. The income tax department and

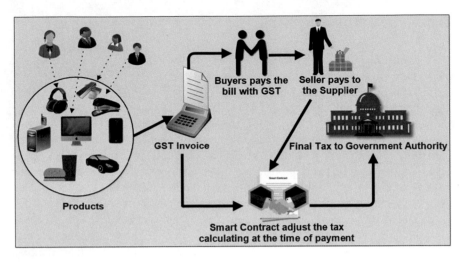

**Fig. 13.7** GST with blockchain

**Table 13.3** Blockchain in income tax process

| Income tax process | Issue | Blockchain-based solution |
|---|---|---|
| Banks are not deducting any tax on their interest income for those who have income below a certain threshold value. Companies and organizations need to fill 15G and 15H forms to show their final income is below the defined threshold value | On the various form of interest income, the bank gives a consolidated view | There is distributed ledger technology that storages interest income from a different bank and income tax department outlook that real-time data |
| All income details of taxpayers are collected from various sources and stored in the one commonplace using form 26AS. After collection, it hangovers to the banker for verification | To intercept the tempering of the information, the banker asks the income tax department to validate the data of the form | The banker validates the information with the hash value on one common platform. The system is not disclosing any personal details of the taxpayers to the banker |
| When the taxpayer displays certificate 197, it permits a lower rate of the tax deducted at the source. | There is a lack of real-time connections between financial organizations, even they are not aware whether the document is submitted by other organizations or not | Income tax department consolidates different nodes and performs real-time validation on the customer data |

bank-enabled user interfaces are connected with the server using bank applications. The server has all the relevant information such as tax credit report (26AS), tax credit reconnection (15GH), proof of the tax deduction (197), and the API layer.

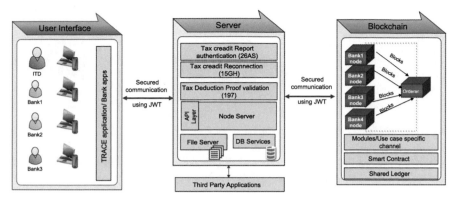

**Fig. 13.8** Architecture of blockchain-based tax process

Securing communication between the bank and third-party applications is enabled using blockchain-based smart contracts and shared ledgers.

The blockchain-enabled taxation system is still not adopted by the Indian government due to cross-border businesses deals. So, how to collect and refund the tax is a major concern. In India, the main tax portion was from indirect taxes like GST and VAT taxes. The goods are transported according to taxes are applied. So, there is a big challenge to adopt new technology in the system. With increased computing power and connectivity to devices, blockchain can be a daily reality to the problems. As of now, there is no application of blockchain in the taxation system anywhere, but soon, blockchain will revolutionize the taxation system.

## 13.7   Case Study: BCT for Land Registry Records

Land is one of the most valuable assets for any individual. Land registry is determined by a land title that protects the individual rights of the landholder also impacts economic, industrial, and social growth. When it comes to ownership distribution, mostly disputes arise. Let's go into the detail of such disputes and how blockchain helps to resolve those disputes will be discussed.

One can easily guess the procedure for land registration, as it is just the same as any other registration. In the abstract, the government appoints an authority that is solely responsible for checking and verifying the documents submitted by the party to claim a particular piece of land. Consider the government as a third party and whenever a third person involves, disputes between the parties always arise. Figure 13.9 shows person A visits government authority for their land registration, but what if any disputes occur at this point only. Trust issues arise when any third party is involved in the land registration process. Another issue in the current flow is the transparency issue. Why does only one authority decide that Person A is the

Person A

Government Authority

Government Database

**Fig. 13.9** Land registry process

rightful owner of the land and why not all the citizens of a nation? Transparency is also a big hurdle in the current system. Scenarios for better understanding are as follows.

1. Person X goes to the government-appointed authority and registers a land on his name. Another person called Y comes on some other day and claims the authority of the same land.

2. Person X goes to the government-appointed authority with the morphed documents and registers a land in his name, and applies for a loan in a bank. Now bank checks the integrity of the documents submitted by person X with the government and it finds that the land is in the name of person X.

3. Person X goes to the government-appointed authority with the morphed documents and register land on his name and tries to sell the land to the other person say person Y. Now, Y person goes to an agent that verifies the document and all and states that the land is of person X only.

### *13.7.1  Blockchain as a Solution for Land Registry*

Let's try to understand how blockchain technology will help us to overcome the aforementioned problems:

- As the land registry record is maintained in the blockchain, anyone can view the data at any time as it is shared publicly.
- No involvement of the third party. Any citizen can directly get connected to the government.
- The system gets transparent to everyone. As all the miners in the system verify the transaction.
- No centralized database is maintained so that no central control will be there.
- Banks can check out the status of the ownership of anyone if it's on the blockchain platform.
- The smart contract will run continuously in the background so that no disputes can occur and the authenticity of the person claiming the land can be verified.

## 13.8   Use Cases: BC Adoptions in India in Tax Payments and Land Registry Records

Blockchain technology offers data integrity, guaranteed data availability, cybersecurity, accountability, and trust. The government is adopting blockchain technology for public sectors like a land registry, banking, voting, e-governance, electronic health records, insurance, and many more sectors in India. Half of the states in India have initiated blockchain projects to handle different challenges of citizen service delivery with the government of Telangana and also the government of Andhra Pradesh rising as two of the leading states in terms of blockchain adoption within the country [40].

According to the Bharat blockchain study 2019, issued in Feb, there are presently over 40 blockchain enterprises being tested by the public sector in India with nearly 92% of those at proof-of-concept or pilot period and 8% at the production section [41]. Now, 8 states of India use the blockchain technology for land registry. Andhra Pradesh a state of India that is the first among all who have adopted blockchain technology for land registry. They have implemented a private blockchain and they have implemented that way so that no landowner can make changes by themselves.

Officials from the Revenue Department, Chief Commissioner of Land Administration, and others would be entrusted with the task of monitoring land records. Officials have keys to access the land records. If any changes are made, the landowner will get a text message informing them of the modification. Based on this, the owner can complain, and the official who made the change can be tracked from the key used [42]. The government has also stated that they are only validating ongoing transactions in the blockchain. Once they are comfortable with the technology, they will move the entire land records to the blockchain network.

### 13.8.1   Problem Description

Many countries face the issue of "Land Mafia", a group of influential people who unethically take away other's land. People think that they own land and when they visit the land registry office to check their land status, they come to know their land now belongs to someone else. The real owner of the land gets frustrated with court hearings by proving that the land belongs to them. The court needs proof, but the land mafia originally tempered it. Digitization in government records are not temper-proof and trackable, so it necessitates the requirement of blockchain technology to resolve the aforementioned issues [43].

At present, everyone, right from the data entry operator, is changing land records, which lead to land disputes. But, once the purification of land records and integrating blockchain technology, it becomes impossible to tamper [44]. Let's understand this issue with some examples: consider three imaginary characters: Joy (a farmer who owns the land), Jack (a government officer and friend of Boby), and Boby (who have personal problems with Joy).

**Fig. 13.10** Flow of
tempering land records

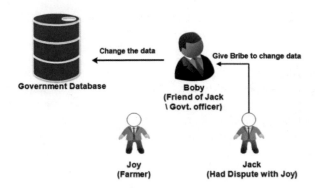

1. Boby was having certain personal issues with the Joy. To take revenge, Boby discovered a good plan. He went to his friend Jack, offered him a bribe, and told him that a farmer called Joy owned land in a certain area. Boby told him to temper the land record and Jack does it for the sake of friendship (How Joy will get to know that his land records are tempered?). Figure 13.10 shows the flow of tempering the data.

2. Joy wanted to sell his land to some other party, so he approached Jack to sell his land to person X with a valid document that states his authenticity. (How many times does Joy have to visit the government office to check the integrity of the process and the status of his land transfer?)

The above-mentioned problem looks simple, but it is huge trouble when it comes to the solution. Let's try to solve this problem in an example form and then formulate a technical way for it in the blockchain. Answers to the above-generated questions:

1. Answer to the first question, which is "How Joy will get to know that his land records are tempered?" is removal of Jack from the process and tracking the land records if any modification found then notify the landowner and take a consense. Figure 13.11 shows the blockchain as a solution to the question 1.

2. Answer to the second question "How many times Joy have to visit the government office to check the integrity of the process and the status of his land transfer?" is: not even a single time if blockchain will be used. He can apply for the land transfer from home and can track the whole procedure from home itself.

Let's formulate the technical perspective of the above solutions and describe the implementation in the next subsection.

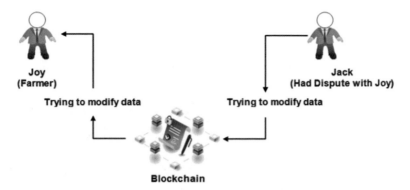

**Fig. 13.11** Blockchain as a solution for land registry

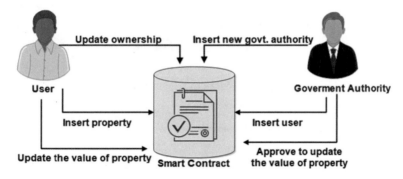

**Fig. 13.12** Process flow of system

## 13.8.2 Implementation

As the core part of the blockchain, a smart contract runs forever that states the rules/consensus by which the data access and storage can be made without the intervention of a third party. Figure 13.12 shows the basic flow of smart contract implemented to resolve the land registry contradictions. We have two entities involved, named user and government authority. The user is a landowner that inserts his/her property, updates its value and updates the land ownership. Whereas the government authority has the right to insert user, insert another authority, and approve updating the property value.

Smart contract implementation has the following functions [45]:

1. Insert/Update property;
2. Approve property adding;
3. Update property value;
4. Approve Update in property value;
5. Update ownership.

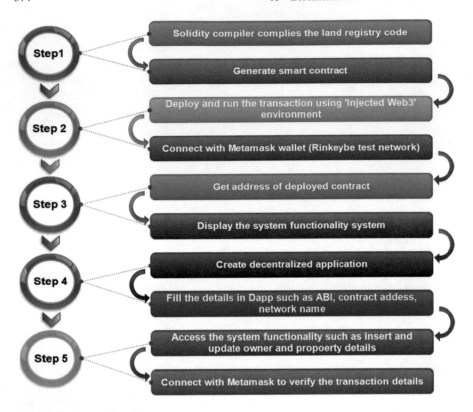

**Fig. 13.13**   Process flowchart

Here, function *InsertProperty* is verified by the administrator and checks whether
that property is in the name of the owner or not. If yes, the submitted proofs are valid
and the inserted property gets approved. Then, the *InsertNewUser* function inserts
the new landowner. Users can update the ownership by using the *UpdateOwnership*
function. Using the *UpdateProperty* function, the government authority can change
the ownership by checking the owner's status and taking their concern. The *Updat-
eValue* function changes the value of the property by the owner and gets verified by
the government authority itself. Updates in ownership are totally on the owner that
will get verified by the smart contract itself. *InsertNewGovtAuthority* functions are
just inserting government authority. Figure 13.13 shows the steps of the land registry
process in the Remix IDE, where we implemented Ethereum-based smart contract
for the land registry.

**Step:1** Figure 13.14 shows the solidity compiler compiles the code and generates a
land registry contract.

**Step:2** After compilation of the code, the next step is to deploy and run (Fig. 13.15)
the transaction using the 'Injected Web3' environment. Using '*Injected Web3*'
we redirected to the MetaMask (Fig. 13.16). It is a wallet that has details of

**Fig. 13.14** Compile the land registry code using solidity compiler

**Fig. 13.15**  Deploy and run

all transactions. It confirms to accept the Ether in the Metamask, we are using the Rinkeby test environment.

**Step:3** Figure 13.17 shows the deployed contract in the deploy and run section.

The deployed contracts show the functionality of the system (as shown in Fig. 13.18) such as insert user, government authority, and property. Update the owner, property, and property value.

**Step:4** Then, open oneclick.com to create the decentralized application (DApp) for the application (as shown in Fig. 13.19). Here, fill in all the mandatory fields such as name, description, Abi, contract address, and network name. After filling in the details, Fig. 13.19 shows we are able successfully able to create DApp.

**Fig. 13.16** Connect with the Metamask

**Fig. 13.17** Deployed contracts

**Step:5** Figures 13.20, 13.21, and 13.22 shows the functionality of the proposed system. Write tab is used to insert and update the owner and property details. whereas read operation is used to retrieve the owner and property details.

## 13.9   Summary

In this chapter, we discovered issues faced by the government and discussed how the revolutionized technology blockchain could handle it. Then, we moved to the

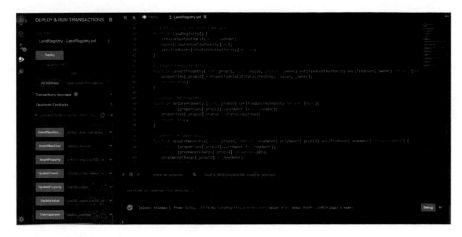

**Fig. 13.18**  System functionality

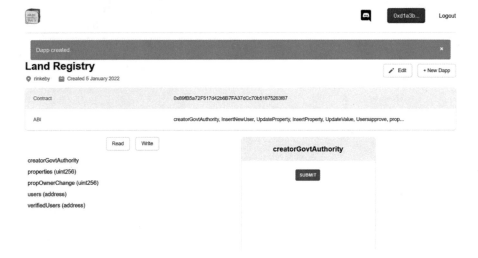

**Fig. 13.19**  Create the DApp for the land registry system

blockchain's advantages and emphasized how it is profitable to the government. In the part of the tax payment process, this chapter discussed how the taxation system works, the role of GST in the taxation system, and the process of GST for the supply chain. After that, this chapter discussed the blockchain technology-based tax implications, where we described the blockchain in the VAT tax, payroll tax, transfer pricing, and income tax. In a case study of GST, this chapter emphasizes the loopholes in the current GST system and how blockchain can be used to overcome those problems. And this chapter discussed the case study on BCT for taxation.

Then, another case study of land registry highlights the land registration issues and challenges along with third party's involvement and its transparency, which is getting

**Fig. 13.20**   Insert new user in the system

**Fig. 13.21**   Connect with Metamask and get the acknowledgment

compromised in the current scenario. Then, we presented a solution for the above-mentioned issues and briefly described the solutions by considering the land registry scenario by implementing a blockchain smart contract. As a concluding remark, blockchain is a trending technology that will revolutionize government services with transparency and tight security.

**Fig. 13.22** Insert the property details in the system

## 13.10 Question and Answer

### 13.10.1 MCQ Questions

1. What are the advantages of using blockchain for GST collection?

(a) Automatic tax refund for the final consumer.
(b) Automatic tax payment to the tax authority.
(c) Tamper-proof invoice generation based on the taxes already levied.
(d) Transactions are done in real time.

2. What data should be part of the blockchain ledger for property transactions?

(a) Property records (all clearance certificates, etc.) along with proof of ownership.
(b) Bank account details of the buyer and seller.
(c) Sale Deed of the property.
(d) None of the above.

3. How will blockchain play provide an advantage in implementing it with GST payments?

(a) Blockchain smart contracts can calculate the invoice based on the tax amount that is already levied during the production process.
(b) Smart contract directly transfers the tax amount to tax authority (CGST or SGST).
(c) It ensures that the return filing process is not required.
(d) The administrative burden for accounting services is drastically reduced.

4. What are the advantages of using blockchain for land registry?

(a) It will allow the government to privately store and modify land record details efficiently and in a tamper-proof manner.
(b) It will provide a mechanism to provide tamper-proof land ownership records which will help in preventing land dispute cases.
(c) Smart contracts can help automate processes such as land distribution through wills of individuals.
(d) It will allow landowners to exchange land multiple times which is currently not possible.

5. Why a blockchain-based solution is suited for managing government data at different levels?

(a) Provide different role-based access policies across different levels of the organization hierarchy in a multi-authoritative setup.
(b) Provide classification of data based on importance even in a decentralized environment.
(c) Provide centralized data management at the national level.
(d) Provide a way to access the data only via the blockchain.

## 13.10.2  Short Questions

1. How is blockchain used for land records?
2. How can blockchain be used in tax?
3. What is Transparent and incorruptible in blockchain?
4. Can blockchain change the nature of land registry in developing countries?
5. What challenges information leak can impose on an organization?

## 13.10.3  Long Questions

1. What area do you think would benefit from blockchain improvements to its current tax compliance methods?
2. What are the Blockchain Use Cases in Government and the Public Sector?
3. How will blockchain streamline the validation of educational and professional qualifications?
4. How will blockchain technology impact the collection of tax?
5. Explain the application of blockchain to land registries.
6. What will be possibilities to connect Blockchain technology with other information technologies in the future?
7. Blockchain and Geospatial industry, how would you combine them together? Use diagrammatic representation for the same.

8. Comments on "Blockchain Governance"—A New Way of Organizing Collaborations?
9. What are the key principles in Blockchain that are helpful in eliminating the security threats that need to be followed?
10. List down any 10 organizations that can use Blockchain technology using their application area?

# References

1. Brogan H (2022) Government. Encyclopedia Britannica, 10 Apr 2020, [Online]. Available https://www.britannica.com/topic/government. Accessed 6 Jan 2022
2. Gupta R, Tanwar S, Kumar N, Tyagi S (2020) Blockchain-based security attack resilience schemes for autonomous vehicles in industry 4.0: a systematic review. Comput Electr Eng 86:106717
3. Kumari A, Gupta R, Tanwar S, Kumar N (2020) A taxonomy of blockchain-enabled softwarization for secure UAV network. Comput Commun 161:304–323
4. Tanwar S, Parekh K, Evans R (2020) Blockchain-based electronic healthcare record system for healthcare 4.0 applications. J Inf Secur Appl 50:102407, ISSN 2214-2126. https://doi.org/10.1016/j.jisa.2019.102407. http://www.sciencedirect.com/science/article/pii/S2214212619306155
5. Akram SV, Malik PK, Singh R, Anita G, Tanwar S (2020) Adoption of blockchain technology in various realms: opportunities and challenges. Secur Priv 3:e109. https://doi.org/10.1002/spy2.109
6. Gupta R, Kumari A, Tanwar S, Kumar N (2021) Blockchain-envisioned softwarized multi-swarming UAVs to tackle COVID-I9 situations. IEEE Netw 35(2):160–167. https://doi.org/10.1109/MNET.011.2000439
7. Portswigger (2020) Canada data breach: 360,000 teachers in Quebec potentially impacted [Online]. Available https://portswigger.net/daily-swig/canada-data-breach-360-000-teachers-in-quebec-potentially-impacted. Accessed 17 Apr 2020
8. Portswigger (2020) Generate data breach impacts 26,000 New Zealand residents [Online]. Available https://portswigger.net/daily-swig/generate-data-breach-impacts-26-000-new-zealand-residents. Accessed 17 Apr 2020
9. Portswigger (2020) Colorado municipality falls victim to Click2Gov software breach [Online]. Available https://portswigger.net/daily-swig/colorado-municipality-falls-victim-to-click2gov-software-breach. Accessed 17 Apr 2020
10. VpnMentor (2020) 'Report: inmates' prescriptions & PII leaked in breach spanning multiple jailhouses [Online]. Available https://www.vpnmentor.com/blog/report-jailcore-leak/. Accessed 17 Apr 2020
11. Forbes (2020) A 'Government Database' of 92 million citizen records for sale to highest bidder [Online]. Available https://www.forbes.com/sites/daveywinder/2019/10/06/a-government-database-of-92-million-citizen-records-for-sale-to-highest-bidder/#3723cbea701b. Accessed 17 Apr 2020
12. Portswigger (2019) Sberbank data breach: Russian lender investigates insider threat [Online]. Available https://portswigger.net/daily-swig/sberbank-data-breach-russian-lender-investigates-insider-threat. Accessed 17 Apr 2020
13. Portswigge (2019) People affected should change their fingers immediately [Online]. Available https://portswigger.net/daily-swig/people-affected-should-change-their-fingers-immediately. Accessed 17 Apr 2020

14. Reuters (2019) In systemic breach, hackers steal millions of Bulgarians financial data [Online]. Available https://www.reuters.com/article/us-bulgaria-cybersecurity/hackers-steal-millions-of-bulgarians-financial-records-tax-agency-idUSKCN1UB0MA. Accessed 17 Apr 2020
15. Portswigger (2019) South African electricity provider said to be exposing customer details online [Online]. Available https://portswigger.net/daily-swig/south-african-electricity-provider-said-to-be-exposing-customer-details-online. Accessed 17 Apr 2020
16. Portswigger (2019) Two years after WannaCry, a million computers remain at risk Online]. Available https://techcrunch.com/2019/05/12/wannacry-two-years-on/. Accessed 06 Jan 2022
17. BBC news (2021) Air India cyber-attack: data of millions of customers compromised [Online]. Available https://www.bbc.com/news/world-asia-india-57210118. Accessed 06 Jan 2022
18. Cleartax (2019) Income tax in India: guide, IT returns, E-filing process [Online]. Available https://cleartax.in/s/income-tax. Accessed 17 Apr 2020
19. Hdfclife.com (2020) Tax structure & taxation system In India [Online]. Available https://www.hdfclife.com/insurance-knowledge-centre/tax-saving-insurance/Tax-Structure-in-India. Accessed 17 Apr 2020
20. Cleartax (2020) Goods & services tax GST (India) what is GST? indirect tax law explained [Online]. Available https://cleartax.in/s/gst-law-goods-and-services-tax. Accessed 17 Apr 2020
21. Kotak (2020) How will GST work - explained with an example [Online]. Available https://www.kotak.com/en/stories-in-focus/how-will-gst-work-explain-with-an-example.html. Accessed 19 Apr 2020
22. Gstindia (2016) GST with examples [Online]. Available https://www.gstindia.com/gst-with-examples/. Accessed 19 Apr 2020
23. Avalara (2020) Will the GST regime adopt blockchain technology? [Online]. Available https://www.avalara.com/in/en/learn/press/will-the-gst-regime-adopt-blockchain-technology.html. Accessed 20 Apr 2020
24. Medium (2018) Introducing blockchain technology to the world of Tax' by Jurgen G. https://medium.com/@jurgeng/an-introduction-to-blockchain-technology-tax-567e536767ec. Accessed 30 Dec 2021
25. Investopedia (2021) Value-added tax (VAT). https://www.gccfintax.com/articles/value-added-tax-and-the-use-of-blockchain-technology-4003.asp. Accessed 30 Dec 2021
26. Setyowati MS, Utami ND, Saragih AH, Hendrawan A (2020) Blockchain technology application for value-added tax systems. J Open Innov: Technology Mark, Complex 6(4):156. https://doi.org/10.3390/joitmc6040156
27. Gccfintax (2021) Value added tax and the use of blockchain technology by Alfredo Collosa. https://www.investopedia.com/terms/t/taxes.asp. Accessed 30 Dec 2021
28. Global Agenda Council on the Future of Software and Society World Economic Forum (2015) Deep shift technology tipping points and societal impact. https://www3.weforum.org. Accessed 30 Dec 2021
29. Microsoft (2019) Price water house Coopers Belasting adviseurs N.V. the Netherlands and Vertex, Two practical cases of blockchain for tax compliance. https://www.pwc.nl/nl/tax/assets/documents/pwc-two-practical-cases-of-blockchain-for-tax-compliance.pdf. Accessed 31 Dec 2021
30. Kim YR (2021) Blockchain initiatives for tax administration. https://dc.law.utah.edu/cgi/viewcontent.cgi?article=1284&context=scholarship. Accessed 01 Jan 2022
31. Thompson AR, Viitasaari V (2017) Payroll tax & the blockchain. Tax notes international, March 13: 1007–1024 https://papers.ssrn.com/sol3/papers.cfm?abstract_id=2970699. Accessed 01 Jan 2022
32. Infosys (2021) Enabling taxation authority as a trusted partner for lending business and tax compliance facilitation. https://www.infosys.com/services/blockchain/case-studies/enabling-taxation-authority.html. Accessed 03 Jan 2022
33. Deloitte (2021) Blockchain and digital asset tax services. https://www2.deloitte.com/us/en/pages/tax/solutions/cryptocurrency-blockchain-taxation.html. Accessed 29 Dec 2021

34. World Economic Forum (2020) Tax implications. https://widgets.weforum.org/blockchain-toolkit/tax-implications/index.html. Accessed 29 Dec 2021
35. Cleartax (2020) Issues in the new GST return system [Online]. Available https://cleartax.in/s/issues-new-gst-returns-system. Accessed 22 Apr 2020
36. Dailypioneer.com (2018) Symphony of blockchain, GST implementation in India [Online]. Available https://www.dailypioneer.com/2018/state-editions/symphony-of-blockchain-gst-implementation-in-india.html. Accessed 22 Apr 2020
37. NPTEL NOC IITM (2019) Blockchain in Government – V (Tax payments and land registry records). https://www.youtube.com/channel/UCYa1WtI-vb_bx-anHdmpNfA. Accessed 07 Jan 2022
38. Inter-American Center of Tax administrations (2021) Blockchain in tax administrations. https://www.ciat.org/blockchain-in-tax-administrations/?lang=en. Accessed 29 Dec 2021
39. Infosys (2020) Case study: India's income tax department uses blockchain to simplify tax processes by Ashutosh Sharma. https://www.infosys.com/services/blockchain/case-studies/blockchain-simplify-tax-processes.html. Accessed 02 Jan 2022
40. Coinjournal (2019) Diana Ngo, public sector driving blockchain adoption in India [Online]. Available https://coinjournal.net/public-sector-driving-blockchain-adoption-in-india/. Accessed 24 Apr 2020
41. Globalcryptonews (2020) Blockchain adoption in India [Online]. Available https://globalcryptonews.io/blockchain-adoption-in-india/. Accessed 24 Apr 2020
42. Ledger Insights (2020) Indian state to implement blockchain for land records [Online]. Available https://www.ledgerinsights.com/indian-blockchain-land-records-registry-andhra-pradesh/. Accessed 24 Apr 2020
43. Analytics India mag (2018) How Andhra Pradesh is emerging as India's blockchain hub [Online]. Available https://analyticsindiamag.com/how-andhra-pradesh-is-emerging-as-indias-blockchain-hub/. Accessed 21 Apr 2020
44. New Indian express (2019) Andhra government to adopt blockchain tech to end land record tampering [Online]. Available https://www.newindianexpress.com/states/andhra-pradesh/2019/dec/15/andhra-government-to-adopt-blockchain-tech-to-end-land-record-tampering-2076359.html. Accessed 21 Apr 2020
45. Github (2020) Property-Registry by Kenrickfong [Online]. Avilable https://github.com/KenrickFong/Property-Registry. Accessed 04 Jan 2022

# Chapter 14
# Impact of Blockchain on Academic Publishing

**Abstract** Blockchain technology creates a distributed environment where the transactions/records are stored securely and immutably. This would be quite useful in the field of publications, especially in open access academic publications. Academic data is huge and dynamic, a contribution of many authors worldwide, which needs to be updated upon publishing any novel research work. So, maintaining its integrity is a bit tedious with the centralized system architecture. Blockchain manages the aforementioned open access data publication issue using digital rights management, consensus mechanism, and smart contract. This chapter discusses the potential of blockchain technology in the academic publishing sector.

**Keywords** Blockchain · Smart contract · Academic publishing

## 14.1 Introduction: Open Access Academic Publishing

Open access is a publication model for scholarly communication. It publishes high-quality research articles and is accessible to all readers (worldwide) without any cost. In contrast, readers have to pay for subscriptions to other publication models to access academic research articles. Figure 14.1 shows the importance of open access in academic publishing. As long as authors try to publish their work on open access platforms, they are more concerned about the misuse of their publication. They try to find a secure and distinct way to publish the content. Suppose the authors use some local word sheets or software to request the publisher. Upon receiving the application, the publisher asks for a raw copy of the content [1]. Then, the author provides content to the publisher. Further, the publisher verifies the authenticity and originality of the content provided by the author. After verification, an agreement is signed by the publisher with the author of the content [1].

Currently, many competing open access models have their pros and cons. Such models are (i) traditional closed access, (ii) gold open access, and (iii) delayed open access. In the traditional closed access model, the author can publish the articles free of cost. Every individual has to pay a certain subscription fee to access the research article. However, the published manuscript is freely available to all readers

**Fig. 14.1** Importance of open access publication

in the gold open access model. But, the authors have to pay a certain charge, i.e., article publishing charge (APC). Delayed open access journals are like traditional subscription-based journals providing free online access after an embargo period. The embargo period may vary from a few months to a few years. Apart from these models, there are certain hybrid journals, which require APC (on voluntary open access publication with specific APC), or free (for subscription-based publication).

The conventional method of content publishing is quite time-consuming and opaque. The main disadvantage of this method is the authors get available only on limited platforms [1]. There is also no transparency in the system for data collection, data analysis, and review of the article. Due to this vague nature, the workflow leads to fewer views and downloads of the article and makes it less noticed. So, it is not a good practice in academic publication [1]. For example, in the case of experimental articles, the journal published only successfully trialed scientific results. Researchers become happy with the publication. The journal does not publish the failed results that explain to the audience the facts and directions for the researchers working in this field. This tendency leads the scientists to be more involved in non-research activities, reducing their productivity, and brainstorming.

In addition to this, the author faces several other challenges for the publication, like inducement of the publishers and publication fees for open access contents [2]. Academic publication is a non-commercial activity; it has become a fast-growing business and is majorly owned by some big publishers. The author must pay the higher charges for the application and subscription. After spending the huge APC, their content is not available in every institution [2]. The author has to pay charges to make the publication available rather than a reader in open access publishing. However, the majority of the authors are not choosing open access publications with the concern of losing their work [1].

## 14.2  The Impact of BCT on Academic Publishing

Sharing knowledge constantly improves the researcher's or readers' experience, beliefs, ability, and skills. In academics, knowledge dissemination impacts the investigation results and explorations that help in reducing the processing and experimental time. We are at a stage where everything is online and the users are learning and teaching via online platforms. They can refer to any material, article, book, or blog to enhance their knowledge. However, on the other side, when the author wants to publish their research work, they face several issues such as a lengthy peer-review process, fewer incentives to publish negative results, and no credit to the contributors [3].

To mitigate the aforementioned issues, technology is required to provide transparency in the peer-review process, handle real-time data, and provide trust and security in the publication process. Blockchain is an emerging technology that can deal with all these issues [4]. It is a distributed ledger system that provides a means for a group to record and exchange information efficiently and securely. Every individual of the group maintains the exact copy of the distributed ledger and group members mutually verify the changes [7]. Any transaction between two entities has to satisfy the contract [6]. A contract is nothing but a set of rules or protocols that need to be followed while transmitting new edits or entries. On every transaction, a new block is added to the chain at the end [7]. Figure 14.2 shows the use of blockchain to resolve market issues, research data storage, peer-review application, and data distribution. It also increases the quality and effectiveness of articles in academic publishing.

Nowadays, the issue of piracy in research is of great concern. People steal others' ideas for their research works and do not give credit to them. This becomes a significant issue and the authors are in a dilemma about selecting relevant journals with strict guidelines on plagiarism and how to protect their contributions from plagiarism. Nowadays, many predatory journals are there in the market, making it challenging for the authors to identify the genuine one where they will publish their

**Fig. 14.2** Impact of blockchain on the academic publishing

work. In this scenario, blockchain delivers data integrity, security, and decentralized ledger to the publication agencies and the authors publishing in them [24]. It offers data storage with guaranteed security. It can allow authors to publish their work in journals only if it satisfies the no plagiarism conditions. It is also useful in identifying high-quality journals among the vast list, including predatory journals. Here, the author's contributions cannot be stolen or reused by any malicious author because of the immutability and timestamp of transactions.

Another vital use case of blockchain in academic publication is the transparency of the peer-review method. Figure 14.3 shows the blockchain-based peer-review process. Here, P indicates publishers, A specifies authors, E designates editors, and R shows the reviewers. In the traditional centralized system, once an author submits an article, the publication house takes their time. The authors are unaware of the detailed process, progress, and steps involved in the publication procedure. The distributed blockchain-based system delivers a stage where the authors will know the progress in the article submitted. It also shares multiple versions of submitted articles to the reviewers for a blind review process. In this way, the blockchain offers secure and anonymous information sharing and the step-by-step progress of the article [5, 26].

**Fig. 14.3** Blockchain-based system for academic publishing

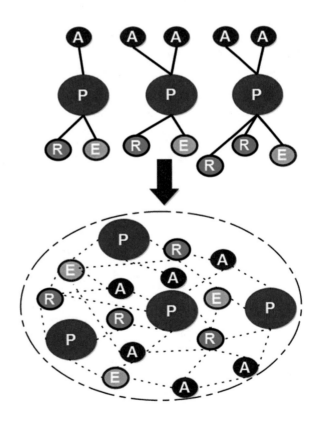

## 14.3  Implications of BCT in Open Access Academic Publishing

In academic publishing, blockchain is quite helpful in controlling content distribution. Nowadays, people find many ways to read or download premium content at no cost via third-party platforms such as sci-hub and many more. While publishing on a blockchain, the digital ledger keeps the publication's complete transaction information that counters copyright theft. A transaction is added to the blockchain only if all members approve it, i.e., authors and publishers [8]. Blockchain's transparency is beneficial, especially for both subscription-based and open access journals to publish their articles on a digital ledger, as it preserves data integrity. It allows publishing platforms and authors to view who is accessing and downloading the article and what changes have been made upon its final release. The verifiability factor of blockchain also helps to provide clarity on article changes, i.e., whether the modified contents are published by the same author or any other person.

Figure 14.4 represents the method of tokenization, where the contents (provided by the author) are divided into tokens. Each token is assigned a unique value when authors own it. An open-source blockchain publication repository can begin by tokenizing the material—that is, describing every segment of publication as a legally equivalent digital object deposited on the blockchain. This can be accompanied by ensuring writers have properties of the connected tokens allotted to them. This is not a minor endeavor as a result of the current mechanism being fairly sluggish thus

**Fig. 14.4**  Tokenization

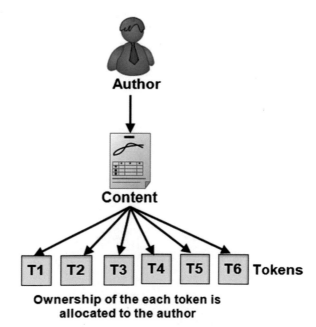

pseudo-transparent, and especially there is a need to be agile whether new publication laws are amended or broken [8].

Various features of blockchain used in academic publishing are

- *Smart contracts*: The exchange of research material obeys explicit instructions encoded as part of the agreement to be automatically executed. After the accepted conditions are met, a smart contract improves the exchange process and performance [10]. Combining a central database with smart contracts could bring enormous benefits. Content ownership is automatically defined through the blockchain, and the use of content and payment of royalties are performed through smart contracts in which the rights are held [11].
- *Cryptocurrencies*: Initial coin offerings, a type of crowdfunding using cryptocurrencies, may eventually be used to finance entire academic publications. In this way, an open access publication represents the value merits of different activities that could develop into a crypto-economic [10].
- *Digital Rights Management*: It is a vital future aspect of blockchain technology. Digital rights are linked to specific problems like reuse, licenses, and royalties currently intermediated by broad organizations and specific goods. Figure 14.5 represents the process flow employed by blockchain technology in providing the digital certificate [10]. Here, the certification is issued by the issuer. The private key provides security for that digital assets. After that recipient can access that document.
- *Integrity Management*: Using blockchain technology, we can maintain the integrity of the research content and keep the content more secure [10].

Cryptocurrency works as an entirely electronic program, with all participants agreeing to the ownership and provenance records. Blockchain implementation in publishing registry is more complicated because of the need to register publication ownership and link the data to the actual world [11]. Although it creates issues in the registry such as accurately representing the publication's significant-world existence and material, legal frameworks must be in place to enforce ownership rights when

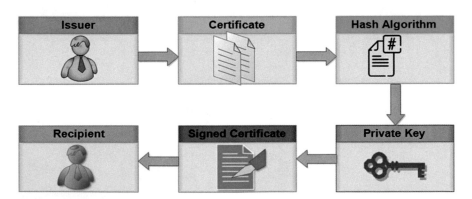

**Fig. 14.5** Digital rights management

blockchain records claim that they are kept, including against publications that are not part of the blockchain or who do not accept it as valid [11].

## 14.4   Smart Contract for Open Access Academic Publishing

Figure 14.6 represents how smart contracts can be implemented between authors and publishers, and each new transaction is added as a block to the chain and whole blockchain gets updated [17, 18]. The benefits of using smart contracts in an open access academic publication are as follows.

1. *Transparency*: It is one of the defining aspects of blockchain and is also provided through smart contracts. As the smart contracts are packed in absolute detail with terms and conditions, which are also reviewed by the parties to the agreement [19].
2. *Time-efficient*: It normally takes more than a couple of days to go ahead with any procedure requiring paperwork. Most intermediaries and needless measures along the way are blamed for the delay in procedures. On the other hand, smart contracts are managed by internet assistance, because they are nothing more than bits of electronic code [19].
3. *Precision*: Smart contracts are coded in a clear and structured manner. It needs to keep all the terms and conditions until it winds up being used. Any provision that is left out of the contract may result in an error during its execution [19].

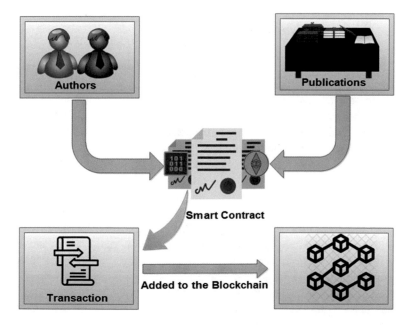

**Fig. 14.6** Pictorial representation of smart contract

4. *Safety and Efficiency*: In modern days, smart contracts with automatic coding functions are the best solutions for data encryption technologies [13]. Since, they follow the highest environmental requirements, the quality of protection they provide allows them to be safe to use for sensitive processes [23].
5. *Savings*: Smart contracts save a good amount of money compared to conventional agreements, as smart contracts involve individuals who are part of the agreement and eliminate the intermediaries. The cost associated with the intermediaries is eliminated [14].
6. *Paperless*: Since smart contracts are computer-coded, paper use is eradicated in the entire process. This, on the one hand, saves costs. At the same time, it is beneficial to businesses internationally as it allows them to save their contracts-related bits of paper use and encourages their contribution to the society [19].

Now, to grasp how blockchain is applicable in the open access academic publishing platform, we discuss various case studies on it in the following section.

## 14.5   Case Study

In earlier sections, we enlightened how blockchain is applied in open access academic publication platforms. In this section, we discuss various real-time use cases on blockchain applicability in academic publishing.

### 14.5.1   Orvium Publication Model

Orvium proposed the publication life cycle model based on blockchain technology. They aimed to encourage open science and research distribution using a token approach for the manuscript submission [15]. They have used proof-of-existence consensus for the peer-review process and authors' transfer licenses and copyright details. They provide easy access to the research information with low publication fees and transparent peer review with better rewards. Orvium offers a model to create decentralized autonomous journals (DAJs). They also used cloud computing, machine learning, and big data analytics with blockchain technology.

Orvium introduced a model that increases transparency, recognizes the researcher's impact, and maximize collaboration in open access publishing. Figure 14.7 shows the proposed Orvium's publication model [15]. Figure 14.8 shows steps followed by Orvium for token mechanics. In their model, user registration and identification is the first step. They use authentication like single sign-on integrated with google and ORCID for a researcher. The platform generates a random key and researchers use it for their transactions. After registration, the next step is the article submission process. The researcher submits their manuscript in a blockchain-based system using a public proof-of-existence and authorship. After submitting the

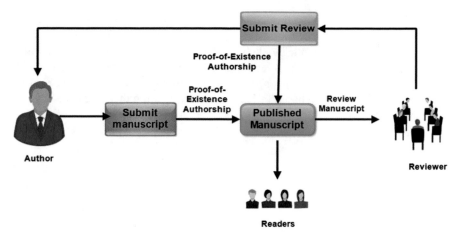

**Fig. 14.7**   Orvium publication model

**Fig. 14.8**   Steps involved in Orvium token mechanics

manuscript, the ORV tokens were staked to the reviewer. The reviewer submits their comment and claims ORV tokens when the review is accepted. The author uses proof-of-existence authorship for publishing papers and researching relevant data. Also, decide user licensing such as paid or free.

### 14.5.2   Eureka: A Scientific Publication Model

Niya et al. [3] presented blockchain-enabled publication solutions such as Eureka's token-based reward system for scientific review and rating platform. It works on open access publishing web front that uses blockchain for peer-review and journal submission. Eureka overcomes the challenges in scientific publishing, no credit for researcher contributions, and reproducibility issues. Figure 14.9 shows Eureka's proposed model for the scientific publishing platform. It is meant for researchers and scientific works. It uses a public blockchain that brings trust and public accessibility. Different entities, such as authors, editors, and reviewers, are involved in the editing and peer-review process. The proposed model used the Eureka token (EKA) for rewards and used Ethereum based infrastructure for creating their digital assets.

Figure 14.10 shows steps for the transactions used by Eureka. In the first step, the author submits the article using EKA. The smart contract informs reviewers of

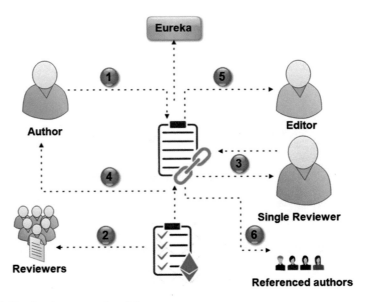

**Fig. 14.9**  Eureka: the proposed model

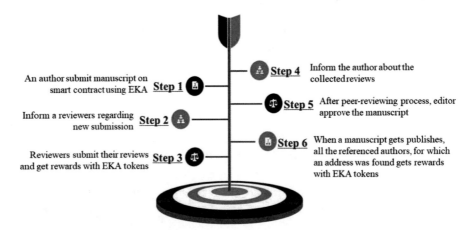

**Fig. 14.10**  Eureka's token-based reward system steps

the new submission when the new event is triggered. Reviewers will thorough the submitted manuscript and submit their comments. After each review submission, they will get a reward with EKA tokens. In the fourth step, the system informs the corresponding authors about the collected comments. When this peer-review process is completed, the editor approves the manuscript. At that movement of manuscript publication, the referenced author gets EKA token in the form of rewards.

### 14.5.3 Blockchain-Enabled Decentralized Scientific Publishing Framework

The proposed system [16] sights to solve incentive issues of conventional systems in scientific publication and communication. It uses a blockchain that fixes loopholes in scientific publications' current practices. The essential part of scientific publishing is both author and reviewer. But authors indeed get rewards from publishing for their reputation and receive incentives from employers for using the publishing model intensively. In contrast, the author pays unreasonably high for the course.

Here, the author considered the ideal open access model based on two key features, such as (i) no economic hurdle for author and reviewer and (ii) both authors and reviewers are fairly incentivized for their contributions. As someone must pay for the services in any business model, the author proposed a blockchain-enabled system wherein everyone is rewarded as per their contribution. Figure 14.11 shows the list of additional features provided by the proposed framework. It offers a low-cost entry environment, prevention against monopoly, secure authorship, and privacy in the blind review process. Figure 14.12 shows the flow of the proposed model. The author submits the manuscript and publishes a pre-print document as per the model. They buy review tokens from the smart contract, get a review for their manuscript, and the reviews are stored in smart contracts. The reviewer gets a review certificate and review tokens from the smart contract.

**Fig. 14.11** Additional feature in scientific publication model

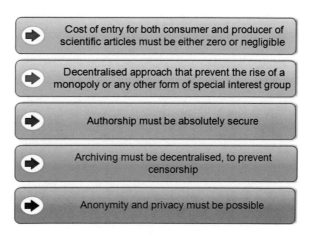

**Fig. 14.12** Proposed
workflow steps

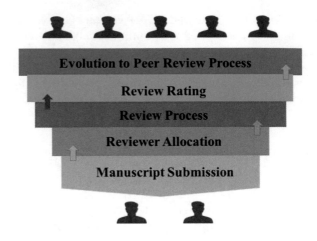

## 14.6   Implementation Details in Open Access Publishing

Here, we present open access publishing implementation details by creating smart contracts between the authors and publishers using a solidity language [20]. The contract holds two primary conditions for the content to get published, such as the authenticity test (plagiarism checking) and the second is the reviewers' comments. Every publisher has an online tool to check the plagiarism of the author's article/content. If the plagiarism is more than the baseline value or the reviewer's comments are negative, the article/content gets rejected or suggests a major revision. Otherwise, the publisher suggests a minor revision before the final publication.

The above mentioned functionality is implemented in the Remix IDE. The following functions are implemented:

a.  Set authors information;
b.  Set publisher information;
c.  Generate book report.

In the solidity, the *AdjustAuthorInfo* function set the author details such as author name, document type, and Aadhar id. The publisher address is set using *AdjustPublisherInfo* function. *ReportofBook* function keeps the plagiarism report and reviewer comments. *useAcademicPublishing* module get unique id as well as Aadhar id from authors and buniqueID from book report. If the plagiarism and comments conditions satisfy then it will get message "Greetings...Your book has been accepted for the publication...." otherwise suggested to revise the book.

Following steps perform the execution of academic publishing code. The first step is open code in the Remix IDE [21, 22].

1.  Figure 14.13 shows the compilation of AcademicPublishing code using the solidity compiler that generates the AcademicPublishing contract.

**Fig. 14.13**  Compiling the academic publishing code using Solidity compiler

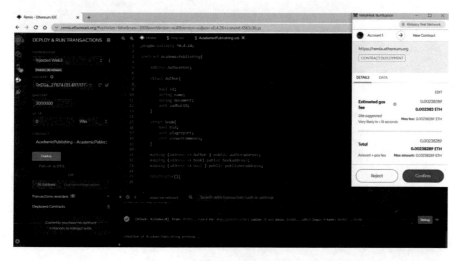

**Fig. 14.14**  Deploy, run, and connect with the Metamask wallet

2. After compilation, we deploy and run the transaction using "*Injected Web3*" environment. Whenever we selected "*Injected Web3*" the tab redirected to the *Metamask* wallet that asks for the transaction confirmation (Fig. 14.14).
3. Contracts are deployed successfully if Metamask confirm the transaction (Fig. 14.15)
4. Figure 14.16 shows the overall system functionalities such as author, publisher, and book details.
5. Deployed contract status is visible on the Metamask wallet (Fig. 14.17).
6. Now, open Oneclick.com to create the decentralized application (Dapp). Here, choose a wallet and select the "connect Metamask" option (Fig. 14.18).

**Fig. 14.15** Successfully deployed the contract

**Fig. 14.16** Functionality of the overall system

**Fig. 14.17** Contract deployment status

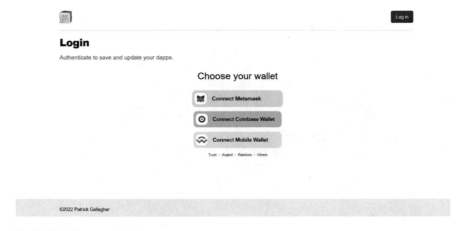

**Fig. 14.18** Choose wallet for DApp

**Fig. 14.19** How to copy ABI

7. To create new Dapp fill in all the required details such as Abi, contract address, and network name (see Fig. 14.21).
8. Figure 14.19 shows how to get Abi address from the Solidity compiler and Fig. 14.20 shows how to copy the deployed contract address from transaction records sections.
9. Now, Dapp is created for the academic publishing application (Fig. 14.22).
10. Figures 14.23, 14.24 and 14.25 shows how to set the author, publisher, and book report details respectively in the dapp.
11. Get academic publishing details. Figure 14.26 displays how to set the plagiarism and reviewer comment value in the report (Fig. 14.27).
12. After, filling value confirm transaction with the Metamask and get acknowledgment (Figs. 14.28 and 14.29).

**Fig. 14.20** Copy the contract address

**Fig. 14.21** Initial stage of Dapp creation

## 14.7   Challenges Faced

Blockchain has a major issue that many people link blockchain with the cryptocurrencies [12]. Crypto has a negative image as it is being used to track criminal activities surrounding hackers and fraudsters. This has imposed bad implications on blockchain technology and made it less favorable and adaptable [25]. Blockchain is acceptable only if people know the distinction between cryptocurrency and blockchain. Such distinction would help in removing the negative impression and can contribute to a greater desire for using this technology [12].

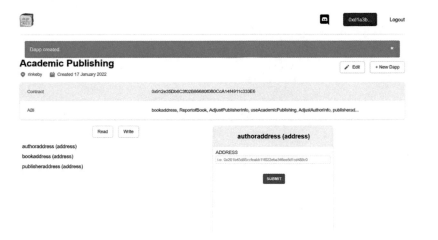

**Fig. 14.22** Dapp for academic publishing

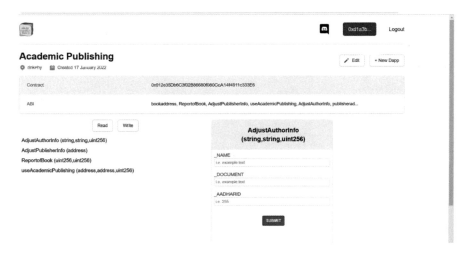

**Fig. 14.23** Set the author details in Dapp

## 14.7.1 Technical Challenges

1. *Immature Technology*: Blockchain is new in the market, so its acceptance is trimmer and need more exploration.
2. Slow and Wearisome: Every detail of the transaction is stored in the chain, so at the end, the chain gets larger and makes the working of blockchain slow.
3. *Scalability*: The blockchain keeps growing and storage requirement becomes an issue that hinders the blockchain scalability.

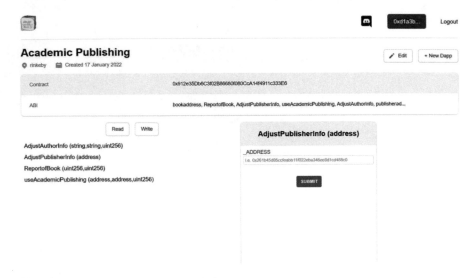

**Fig. 14.24** Setting up of publisher details in Dapp

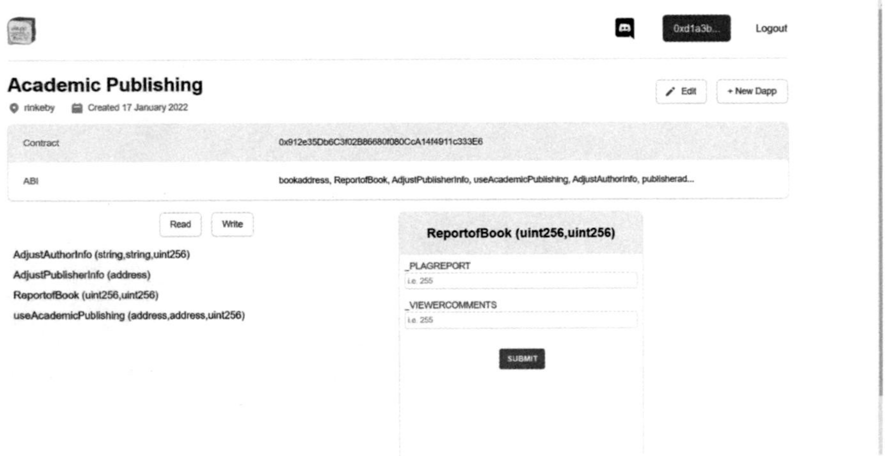

**Fig. 14.25** Setting up of book report in Dapp

4. *Interoperability*: Exchange of information with other types of chains such as public, private, and the consortium is an issue.
5. *Autonomous*: As no human interaction is required, the system becomes autonomous.
6. *Integration with legacy systems*: Blockchain technology is not much familiar to the people and government. This is why the blockchain is facing legal issues.

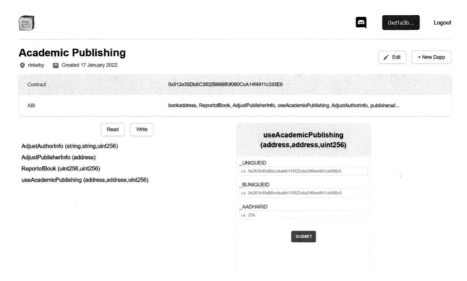

**Fig. 14.26** Get academic pushing details in Dapp

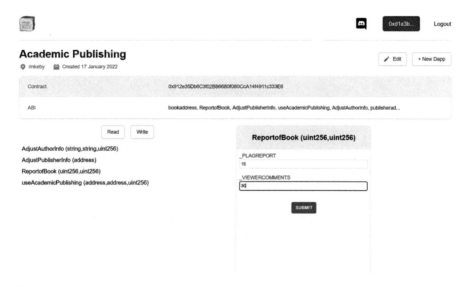

**Fig. 14.27** Submit plagiarism and comments of book report

## 14.7.2 Organizational Challenges

1. *Lack awareness and understanding*: As mentioned earlier, being a new technology, very few people are comfortable to work with it.
2. *Productivity Paradox*: Blockchain is having image issues of acceptance of this technology in people is less.

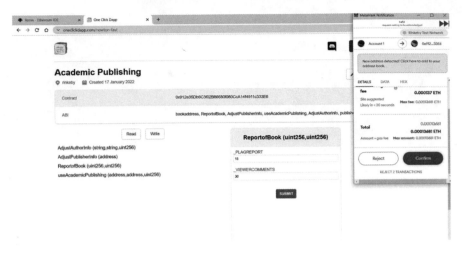

**Fig. 14.28**  Confirm submission to connect with the Metamask wallet

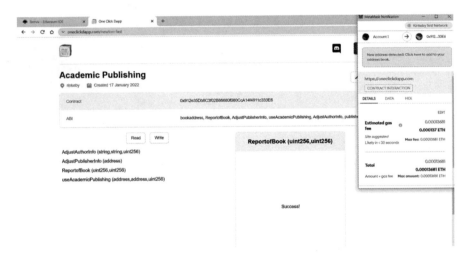

**Fig. 14.29**  Get an acknowledgment from Metamask wallet

3. *Lack of cooperation*: Clients are not willing to shift their data onto the blockchain, so there is a lack of cooperation.
4. *Security and privacy*: As scalability is an issue in blockchain, the cost of implementation keeps increasing.
5. *Lack of regulatory clarity good governance*: Bitcoin, the main implementation of blockchain is not legal yet, so there is a lack of clarity and good governance.

## 14.8 Summary

Blockchain technology gave feasible solutions that overcome data integrity issues, give fair credit contribution, and provide transparency in the open access academic publishing. So, assuming that the above problems can be resolved, open access blockchain academic publishing can be used to publish the content by treating them as blockchain transactions, providing a verifiable and permanent record. The distributed existence of the ledger would ensure no downtime and no server failure would ever affect the service's availability [17].

An additional benefit of disseminating content through the blockchain is its security and transparency. Content is currently downloaded and shared via various platforms such as publisher platforms, ResearchGate, PubMed Central, etc., which makes it difficult for the publisher/author to track its usage. Blockchain resolves this issue [9]. Technological developments usually take a long time to develop and enter into a stable shape. As with every technical breakthrough, blockchain will follow the same sluggish trajectory of adoption in the coming years. However, blockchain offers incentives to the miners, but it is still not well accepted.

The list of the above-listed barriers to blockchain adoption strongly underlines the need for technical improvements. Blockchain's ability to bring significant improvements to other markets and sectors, including the supply chain sector, has been unveiled. Product development is also one of blockchain technology's most evident and practical implementations, and we can expect it to develop rapidly soon.

## 14.9 Question & Answer

### 14.9.1 MCQ Questions

1. Blockchain systems could enable secure sharing with the benefit of certifying the results of ...

(a) peer-review process.
(b) research paper selection.
(c) research paper submission.
(d) none of the above.

2. An academic publisher's role is ...

(a) distributes academic research and scholarship.
(b) to promote proper validation of scientific findings.
(c) to a group of people for purposes of further distribution, public performance, or public display.
(d) all of the above.

3. What is the importance of open access publishing?

(a)  access for the public.
(b)  to get more exposure of work.
(c)  higher citation rates.
(d)  all of the above.

4. What are the different features of blockchain open access academic publishing?

(a)  smart contract.
(b)  integrity management.
(c)  digital rights management.
(d)  all of the above.

5. What are the benefits of smart contract in open access academic publishing?

(a)  space efficient.
(b)  transparency and trust.
(c)  reduce workload.
(d)  none of the above.

### 14.9.2   Fill in the Blanks

1. Open access is a publication model for academic communication that produces research-relevant data and is accessible to readers ...... .
2. Gold Open access also known as ...... .
3. An open-source blockchain publication repository can have to begin by tokenizing the material.
4. Ownership of each token is allocated to the ...... .
5. Blockchain provides ..... in the peer-review process.

### 14.9.3   Short Questions

1. Blockchain-based Recommender system for Scientific Publishing: A Case Study. Use diagrammatic representation for your explanation.
2. How blockchain helps to avoid publishing in predatory journals?.
3. Challenges Associated to Blockchain Technology when it applies to the academic publishing.
4. How can blockchain be used in academic publishing?

### 14.9.4   Long Questions

5. How blockchain simplifies the peer-review process?
6. What, in your opinion, are the best and the worst aspects of the peer-review and publication process using blockchain?

7. Adoption of Blockchain Technology in Fake News Identification: A Case Study. Use diagrammatic reorientation for your explanation
8. What are the differences between open access and standard subscription-based publication?

# References

1. Grech A, Camilleri AF (2017) Blockchain in education. Publications Office of the European Union, 132 S
2. van Rossum J (2018) The blockchain and its potential for science and academic publishing. Inf Serv Use 38:95–98. https://doi.org/10.3233/ISU-180003
3. Niya SR, Pelloni L, Wullschleger S, Schaufelbühl A, Bocek T, Rajendran L, Stiller B (2019) A blockchain-based scientific publishing platform. In: 2019 IEEE international conference on blockchain and cryptocurrency (ICBC), pp 329–336. IEEE
4. Gupta R, Kumari A, Tanwar S, Kumar N (2020) Blockchain-envisioned Softwarized multi-swarming UAVs to tackle COVID-19 situations. IEEE Netw. https://doi.org/10.1109/MNET.011.2000439
5. Cabells Blog (2021) The impact of blockchain tech on academic publishing by JBran-nam. https://blog.cabells.com/2021/01/27/the-impact-of-blockchain-tech-on-academic-publishing/. Accessed 10 Jan 2022
6. Kakkar R, Gupta R, Tanwar S, Rodrigues JJPC. Coalition game and blockchain-based optimal data pricing scheme for ride sharing beyond 5G. IEEE Syst J. https://doi.org/10.1109/JSYST.2021.3126620
7. Phung SP, Raju V. Utility of blockchains in publishing sector: focus on Academic Publishing. Int J Psychosoc Rehabil
8. Lizcano D, Lara JA, White B et al (2020) Blockchain-based approach to create a model of trust in open and ubiquitous higher education. J Comput High Educ 32:109–134. https://doi.org/10.1007/s12528-019-09209-y
9. The Institute of Chartered Accountants in England and Wales, incorporated by Royal Charter RC000246 with registered office at Chartered Accountants' Hall, Moorgate Place, London EC2R 6EA. https://www.icaew.com/technical/technology/blockchain/blockchain-articles/blockchain-case-studies
10. HASIB ANWAR. https://101blockchains.com/introduction-to-blockchain-features
11. Pratap M (2018) https://hackernoon.com/how-is-blockchain-disrupting-the-supply-chain-industry-f3a1c599daef. 8th Aug 2018
12. Carlo RW (2020). https://www.finextra.com/blogposting/18496/remaining-challenges-of-blockchain-adoption-and-possible-solutions. 29 Feb 2020
13. Tam KC. https://blockgeeks.com/guides/smart-contract-development
14. Benjamin (2019). https://hackernoon.com/a-brief-introduction-to-smart-contracts-53173x9g. 23rd Nov 2019
15. Orvium (2019) Whitepaper: accelerated Scientific Publishing (v1.7). https://docs.orvium.io/Orvium-WP.pdf. Last accessed 16 Jan 2022
16. Coelho FC, Brandao A (2019) Decentralising scientific publishing: can the blockchain improve science communication?. Memorias do Instituto Oswaldo Cruz vol. 114:e190257. https://doi.org/10.1590/0074-02760190257
17. Wikipedia Smart Contract. https://en.wikipedia.org/wiki/Smartcontract
18. Gupta R, Tanwar S, Al-Turjman F, Italiya P, Nauman A, Kim SW (2020) Smart contract privacy protection using AI in cyber-physical systems: tools, techniques and challenges. IEEE Access 8:24746–24772. https://doi.org/10.1109/ACCESS.2020.2970576
19. Mayank Pratap's Smart contract basics. https://hackernoon.com/everything-you-need-to-know-about-smart-contracts-a-beginners-guide-c13cc138378a. 27th Aug 2018

20. Smart Contract Reference. https://ethereum.org/
21. Solidity Reference. https://solidity.readthedocs.io/en/v0
22. Remix Reference. https://remix.ethereum.org
23. Vian K (2016) Own your achievements: three ways blockchain tech is disrupting education. https://Blockchainfutureslab.wordpress.com/2016/03/16/ownyour-achievements-three-ways-Blockchain-tech-is-disrupting-education
24. Watterson A (2016) The blockchain for education: an introduction. http://hackeducation.com/2016/04/07/Blockchain-education-guide
25. Ryan P (2017) We need to figure out how to use the Blockchain properly. https://www.weforum.org/agenda/2017/06/Blockchain-is-stalling-butwhats-holding-it-up
26. Cheng S, Daub M, Domeyer A, Lundqvist M (2016) Using Blockchain to improve data management in the public sector. http://www.mckinsey.com/business-functions/digital-mckinsey/our-insights/using-Blockchain-to-improve-data-management-in-the-public-sector

Printed in the United States
by Baker & Taylor Publisher Services